NEW DEVELOPMENTS IN OBESITY RESEARCH

NEW DEVELOPMENTS IN OBESITY RESEARCH

LAWRENCE F. DITMIER
EDITOR

Nova Science Publishers, Inc.
New York

For permission to use material from this book please contact us:
Telephone 631-231-7269; Fax 631-231-8175
Web Site: http://www.novapublishers.com

NOTICE TO THE READER

Library of Congress Cataloging-in-Publication Data
New developments in obesity research / Lawrence F. Ditmier (editor).
p. ; cm.
Includes bibliographical references and index.
ISBN 1-60021-296-4
1. Obesity--Research. I. Ditmier, Lawrence F.
[DNLM: 1. Obesity. 2. Biomedical Research. 3. Nutrition. 4. Socioeconomic Factors. WD 210 N532 2006]
RC628.N49 2006
616.3'980072--dc22 2006018905

Published by Nova Science Publishers, Inc. ✢New York

CONTENTS

PREFACE

Far from a sign of healthy prosperity and contentment, overweight and obesity are now considered high risk factors for a wide range of diseases including early death and disability, heart disease, diabetes, reproductive problems, cancer, breathing problems and arthritis. Obesity, now at epidemic levels in many countries, is defined as an excessively high amount of body fat or adipose tissue in relation to lean body mass. The amount of body fat (or adiposity) includes concern for both the distribution of fat throughout the body and the size of the adipose tissue deposits. This book includes within its scope the causal connection of obesity to diseases as well as the prevention and treatment of obesity. Leading-edge scientific research from throughout the world is presented.

Chapter one explores the ramifications of nutrition, rapid growth, early puberty, obesity, increased height, and several biochemical factors on health and longevity. In particular, it presents new findings indicating that many popular beliefs require reevaluation, such as the role of birthweight and early puberty on childhood and adult health. New reports indicate that commonly accepted body mass index values for normal, overweight, and obese designations are too high and need to be reduced. It is shown that only a ~ 2% excess in daily energy consumption can increase body weight by 22.7 kg (50 lb) within 10 years. Additional findings are presented indicating that rapid growth promotes increased cardiovascular disease, type 2 diabetes, and cancer in adulthood. Extensive research also indicates that promoting maximum attainment of one's genetic potential for height increases the risk for cancer and cardiovascular disease in adulthood and reduces longevity. Early puberty is also shown to increase the risk of cancer and obesity in adulthood. Also discussed are the role of nutrition and rapid growth in producing an early adiposity rebound and the impact of body size and nutrition on the concentrations of several biochemical factors. These biochemical factors include growth hormone, insulin-like growth factor-1, insulin-like growth factor binding proteins, sex hormone binding globulin, C-reactive protein, homocysteine, and creatinine. Current findings on the optimum diet for maintaining health and extending longevity are given and support a near vegetarian diet, similar to the traditional Mediterranean practice.

In chapter two the metabolic syndrome (MS) describes a cluster of cardiovascular risk factors, including central adiposity, insulin resistance or overt type 2 diabetes mellitus (DM2), atherogenic dyslipidemia, and elevated blood pressure or overt hypertension. Obesity, especially central adiposity, is associated with insulin resistance, the central feature of MS. Predictably, the incidence of MS increases with increasing body mass index (BMI) but also occurs with weight gain, even at normal weight. MS predisposes individuals to the development of DM2 and is responsible for increased all cause and coronary artery disease (CAD) mortality. MS, therefore, likely represents a latent disease process that may be viewed as preclinical diabetes or preclinical atherosclerosis. Because transitional conditions are more amenable to therapeutic intervention than late-stage disease, it is becoming increasingly recognized that early identification and treatment of MS is necessary to prevent CAD mortality. Recent clinical trials have demonstrated that weight loss through lifestyle modification with or without pharmacologic intervention can improve all components of MS. Although the exact mechanisms underlying MS remain undefined, physicians can contribute to improved long-tem clinical outcome by endorsing a healthy lifestyle that promotes weight loss or maintenance and physical activity. This may lead to the reduction in morbidity and mortality associated with obesity, DM2, and MS.

The objective of chapter three and purpose of this study was to compare the maximal exercise performance of overweight subjects during friction braked cycle ergometry of 10 s duration when resistive forces reflected total-body mass (TBM) or fat-free mass (FFM). Data collection was administered using a single blind, cross - over design. Subjects were assigned in a random fashion to either the TBM or FFM protocol. 11 healthy, overweight, male university students (age 22.3 ± 2 yrs, body fat 27.1 ± 2 %) participated in the study. Body composition was assessed using underwater weighing procedures. Fat mass was calculated from body density values. Power outputs were calculated using high intensity cycle ergometry. Increases ($P < 0.01$) in peak power output (PPO) were found between TBM and FFM (1029 ± 98 W TBM vs. 1397 ± 146 W FFM). Significant decreases ($P < 0.05$) were observed for the time taken to reach PPO (4 ± 1.4 s TBM vs 2 ± 1 s FFM). Pedal velocity significantly increased ($P < 0.01$) during the FFM protocol (98.2 ± 9.7 rpm TBM vs. 139.6 ± 6.6 rpm FFM) and resistive forces significantly decreased (7.5 ± 0.4 kg TBM vs. 5.5 ± 0.4 kg FFM). The findings of this study suggest that greater peak power outputs are obtainable when resistive forces reflect FFM as opposed to TBM. The results may have implications that relate to both the pathology of disease, related clinical examinations and exercise prescription in overweight populations.

Chapter four states that the prevalence of overweight and obesity is disproportionately high among Black, Latina, and Native American women compared with non-Hispanic White women in the United States. The basis for this disparity is likely multifactorial with genetic, environmental, and behavioral contributions, yet remains incompletely understood. This chapter examines the evidence for culture-specific attitudes and behaviors that may enhance risk for obesity with a special focus on differences in body image and disordered eating across ethnic minority groups in the United States. In addition, differences in patterns of physical activity, culturally informed explanatory models of health, and dietary practices across ethnic groups are reviewed. Clarification of the socio-cultural contributions to differential risk for disordered eating and obesity will be essential to the development of effective and culturally sensitive interventions across ethnically diverse populations.

In chapter we see that since 1980 the prevalence of obesity in many developed countries has doubled or tripled, and many developing countries are following the same trend. The World Health Organisation describes this as a serious global epidemic and a major contributor to the global burden of chronic disease and disability. In the last 25 years there has been a widespread environmental change resulting in a small positive disturbance in energy balance for a large proportion of the global population.

This chapter reviews current evidence regarding the underlying determinants of this epidemic. The mechanisms of energy balance are described, and limitations in interpretation of energy intake data from dietary surveys and energy expenditure data from physical activity surveys are discussed.

There is no evidence of a decrease in population energy expenditure as a factor in the obesity epidemic. On the contrary, both theoretical considerations and objective evidence demonstrate that per capita energy expenditure increases as mean BMI increases.

Data from Europe, the United States and Australia show an increase in self-reported energy intake and an increase in population food supply energy concurrently with the dramatic increase in the prevalence of overweight and obesity in children and adults since 1980. Increased consumption of sweetened drinks, snack foods and fast food with an increase in portion sizes contribute to increased energy intake. Absolute fat intake has changed little, but fat as a percentage of energy has decreased due to an increase in carbohydrate intake and total energy intake. Economic evidence also supports the role of low-cost, energy-dense but nutrient-poor foods as major contributors to this epidemic.

Population physical activity level has also decreased, but over a much longer period than the recent epidemic of obesity and a large proportion of adults and children in developed countries are insufficiently active for good health. Physical inactivity increases the susceptibility of individuals to obesogenic changes in the food supply. Increasing physical activity is associated with many health benefits apart from its role in obesity prevention.

The increased prevalence of ex-smokers in developed countries may have contributed to the increased prevalence of adult obesity, but evidence for this is conflicting, and it is unlikely to have contributed to the epidemic in children or in developing countries.

Comprehensive efforts are required by governments, industry and communities working together to address these determinants and create the economic, physical and social environments that support individuals to choose healthy foods and an active lifestyle.

In chapter six we see that at the same time the incidence of obesity has been increasing dramatically world-wide, new evidence has begun to shed light on obesity as an immunologic disorder. Like macrophages, adipocytes secrete a wide array of cytokines that mediate the inflammatory response, and the increased risk obesity poses for developing heart disease, stroke, and diabetes are closely linked to elevated levels of adipocyte-secreted immune modulators. New scientific developments have also revealed that high levels of several immunologic factors are precursors to weight gain, and point to chronic-low grade inflammation as a contributing factor in the onset and progression of obesity. Both infectious organisms and environmental chemical agents are capable of inducing lipogenesis and adipogenesis by stimulating pro-inflammatory cytokine signaling, and theoretical models of how obesity and inflammation intersect are being developed. These models offer promising approaches for research to extend our understanding of the current obesity epidemic.

In chapter seven we are informed that Alström syndrome (ALMS) (OMIM 203800) is a rare autosomal recessive genetic disorder with approximately 350 cases known worldwide. ALMS patients typically present in the first months of life with an ocular disorder characterized by progressive retinal dystrophy that leads to blindness by the end of the second decade. Other early features of ALMS include obesity, beginning in early childhood, hyperinsulinemia or insulin-resistance, type 2 diabetes, and hyperlipidemia. Bilateral neurosensory hearing loss develops in childhood and usually progressively worsens. Impairment of heart, kidney, or liver function at different stages of life may severely affect patients' prognosis and survival. Other less frequently observed features involve the pulmonary, bone-skeletal, neuromuscular, hemopoietic and endocrine systems. In most of the patients there is normal intelligence, although there are some reports that showed a possible delayed intellectual development.

In 2002, the gene responsible for ALMS, *ALMS1*, on chromosome 2p13 was identified. *ALMS1,* a novel gene of unknown function, is comprised of 23 exons and encodes a protein of 4,169 amino acids. *ALMS1* is ubiquitously expressed at low levels in all tissues reported to be pathologically affected in ALMS patients and thus far does not share significant sequence homology with other known genes. The diagnosis of Alström syndrome remains a difficult task both for clinicians and geneticists, considering the overlap between ALMS and other genetic disorders with retinopathy, obesity, or diabetes. Initially, many patients are diagnosed clinically without genetic confirmation, which often results in multiple misdiagnoses and delayed diagnosis. No specific treatment of ALMS currently exists. The ALMS syndrome deeply impacts the life of patients and families. The formation of various pro-active groups, including Alström Syndrome International, Alström Syndrome-Canada, and Alström Syndrome-UK have had a fundamental role, in raising interest in the syndrome, organizing scientific meetings, and giving support to patients and their families. Moreover, an International Scientific Advisory Board on ALMS has been established which has the role of translating scientific achievements into everyday practice.

Chapter eight concerns itself with nutrition research and the study of those processes whereby cells, tissues and the whole body obtain and utilise substances contained in foods and drinks in order to maintain their structures and functions. Thus, nutrition oriented scientists are interested in understanding the balance between the supply and demand of substrates to ensure the functional integrity of the organisms, including homeostasis, growth and reproduction. Therefore, nutrition research involves the advances in knowledge concerning not only food composition, dietary intake and food/nutrient utilisation by the organism, but also and the short or long-term influences of food/nutrient consumption on health and nutrient functions, which can be achieved through "in vitro" and "in vivo" studies with different experimental models. Thus, processes involved in nutrient utilisation such as digestion, absorption, metabolism and excretion as well as nutrient functions can be evaluated and measured by using "in vitro" approaches. These tests involve a laboratory replication of an "in vivo" function. On the other hand, whole animal systems have been used in measuring nutrient utilization and fate as well as to assess nutrient functions. The selection of laboratory tests to be used in different models should take into consideration their precision, accuracy, analytical specificity and sensitivity, predictive value and validity. Most of the nutritional experiments and research methodology described for animals can be also applied to human beings. However, in some circumstances animal studies are preferred for practical reasons, but in others it is necessary to carry out the experimentation directly in man due to difficulties for data extrapolation. The impact of food intake on health and nutritional outcome is conveniently assessed thought dietary intake estimations, anthropometry, biochemical markers and clinical indications of the nutritional states. The selection of research tests to be used in different models should take into consideration their precision, accuracy, analytical specificity and sensitivity, predictive value and validity. In any case, adequate ethical guidelines and regulations must be followed and accomplished in all experiments carried out in animal and human beings.

In: New Developments in Obesity Research
Editor: Lawrence F. Ditmier, pp. 1-40

ISBN 1-60021-296-4

Chapter 1

NUTRITION, OBESITY, GROWTH AND LONGEVITY

Thomas T. Samaras
Reventropy Associates, San Diego, California

ABSTRACT

This chapter explores the ramifications of nutrition, rapid growth, early puberty, obesity, increased height, and several biochemical factors on health and longevity. In particular, it presents new findings indicating that many popular beliefs require reevaluation, such as the role of birthweight and early puberty on childhood and adult health. New reports indicate that commonly accepted body mass index values for normal, overweight, and obese designations are too high and need to be reduced. It is shown that only a ~ 2% excess in daily energy consumption can increase body weight by 22.7 kg (50 lb) within 10 years. Additional findings are presented indicating that rapid growth promotes increased cardiovascular disease, type 2 diabetes, and cancer in adulthood. Extensive research also indicates that promoting maximum attainment of one's genetic potential for height increases the risk for cancer and cardiovascular disease in adulthood and reduces longevity. Early puberty is also shown to increase the risk of cancer and obesity in adulthood. Also discussed are the role of nutrition and rapid growth in producing an early adiposity rebound and the impact of body size and nutrition on the concentrations of several biochemical factors. These biochemical factors include growth hormone, insulin-like growth factor-1, insulin-like growth factor binding proteins, sex hormone binding globulin, C-reactive protein, homocysteine, and creatinine. Current findings on the optimum diet for maintaining health and extending longevity are given and support a near vegetarian diet, similar to the traditional Mediterranean practice.

Introduction

Nutrition has both good and bad aspects for human health and longevity. Too little nutrition undermines the immune system, stunts growth, lowers energy capacity and reduces well-being. Too much food raises several hormonal levels which can lead to accelerated growth, obesity, early puberty, chronic disease and reduced longevity.

Excessive nutrition leads to increased levels of insulin and insulin-like growth factor-1 (IGF-1), which in turn affect levels of IGF binding protein-1 (IGFBP-1), sex hormone binding globulin (SHBG), creatinine, and C-reactive protein. High levels of insulin, IGF-1, creatinine and C-reactive protein are associated with cardiovascular disease (CVD), cancer, and all-cause mortality. Low levels of SHBG and IGFBP-1 are also tied to increased insulin resistance, CVD, cancer, and all-cause mortality. It should be noted that the term "excessive nutrition" used in the introduction of this paragraph may actually be "normal nutrition" in the developed world.

Obesity has reached an epidemic level with virtually no new approaches for combating it. Therefore, a new paradigm is needed which addresses the entire life cycle of humans with new nutrition guidelines starting after birth and proceeding through old age. This paradigm involves maternal nutrition, birthweight, growth rate, obesity and body size (height and weight).

Attainment of one's genetic potential for body height is another problem related to obesity and longevity. We need to let go of this false premise and emphasize a healthful and non-obesity promoting diet that starts with preconception and proceeds through the growth cycle and adulthood.

Role of Nutrition in Childhood Growth and Obesity

Rapid childhood growth is an important aspect related to increased obesity and chronic disease in adulthood. An excessive growth rate promotes CVD and cancer and can be controlled by nutritional changes; e.g., reducing the intake of animal protein, refined carbohydrates, and animal/total fat.

Nutrition is fundamental to life. It enables infants to grow into adults, it provides various chemical and biological agents to fight off disease, it enables the repair or replacement of damaged cells and tissues, and it allows the performance of various physical activities. Unfortunately it can also create biochemical changes that promote excessive growth and chronic disease and shorten longevity. The beneficial or harmful role of nutrition starts with the mother and continues through infancy, childhood and adulthood.

Today's greatest threat to public health lies with increasing obesity of our population. In developed countries, the trend towards greater weight originates with an over abundance of energy-rich food choices which can be consumed in quantities that earlier mankind rarely experienced. In addition, a false impression exits that overweight or obesity are caused by large amounts of increased consumption. Unfortunately, this is not true. For example, if a child on a 10,000 kilojoule (kJ) diet consumes only ~ 2% more energy [200 kJ (48 kcal)] per day (d) than he or she needs and stores this energy in the form of fat, he or she will exceed normal body weight by 22.7 kg (50 lb) in 10 years (y) [200 kJ x 365 d x 10 y = 730,000 kJ/32,186 kJ/kg = 22.7 kg (50 lb); 1 kg of fat = 32,186 kJ].

Unfortunately, the trend towards increasing height, which is generally viewed as a positive value, is also related to obesity. In fact, Amador et al (1992) reported that the same factors that increase lean body mass and promote linear growth also increase fat accumulation.

Early excess nutrition affects our physical characteristics and promotes obesity. In contrast, during childhood and adolescence, shortage of food reduces body size in terms of height and weight as found in Europe during WW II (Tretli, 1989).

While it is well known that nutrition has had a major role in the secular trend towards larger human size, the exact impact is unknown because formal experiments cannot be done on humans. However, Hoppe et al (2004A) reported a .51 cm (.2") increase in height and a .44 kg (~1 lb) increase in weight at 10 y of age for a 1% increase in energy from protein at 9 months. Animal studies have also shown that caloric restriction of about 30% can reduce the weight of monkeys by 33% compared to normally fed animals (Lane et al,1995). A study involving horses found that a 30% reduction from recommended calories and protein resulted in horses that were 25% shorter in length (Loveridge and Noble 1994). Thus, substantial reduction in nutrition sustained over the growth period can have a major impact on final adult size. Of course, other factors, such as genes, illness and environment can affect growth during childhood and adolescence. In Scotland, socioeconomic class also played a role in growth and height. Researchers found that adults whose fathers were from lower socioeconomic classes averaged 4.6 cm shorter than adults whose fathers were from the upper classes (Okasha et al, 2002). This relation between height and socioeconomic status exists throughout most of the world (Komlos and Baur, 2004).

Small infant or childhood size can result from low BW, under nutrition, maternal illness or childhood illness. However, subsequent catch-up growth can offset much of this reduced growth although promotion of greater height and weight may not be healthful. The ramifications of increasing height on health are discussed in another section.

Excess nutrition during the growth period can also promote obesity and cancer during childhood, adolescence and adulthood (Samaras et al, 2003A). Even if children grow up without excess adiposity, over nutrition during adulthood can result in overweight or obesity along with its various comorbidities and reduced longevity. In addition, overweight children who grow up to be normal weight adults are still at increased risk of disease in adulthood (Alexander, 2001).

Body Mass Index and its Relation to Obesity and Health

The body mass index (BMI) or Quetelet Index is the most common index used for measuring whether someone is overweight or obese. It is often linearly related to chronic disease, such as coronary heart disease (CHD) and cancer (Oken and Gillman, 2003; Calle et al, 2003). For example, based on > 900,000 subjects, Calle et al (2003) found cancer rose linearly with increasing BMI. The BMI = weight (kg)/height (m)^2 (where kg = kilograms and m = meters). The BMI has several key values for identification of the obesity state. According to the World Health Organization (WHO), an adult BMI of < 18.5 kg/m^2 is designated as underweight, 18.5-24.9 kg/m^2 is considered normal weight, 25-29.9 kg/m^2 is defined as pre-obese, and $\geq$ 30 kg/m^2 defines obesity (James, 2004). Although widely accepted, the upper limit of a normal BMI (24.9 kg/m^2) appears to be too high as described in the next section.

Some researchers argue that one can have a highly muscular build, low adipose tissue and a high BMI, and thus the index falsely indicates adiposity when it is not present. In spite of this limitation, the BMI appears to work well for most studies. Crude body weight also negatively correlates with longevity; e.g., it was found that for professional baseball players (presumably with low body fat), body weight during playing years was a predictor of shorter longevity in later years (Samaras and Storms,1992; Samaras et al, 2002A). Since young baseball player weights reflected mostly fat-free skeletal, organ and muscle mass, a high lean body weight is not necessarily a healthful physical feature, at least during young adulthood.

A problem with the BMI is that it does not provide an equitable comparison in epidemiological studies because taller subjects tend to be more linear in build compared to shorter ones with the same BMI. While the BMI presents a potential problem in identifying someone's obesity status, most researchers accept the BMI as a reasonably reliable indicator of relative body fat. In addition, it is a good guide for predicting one's future morbidity and mortality, regardless of height.

It should be noted that the BMI varies dramatically as an infant proceeds into childhood, adolescence and adulthood. The BMI increases by ~ 5 kg/m^2 between birth and the first year of life and then decreases to about 15 kg/m^2 by 4-8 years (y) of age (Freedman et al, 2001). It then increases progressively until young adulthood where it reaches a level of 20-25 kg/m^2. The age at which the child reaches its minimum BMI before starting its subsequent rise is called "adiposity rebound" (AR), and early AR predicts future obesity. Adiposity rebound is discussed further in a subsequent section.

Galileo Galilei was the first scientist to show that the skeleton of a large animal increases as the third power of its linear dimensions (Loveridge and Noble, 1994). This finding is reflected in another important index, the Ponderal Index (also called the Rohrer Index). The Ponderal Index is obtained by dividing weight (kg) by height $(m)^3$. This index is used because it reflects a proportional weight change with height; e.g., a study of four generations of Harvard students showed a secular increase in height and weight that closely followed the Ponderal Index (Damon, 1968). More recently, Freedman et al (2001) found that each 5 cm increase in childhood height resulted in a ~ 2 kg/m^2 rise in adult BMI. (This value reflects excess adiposity since the increase in the BMI should be ~ .7 kg/m^2 for a proportionately larger body.) In addition, this index follows the BMI up to the AR and then stays flat through adolescence while the BMI continues to increase with age (Freedman et al, 2001). Bellizzi and Dietz (1999) also reported that adult BMI increases in step with increasing height. In contrast, Leon et al (1998) tracked about 13,000 adults from birth through ages 1,15, and 65 y and found the Ponderal Index was the same (~ 26 kg/m^3) for each of these three ages.

As mentioned, many researchers prefer the BMI for establishing relative weight for height, and a number of studies have found this index applies to epidemiological studies. However, Samaras et al (2002B) found over 80 population samples from the US and other parts of the world which have followed the Ponderal Index as height increased with an error of <10%. Many were fairly large studies and one involved about 1/2 million subjects.

Epidemiological studies comparing the morbidity or mortality of short people in comparison to tall ones need to consider whether life long socioeconomic status and body type are important health factors; e.g., stocky vs linear. It is obvious that we cannot compare the health of a 152 cm, 91 kg (5',200 lb) person to a 183 cm, 77 kg (6', 170 lb) person to study the impact of height on health. While epidemiological studies compare much smaller height differences, short and tall people of the same body proportions should be compared to have a fair comparison. Thus, if body type can affect results (Krakauer et al, 2004), comparing short and tall people with the same BMI produces an unfair comparison and may invalidate study findings. However, if short and tall groups with the same body type are to be compared equitably, the Ponderal Index should be the same for both groups.

The BMI can also be used in height-health studies, but it has to be increased in proportion to height to maintain equitability. That is, a 10% taller person (or cohort) should have a 10% greater BMI compared to the shorter person (or cohort) to minimize confounding due to body type. For example, a 152 cm, 50 kg (5' 110 lb) person with a BMI of 21.4 kg/m^2 should be compared to a 183 cm, 86 kg (6' 190 lb) person with a BMI = 25.8 kg/m^2 to approximate the same body type. In this case, a 20% taller person has a 20% higher BMI.

One final point on the BMI. Higher socioeconomic people in developed countries are generally taller and leaner than people in lower classes. One reason for their leaness is that they are better informed on healthful eating practices. They also exercise more during their free time and are more conscious of their body weight. In addition, since most foods are standardized, taller people consume fewer calories in proportion to body weight when they have a banana, glass of milk, or a restaurant meal. As mentioned, a small change in caloric intake can have a big effect over 10 or 20 y. Thus, many tall upper socioeconomic class people are likely to be relatively lean, especially in middle years and beyond.

ARE OVERWEIGHT AND OBESITY PROPERLY DEFINED?

The conventional BMI limits defining overweight and obesity are being redefined based on new findings (James, 2004). For example, the WHO recently reported that the desirable BMI has been reduced to between 21 and 23 instead of the usual upper limit of 24.9 kg/m^3 previously described. This lower BMI reflects the fact that diabetes, gallstones, hypertension, and coronary heart disease (CHD) increase linearly starting with a BMI of 19-20 kg/m^2 (James, 2004). Cancer is also linearly related to BMI as well. In addition, the desirable CHD protective factor, high density lipoprotein cholesterol, declines with increasing BMI starting below a BMI of 20 kg/m^2. Interestingly, a recent study by Komlos and Baur (2004) found that people who report that they are in excellent health have lower BMIs than people with lower health self- appraisals.

A Korean study found a linear relation between blood pressure and a BMI range of 16 to 31 kg/m^2 (Jones et al, 1994). Based on 22,354 subjects, the researchers concluded that a strong positive correlation exists between BMI and blood pressure for lean as well as obese subjects.

While Western society is not likely to move towards an average BMI of 20 kg/m^2 or less, it is important not to accept the conventional BMI designation of $<$ 24.9 kg/m^2 as optimum for health. Most traditional societies that are free of Western diseases have a BMI of about 20-22 kg/m^2. Unfortunately in the overfed Western world, people who are this thin are often suffering from clinical or subclinical health problems, are smokers, had pathological conditions during childhood, or in the case of the elderly suffer from lack of money to eat adequately. However, it appears that very lean people who eat a well balanced diet, exercise regularly, don't smoke, have satisfying lives and receive good medical care tend to be the healthiest (Kagawa, 1978; Willett, 2003; Willcox et al, 2004; Samaras and Elrick, 1999).

Realistically, many Westerners will have a very difficult time trying to attain a BMI of 22 kg/m^2 or less due to having heavy frames and high musculature or low metabolic rates. The goal for these people should be to attain the lowest BMI feasible by adopting a different dietary regime which emphasizes high fiber, low animal food, and low calories and by exercising regularly. While any weight loss diet needs to be more spartan than the standard American diet, it still needs to be designed to provide an enjoyable eating experience. If meat and dessert are important to the quality of life for an individual, then special days should be set aside for occasional deviation from a low calorie vegetarian diet. In addition, regular exercise and a predominantly plant-based diet can reduce the risks of chronic diseases even in relatively high BMI people (Chrysohoou et al, 2004).

Birthweight, Overweight, Height and Nutrition

Some of the findings reported in this section represent unique or different paradigms in relation to birthweight, overweight, height, and nutrition. However, they are based on both human and animal findings involving hundreds of research papers.

Birthweight, Mortality and Overweight

Most medical professionals believe that high birthweight (BW) is a positive health indicator (Friedlander et al, 2003). This belief is based on many epidemiological studies that find higher BW correlates with reduced perinatal and infant mortality and lower adult chronic disease (Samaras and Elrick, 2003A, Melve et al, 2000; Elrick et al, 2002). However, the following research conflicts with this generally accepted paradigm.

Eriksson et al (2004) conducted a long-term longitudinal study of CVD mortality in 1319 men who weighed ≤ 3000 g at birth. Tracked from 20 y to 85 y of age, the researchers found BW did not significantly affect CVD morbidity or mortality. In addition, a review of BW and nutrition by Rush (2001) found that increased BW due to excessive maternal nutrition does not necessarily improve perinatal survival because greater prepregnant weight is positively tied to increased BW and higher perinatal mortality.

The mean BW of children has increased in many developed countries in spite of a substantial rise in low BW preterm births (Wen, 2003). For example, the UK has experienced an increase in BW in recent years due to increasing size of mothers, and it is common for tall (>165 cm) mothers to produce macrosomic (≥ 4000 g) infants (Bergmann, 2003). In addition, high maternal weight (≥ 26 kg/m^2) before pregnancy is a strong predictor of macrosomia. Furthermore, a gain of ≥ 16 kg vs < 10 kg increases the risk of having macrosomic infants and maternal diabetes. Macrosomic female infants have a higher risk of becoming obese or developing type 2 diabetes when they are born to obese or diabetic mothers. In addition, these girls then tend to give birth to macrosomic neonates. Unfortunately high BW or large for gestational age infants have higher risk of many ailments including increased heart defects and stillbirth deaths and are more likely to become obese in early adulthood (Lapunzina et al, 2002; Bergmann et al, 2003).

The correlation between BW and later BMI is well established. Oken and Gillman (2003) reported that over 24 studies have evaluated BW and later BMI. Most of them found that BW is positively correlated with later BMI and none found a negative relationship. The usual increase in BMI was .5 to .7 kg/m^2 per 1 kg increase in BW. James (2004) noted that obesity (high BMI) was intrinsically connected to gallstones, diabetes, hypertension, and CHD. However, while higher BW leads to higher BMI and chronic disease in adulthood, low BW leads to central obesity which is strongly tied to adult disease as well.

It appears that a small increase in BW can produce large weight changes in adulthood. For example, Sorensen et al (1997) examined several thousand adult males and found that normal weight men had an average BW of 3445 g, overweight men had a BW of 3547 g and obese men had a BW of 3571 g. Thus, the BW difference between normal adult weight and obesity was only 126 g.

One of the most important purported benefits supporting the advantages of higher BW infants is lower adult blood pressure. Many studies have supported this finding, usually after adjusting for current height and weight. However, Huxley et al (2002) evaluated over 55 studies and concluded that the connection between blood pressure and BW was of minor importance compared to the adult height and weight attained by the infant. Thus both low and high BW infants can maintain normal blood pressure if they do not become overweight in adolescence and adulthood.

In recent years, a number of studies have found that the optimum BW and birthlength were not the heaviest and longest infants. For example, Melve et al (2000) in a large study of >1.7 million subjects found infants 49 cm long (vs > 50 cm) and weighing between 3250 g and 3500 g had the lowest perinatal mortality (.6/1000) for 1992-1971 birth year. Below 49 cm, mortality increased to 1/1000 and above 49 cm, mortality increased to about 38/1000 for a birthlength of 54 cm. Based on 14 BW groupings (range: 1500-4999 g), longer infants had a higher risk of perinatal mortality in each group.

An Italian study involving more than 2 million male and female subjects over a 40-y period evaluated BW and infant stillbirth mortality and found that BW declined substantially over the last 40 y and paralleled declines in stillbirth mortality from 21/1000 to 1.8/1000 during the same period (Ulizzi, 2002). For male infants born alive, BW declined from 3526 g in 1954 to 3414 g in 1994. Females saw a decline from 3410 g to 3275 g for the same period. In addition, birthweight in Japan has declined in recent years and infant mortality is one of the lowest in the world at ~ 3/1000 compared to ~ 8/1000 in the US (Sawaguchi and Sawaguchi, 2002).

Another problem with greater BW is its positive correlation with adult cancer and height. Andersson (2001) found that the highest BW quintile had 1.71 times the cancer risk compared to the lowest BW quintile for a range of 2745 g to 4241 g. Recently, a large study found BW to be positively correlated with breast cancer (Ahlgren, 2004). Okasha et al (2002) conducted an extensive review of BW and its relation to cancer. They found high BW and childhood and adult tallness were positively correlated with various cancers, including breast, prostate, colo-rectum, haematopoietic system and endometrium.

Researchers have reported that traditional societies tend to have low BW infants but have exceptionally low CVD, cancer and diabetes (Bradley, 2001; Saweri, 2001). Almost 90% of the world's low BW infants are born in developing world. If low BW per se promoted future chronic disease, it should be evident in these societies. Surprisingly, even the elderly rarely (< 5%) die from Western diseases. Instead death occurs primarily from infections, traumas, malnutrition, and neglect (Drury, 1972). (However, cancer mortality, especially from prostate cancer, is higher for the elderly compared to their youth and may be related to adoption of Western dietary patterns during the 20th century.)

A gestational age of 37 to 42 weeks is considered normal and BW, birthlength and BMI correlate with the length of gestation as well as with each other (Leon et al, 2000). A BW below 2300 g is not desirable because it is also a risk factor for later obesity (Leong et al, 2003). However, this could be due to excessive nutrition to promote growth in small children. In addition, infants born after 42 weeks were found to have the highest mortality from CHD in adulthood when compared to all gestation periods (Forsen et al, 1997). Although their BWs were not given in the study, later gestation infants tend to be heavier and have greater heights and BMIs as adults (Leon et al, 2000; Leong et al, 2003).

The importance of BW to adult health is also challenged by three studies involving adults exposed to famine during their conception or gestation period. These studies involved Finland, Russia, and Holland (Roseboom et al, 2001; Stanner et al, 1997). A substantial drop of a few hundred grams in BW occurred during famine conditions. As expected, infant mortality was higher but once adulthood was reached, no differences in mortality were found. These findings, based on over a million subjects, conflict with the general belief that low BW has harmful effects in adulthood. This example is not given to indicate that famine exposure is harmless, but that low nutrition in infancy may not have serious health consequences in adulthood, especially if overweight and other poor health habits are avoided.

Another point on BW is related to twins who tend to be about 1000 g lighter than singleton infants. Zaretsky (2003) reported on a longevity study of about 27,000 twins who were WW II US veterans born between 1917 and 1927. The study found that these twins averaged 80.5 y, a respectable longevity for lower BW babies which are supposed to be at a health disadvantage. [In 1950, 20 y old males from the general population had a life expectancy of 77.3 y (Statistical Bulletin, 1964).] Even more striking was the finding that identical twins lived to 82 y. Identical twins are a few hundred grams lighter than fraternal twins but lived 1.5 y longer in spite of their lower BW. The longevity of twins is even more impressive because they tend to experience catch-up growth due to low BW and this has been found to increase the risk of CHD in singleton adults (Silventoinen et al, 2003) and presumably would apply to twins as well.

Type 2 diabetes has been found to be related to low BW. However, Hypponen (2003) found that small infants did not grow up to have type 2 diabetes if they avoided excess postnatal weight gain. Wilkin et al (2002) and Murtaugh et al (2003) also found that BW was not important in relation to insulin resistance, type 2 diabetes and CVD risk factors when excess weight was avoided in childhood and adolescence. However, both low and high BW were correlated with increased CVD and type 2 diabetes (Murtaugh et al, 2003).

Birthweight is strongly correlated with height and BMI in adulthood, and both height and BMI are positively correlated with blood pressure (Williams and Poulton, 1999; Oken and Gillman, 2003; Seidman et al, 1993; Samaras et al, 2003A). Seidman et al (1993) in an Israel study of 30,083 17 y old males and females found that height increased linearly with BW from 1000 g to 5000 g. For males, height increased from ~ 165 cm for a BW of 1000 g to ~179 cm for a BW of 5000 g. Females averaged ~156 cm for a BW of 1000 g and ~168 cm for a BW of 5000 g. A similar finding occurred in Denmark but the lower and upper limits of height were higher by several centimeters compared to the Israel study (Sorensen et al, 1999).

In conclusion, based on a study of 165,136 Swedish men, Leon et al (2000) concluded that whether size at birth is a good or poor proxy for fetal growth and development problems is subject to speculation and uncertainty in terms of its impact on clinical or public health. There is no doubt that normal fetal growth and BW are important, but it is debatable whether BW per se is important independent of the size of the mother, her diet, gestational period, and socioeconomic and medical conditions (Wilcox, 2001). However, substantial research indicates that higher BW children are at greater risk for adult obesity and various chronic diseases, and low BW children are at high risk for chronic disease if they become overweight or obese.

Height, Weight and Health

While our society views tallness as reflecting better health (Komlos and Baur, 2004; Okasha et al, 2002), research over the last 30 y has found that lifespan is reduced by .5 y with every cm increment of height (Miller, 1990; Samaras and Storms, 1992; Samaras, 1996; Samaras and Elrick, 1999; Samaras et al, 2003B). In addition, a longitudinal study found that children that were in the ≥ 90th percentile of height were 5 times as likely to have an adult BMI ≥ 30 kg/m^2 (obese) as children who were in < 50th height percentile. Skinfold thickness, a measure of obesity, was also substantially greater for taller children (Freedman et al, 2002).

Studies showing increasing body size and decreasing longevity conflict with the fact that life expectancy has increased sharply over the last 150 years. The reason for this paradox is that improvements in socioeconomic factors, sanitation, and medical care have more than offset the negative aspects of growth in body size. This is especially true for reduced infant and early childhood mortality, lower maternal mortality during the birth process, reduced infectious diseases and epidemics, reduced malnutrition and the ability of the medical profession to keep elderly people alive with illnesses that would have killed them in earlier decades.

Gunnell et al (2001) reviewed about 300 studies involving body height and cancer and found that there is a 20 to 60% increased risk for cancer of all types with increasing height. More recently, Ahlgren et al (2004) reported that most studies on body size and cancer found that taller women have more breast cancer regardless of menopausal status. The tallest quintile of women had 1.51 times the risk of cancer compared to the shortest quintile. These human findings are consistent with animal findings; e.g., monkeys subjected to caloric restriction were smaller than those normally fed and had substantially lower cancer (Bodkin et al, 2003). A study involving 47,690 men (57 to 92 y of age) found that the relative risk (RR) for Western type cancers increased from RR =1 for a height of < 167.64 cm to RR = 1.72 for a height of ≥ 183 cm (Giovannucci et al, 2004). Engeland (2003A) also found in a 21-y study involving about 1 million men that the tallest had 1.72 times the risk of prostate cancer as the shortest men. (While most studies have found a positive correlation between height and prostate cancer some have not found a substantial difference; e.g., Habel et al, 2000.)

Findings on the relation between height and ovarian cancer are not entirely consistent but strongly favor a correlation between height and ovarian cancer. For example, initially, the

Iowa Woman's Heath Study (Anderson et al, 2004) did not find an overall relation between height and ovarian cancer, but a supplemental analysis found an 86% greater incidence for serous ovarian cancer for the highest height quartile (> 165 cm) vs the shortest quartile (≤ 155 cm). (Serous cancer is the most widespread form of ovarian cancer and accounts for 40% of epithelial tumors.) In addition, a very large study found a 48% increased risk of ovarian cancer for women ≥ 175 cm in height compared to 150-154 cm (Engeland et al, 2003B). This study involved 1.1 million Norwegian women tracked for an average of 25 y. Another study by Schouten et al (2003) tracked 62,573 Dutch women for 7.3 y and found that women ≥ 175 cm had 2.17 times the risk for ovarian cancer compared to women < 160 cm in height.

Holzenberger et al (1991) evaluated the height and longevity of 43 provinces in Spain involving > 1.3 million men of military age in the mid 19th century. Based on a tracking period of > 50 y, they found that survival to age 70 y and over was substantially greater for shorter males; i.e., regression analysis resulted in a negative correlation coefficient (r) = -.58, p < .001 between youthful height and later survival over a height range of ≤ 157.5 to ≥ 164.5 cm among 43 provinces. The researchers concluded: The shorter the average height at 18 y of age in a province, the greater the probability of men reaching advanced ages in that province. Confounding due to selection procedures, migration to other counties or countries, mortality due to epidemics, or population census errors was essentially ruled out by the researchers. They concluded that longevity in the 20th century is reduced due to accelerated growth, but this reduction is more than offset in developed countries by the life prolonging influences of improved economic conditions and advances in medical technology.

More recently, Krakauer et al (2004) in a 10-y longitudinal study of 116 male and female Swedes 50-74 y of age at baseline found that age at death was negatively correlated with height. They found an r = -.26, p = .006 between height and age at death. In 1991, Miller (1990), based on an Ohio population of 1679 deceased men and women, found that age at death declined by .47 y/cm as height increased. In the previously mentioned Swedish study, Krakauer et al also found a decline in longevity of - .52 y/cm increase in height (personal communication, July 11, 2002), which was identical to the average value reported ~ 3 y before by Samaras and Elrick (1999). In addition, the slope between white California males and females was also -.47 y/cm. The closeness of the slope values for a variety of unrelated populations is remarkable as shown next.

1.	Ohio study (men and women combined):	.47 y/cm
2.	Swedish study (men and women combined):	.52 y/cm
3.	Finnish athletes (men only):	.49 y/cm
4.	Variety of groups (men vs men and men vs women):	.52 y/cm
5.	Mid 19th C French data (average for men and women):	.51 y/cm
6.	California male vs female life expectancy:	.47 y/cm

Items 4-6 are from Table 2 in Samaras and Elrick (1999) which shows an individual population variation from - .35 y/cm to - .81 y/cm for 13 population samples.

In 2004, a review paper (Samaras et al, 2004) on CVD and height reported a strong positive correlation between height and CHD and a moderate increase in stroke. While a number of epidemiological studies contradict these findings, the differences they find range from virtually none to relatively small decreases with increasing height. For example, Davey Smith et al (2000) found short people had a 12% higher mortality rate vs 10 cm taller men, but Samaras et al (2004) found taller (10-12 cm) men and women had higher mortalities from 100% to over 400%. An analysis of this paper's data is presented in Table 1. (A relative risk (RR) of < 1 indicates lower CHD risk compared to an RR = 1 for the taller cohort.) The findings of the 2004 review were based on a variety of populations throughout the world and involved millions of deaths.

Table 1. Summary CHD mortality and height findings

Source[a]	CHD Mortality Relative Risk (RR)	Remarks/comments
Fig 1	.42	Men 45-65 y of age; 11 European countries-shortest vs tallest
Fig 2	.25	Men 65-84y of age;16 populations (US, Europe, Japan)-shortest vs tallest
Fig 3	.46	California ethnic males 25-84 y of age-shortest vs tallest males
Fig. 3	.46	Calif ethnic: shorter females vs taller females 25-84 y of age
Fig. 3	.42	Calif shorter white females vs taller white males 25-84 y of age
Fig. 4	.68 .34 (adjusted for 10 cm height difference.)	Native Americans 65-74 y of age.Trend based on a 5 cm difference in height between shortest and tallest instead of 10-14 cm found in other populations.
Fig. 5	.40	Japanese males 60-64 y-shorter vs 3 cm taller.
	.12	Adjusted for a 10 cm height difference.
p RA66	.26	Shorter Muslin vs Hindu males 20->60 y of age. Incidence of CHD and hypertension
Health, United States, 1999	.34	White females vs males 65-74 y of age. Year of mortality data: 1985
	.50	Year of mortality data: 1997 (The RR change in1997 vs 1985 may be due to changes in prevention and treatment of CHD risk factors for men.)

[a] Figure numbers and page numbers refer to Samaras et al (2004) except for the publication. Health, United States 1999, NCHS, pp 164-5.

Interestingly, the rate of increase in CHD with increasing height was similar for both men and women (Samaras et al, 2004). That is, the increase in CHD mortality per centimeter increase in height for taller vs shorter women was similar to the rate of increase in mortality for taller men vs shorter men. Traditionally, the lower CVD of women compared to men has been attributed to their hormones. However, this argument is weakened based on recent findings that hormone replacement therapy for menopausal women promotes heart attacks, stroke and breast cancer rather than reducing them (Mayo Clinic, 2005).

Developing societies have very low CVD compared to the West and their people are smaller in body size. About 70% of the world's elderly (> 59 y) live in the developing nations and have relatively little Western disease (Brown et al, 1997; Eaton, et al, 1988). For example, Eaton reported that the following populations have little or no CHD:

Eskimos
Kenyan Kikuyu
Solomon Islanders
Navajo Indians
Masai pastoralists
Australian aborigines
Bushmen
New Guinea highland natives
Congo pygmies

All these populations are substantially shorter than the US whites and blacks and Northern Europeans and have low BMIs. While the Masai average 173 cm (~ 5'8"), they have traditionally been of low weight (about 60 kg or 132 lb). However, with economic growth, height and CVD have been increasing in the developing world, especially in urban areas. For example, South Korea has seen a 2000% increase in cancer and CHD over the last 50 y (Oken and Gillman, 2003). Along with rapid economic development, height and weight have also increased substantially during this period. Type 2 diabetes has also increased by over 1400%.

Additional evidence comes from a Chinese study. For example, during the study period, American males, who were about 10 cm taller than Chinese males and had 16.7 times the CHD mortality of shorter Chinese males (Campbell et al, 1998). Within China, height was also positively correlated with CHD and stroke among Chinese of the same ethnic background (Chen et al, 1990). (These data are supported by animal research as well, including dogs (Miller and Austad, in press), monkeys (Bodkin, et al, 2003), and rodents (Bartke, 2003; Rollo, 2002).

Rhesus monkeys on caloric restriction have also been found to have much lower CVD disease mortality compared to larger ad lib fed monkeys (Bodkin et al, 2003). This study also found that calorie restricted males and females, whose weights were kept between 10 and 11 kg, had about the same average longevity. This and other findings indicate that life span differences between men and women may be due mainly to differences in body size (Miller,1990; Rollo, 2002; Samaras et al, 2003B; Stindl, 2004).

A higher BMI is also important in promoting CHD. Eriksson et al (1999) found that men who died from CHD were thin at birth and experienced increases above average in BMI from 7 to 15 y. In addition, children who were tall at 11 y of age had higher risk from CHD mortality in adulthood compared to the shortest children. During childhood or adolescence, a reduction in the rate of increase in weight and BMI reduced the risk of CVD in young adulthood (Sinaiko et al, 1999). And this reduction in risk was independent of body weight during early childhood.

One mechanism for increased CHD through increased body size deals with the left ventricle of the heart which provides oxygenated blood to the body. Researchers have reported that increased body size (weight, height and BMI) correlates with left ventricle mass and left ventricular hypertrophy (enlargement due to increase in cell size rather than cell number) which reduce pumping efficiency and promote CHD mortality (de Simone et al, 1997; Levy et al, 1988). In addition, Falkner et al (1998) reported height was the strongest factor involved in higher blood pressure for children and adolescents. The study was prepared by the National High Blood Pressure Education Program Working Group and involved >

60,000 subjects. Of course, increases in BMI or crude weight also tend to raise blood pressure and CVD.

Another study involving very lean Chinese subjects found that hypertension increased progressively for BMIs from 18 to 24 kg/m^2 (Hu et al, 2000). In addition, a higher BMI was directly correlated with increased levels of total cholesterol, triglycerides, and fasting glucose. In contrast, beneficial high density lipoprotein cholesterol (HDL-C) declined with increasing BMI. Higher blood pressure and a poor lipid profile are important factors in promoting CVD morbidity and mortality.

Sinaiko et al (1999) conducted a longitudinal study of various childhood parameters and their relationship to young adulthood. They found that initial childhood (7.7 y) weight, BMI and height were correlated with young adult weight, BMI, and height. In addition, fasting insulin, lipids and systolic blood pressure were positively related to all body size parameters in children and adults. They concluded that weight gain in excess of normal growth during childhood increases adult CVD risk.

Another possible reason that shorter people have greater life spans is their relatively larger organ size. Women also have relatively larger organs compared to men. Exceptions include the heart and lungs which are almost directly proportional to body weight. Censi et al (1998) reported that shorter people have about 13% higher organ weight in relation to fat-free mass. As people age, organ cells die or lose functional capacity. However, since shorter people have relatively larger organs and lower body demands, they may retain greater functional capability in terms of their body mass as they age if the rate of cell loss is similar to that of taller people. In addition, since the absolute organ masses are generally lower due to fewer cells, the opportunity for cancer incidence with age appears to be smaller.

In conclusion, promoting higher BW and subsequent larger body size (height and weight) through increased nutrition does not necessarily lead to better health prospects but can lead to overweight and obesity in childhood, adolescence and adulthood. The belief that fat children will outgrow their condition is not validated by experience: overweight or obese children tend become overweight adults (Alexander, 2001). In addition, early food habits are carried forward through one's lifetime unless concerted corrective actions are taken (Hoppe, et al 2004B; Knoops et al, 2004).

Dangers of Excessive Protein Intake

Protein is an essential macro nutrient and ~ .8 g/kg (.36 g/lb) of body weight per day (d) is needed by the average child and adult although athletes may require 1.2 g/kg/d (Samaras, 1994). However, a number of studies have found a correlation between protein intake and obesity and chronic disease. One study found that a high intake of protein up to 8 y of age resulted in obesity at 8 y (Metges 2001). Skinner et al (2004) also reported that the average intake of protein and fat between 2 and 8 y of age positively correlated with BMI at 8 y. A similar finding was made earlier by Rolland-Cachera et al (1995) who reported that protein at 2 y of age was the only nutrient tied to obesity at 8 y of age. In contrast, Skinner et al (2004) reported that the average carbohydrate intake between 2 and 8 y of age was inversely related to BMI at 8 y.

An obesity study in Greece involved over 27,000 men and women 25-82 y of age (Trichopoulou et al, 2002). The researchers found that protein was the only macro nutrient responsible for increasing BMI. Fat and carbohydrates had no apparent effect unless they were consumed in high amounts thereby increasing total energy intake. The results of the study supported previous studies showing that moderate carbohydrate intake does not promote obesity. Of course, excessive high-glycemic or refined carbohydrates can promote obesity especially in sedentary populations.

Grant (1998) described a number of ecologic studies that have found a positive correlation between protein intake and increased mortality. He reported that one of the earliest ecologic studies by Yudkin found ischemic heart disease was positively correlated with animal protein and fat. Yerushalmy and Hilleboe also found animal protein, total calories, animal fat and total fat had high positive correlations with CHD mortality. In addition, carbohydrates had a positive but low correlation with CHD, but vegetable fat and vegetable protein had small to moderate negative correlations.

In a comprehensive review of maternal nutrition and perinatal survival, Rush (2001) reported that "There is repeated and consistent evidence that high density protein supplements....are toxic." This finding applies to protein levels substantially exceeding 20% of total energy intake for pregnant women. (Normal levels of protein intake as a percentage of total caloric intake range from 10 to 20%, except for calorie restricted diets, which may require higher intakes of protein to satisfy lean body mass needs of ~ .8 g/kg/d.)

In addition, a large Chinese study found that there is a direct correlation between the amount of animal protein eaten and the increased incidence of CHD, cancer and diabetes (Campbell et al, 1998).

Hoppe et al (2004A) found that among well-nourished adults, protein intake was positively correlated with IGF-1 which also relates to increased cancer. In addition, milk intake in children was found to be related to IGF-1 concentrations, increased growth velocity and greater height (Hoppe et al, 2004B). And high growth velocity early in life was related to increases in some chronic diseases in adulthood. Based on a review of 21 studies, IGF-1 was found to be correlated with increased risk of prostate cancer and pre-menopausal breast cancer (Renehan et al, 2004). In contrast, Miller and Austad (in press) found low IGF-1 and small size extended longevity in animals.

The lowered ability of the body to oxidize fat promotes weight gain and is tied to greater protein intake and its relation to increased insulin secretion (McCarty, 2000). By itself dietary protein results in modest rises in insulin levels but when combined with co-ingested carbohydrates, protein intake potentiates insulin response reducing fat oxidation and promoting fat storage. Insulin levels can be reduced by lower protein, fat, and calorie intake and increased regular exercise.

Singh et al (2003) reviewed the findings of six studies involving very low protein intake and concluded that a very low meat intake is correlated with greater longevity. It was found that following a low-meat diet for more than 20 y decreased mortality substantially and increased longevity by 3.6 y. However, low meat consumption did not appear to reduce mortality substantially after 90 y of age. In addition, both the Adventist Health Study and the Health Professionals Follow-up Study found beef or red meat consumption correlated with higher risk of coronary disease, especially for men (Willett, 2003).

Menotti et al (1999) reported on food intake patterns and 25-y mortality from coronary heart disease in the Seven Countries Study. They found high protein and high fat foods, such as meat, butter, milk, and cheese, were positively correlated with CHD, but fish and oils were negatively correlated with heart disease.

Age at adiposity rebound was tied to a high percentage of energy from protein at 2 y of age; e.g., the higher the per cent protein ingested, the earlier the AR (Rolland-Cachera et al, 1995). In developed countries, by age 1 y the infant diet is characterized by high protein and low fat with up to 4 times the protein needed for normal nutritional needs (Rolland-Cachera et al, 1999). In contrast, mother's milk provides only 7 to 10% of its energy content in the form of protein and 50% from fat. Thus, a high protein, low-fat diet is not adapted to the infant's needs and appears to conflict with nature.

In recent years, the recommended protein intake during the first year of life has declined substantially as follows: for 1-2 mo old infants: from 2.25 to 1.99 g/kg/d; for 5-6 mo infants: from 1.30 to .92 g/kg/d; and for 9-12 mo infants: from 1.15 to .78 g/kg/d (Dupont, 2003). It is interesting to note that protein in human milk is slightly higher than recommended. (A graphical plot for protein intake from birth to 12 months of age, based on recent research, is available from Dupont's paper.)

Childhood obesity is tied to increased height and muscle mass. An early high-protein diet may account for these characteristics and for the secular trend in body parameters (increased height, muscle and fat mass over the last 100 y). Physical changes in body parameters are due to changes in hormonal status; e.g., high IGF-1 levels (concentrations), and reduced growth hormone levels are features of children with simple obesity.

Studies of mice support human findings; e.g., Rollo et al (1996) found that increasing protein intake from 23% to 38% of total calories decreased longevity by 50%.

Tallness, which is related to protein intake (Hoppe et al, 2004B; Chen et al, 1990) has also been found to increase the risk of hip fracture. Hemenway et al (1994 and 1995) found hip fracture risk increased to over 2 times for men ≥ 183 cm (6') vs < 175 cm (5'9") based on 50,000 men 40-75 y old. Women ≥ 173 cm (5'8") were also twice as likely to have a hip fracture compared to women < 157 cm (5'2"). This study was based on 90,000 women 35-59 y of age and tracked for 10 y.

Conflicting research on the relation between protein intake and calcium loss exists (Tufts University Health and Nutrition Letter, 2004A). However, Frassetto et al (2000) found a strong positive relationship between animal protein intake and the risk of hip fracture. They studied the hip fracture rates of elderly women in 23 countries based on their protein intake and found that the countries with the lowest protein intake, such as Nigeria and Papua New Guinea, had much lower hip fracture rates compared to the highest--Germany and the Scandinavian countries. For example, Germany had 200 times the risk of hip fracture compared to Nigeria. The reason for this condition appears to be the leaching of calcium from the bone when a high- protein diet produces an acidic condition in the body. However, vegetable protein appeared to have a negative correlation ($r = -.37$, $p < .04$) between its consumption and hip fracture incidence. (The correlation for animal protein and hip fracture was .82, $p < .001$.) An earlier report also noted that osteoporosis is low in the developing world although calcium intake is low as well (Loveridge and Noble, 1994).

In general, short people in the West are also less likely to break a bone on falling. Thus, it may be that small body size is a key factor in hip fracture risk with low protein being the instrument for producing shorter height and lower bone loss.

Rapid Growth does not Promote Long-Term Health

High growth rate has generally been viewed as beneficial (Hoope et al, 2004B). However, animal studies over the last 70 y and recent human findings are challenging this viewpoint. In addition, Singhal et al (2003 and 2004) recently reported that accelerated or catch-up growth has long-term negative health consequences and challenge current public health advice that promotion of rapid infant growth is desirable. For example, rapid childhood growth is tied to increased risk of CVD and diabetes in adulthood, especially for low BW children. They also found that greater than normal weight gain and linear growth during the first 2 weeks after birth correlated with increased long-term CVD independent of BW (Singhal et al, 2004). They and others also observed that catch-up growth may simply be due to over nutrition and that CVD, obesity and glucose intolerance are the result of an overweight population (Singhal et al, 2004; Ong and Dunger, 2004; Wilkin et al, 2002).

Stettler et al (2002) found rapid growth in the first 4 months of life was tied to increased adiposity at 7 y of age. Ong et al (2000) have shown rapid growth in the first 2 y of life leads to obesity at 5 y of age. And Monteiro et al (2003) found that rapid growth in weight during the first 43 months (mo) of life increased the risk of overweight and obesity in adolescence. In addition, between 20 and 43 mo, rapid height gain correlated with increased overweight in adolescence. Another study (Parsons et al, 2001) found that 33 y olds were more likely to be obese when they experienced rapid childhood growth in height during the first 7 y of life. Kimm and Obarzanek (2002) reported that obesity is often associated with accelerated increases in height in prepubertal children. Interestingly, when overweight children lose weight, they also see a decline in the rate of linear growth.

Alexander (2001) found that 70% of children who experienced rapid growth in weight and became obese as preadolescent children remained obese as adults. In addition, Hoope et al (2004A) found rapid childhood growth in height was positively correlated with IGF-1, total energy intake, total protein intake and percentage of calories from protein. Rapid growth in height was also correlated with obesity in adulthood (Kimm and Obarzanek 2001; Parsons et al, 2001). Pietilainen et al (2001) found that body size (height and weight) tracked from birth to late adolescence and was strongly influenced by the parent's body size. In addition, height at 16 y had a very strong correlation coefficient ($r = .92$) with height at 33 y of age (Ahlgren et al, 2004).

In Spain a comprehensive study of 43 provinces found that slower growth up to ~ 20 y was related to increased longevity (Holzenberger et al, 1991). This study was based on 1.3 million young male subjects and tracked survivors beyond 70 y of age. Consideration of confounding factors did not alter the results.

Must and Strauss (1999) reported that being overweight in childhood leads to substantially increased all-cause and CHD mortality in adulthood. This increased risk due to childhood or adolescent overweight was independent of adult weight, socioeconomic class, and smoking.

Normally in the West, low BW infants experience catch-up growth and large infants experience catch-down growth (Ong et al, 2000). This is probably due to overfeeding lower weight children and somewhat underfeeding larger infants. However, catch-up growth is not a universal experience because it was found that poor low BW children in Hong Kong actually experienced slower than normal growth (Luo et al, 2000). This paradox probably reflects dietary factors related to the subjects' Chinese ethnicity and socioeconomic status.

A large study on the relation of rapid childhood growth and adult cancer revealed a strong increase in incidence with above average increase in leg length during adolescence (Gunnell et al, 1998). Non-smoking cancer risk increased by 78% for each 3-4 mm (.12-.16") increase in leg length over average. This study involved 1167 children 2-8 y of age at baseline who were tracked for 60 years. Another study (Ahlgren et al, 2004) involving 117,415 Danish women found breast cancer risk was higher due to three factors: (1) high peak growth rate early in childhood, (2) tallness from 8 to 14 y of age, and (3) rapid height increase during puberty (8 to14 y).

Giovannucci et al (2004) evaluated the impact of excessive energy intake on Western cancers. Based on 47,690 male health professionals, cancer risk increased with maximal growth during childhood and adolescence.

It is well established that obesity acquired during childhood is harmful to future health and well-being (Cameron and Demerath, 2002). Therefore, rapid growth leading to infant, childhood, and adolescent obesity is a critical public health issue that deserves intensive and substantial nutritional changes.

EARLY ADIPOSITY REBOUND AND FUTURE OBESITY

Adiposity rebound (AR) occurs during childhood following a decline in BMI from 1 y of age to 4 to 8 y (Freedman et al, 2001; Freedman et al, 2002). The BMI then increases in a linear manner until nearly full growth is obtained. The AR has been used as a marker to identify the potential for later obesity because an early AR generally predicts overweight or obesity in adolescence and early adulthood. In addition, the magnitude of BMI at AR is correlated with height and BMI at age 3 y and the BMIs of the parents. Williams and Dickson (2002) also reported that earlier AR is correlated with tallness at 3 y of age. Freedman et al (2001) found that adults who reached AR before 5 y of age had an adult BMI 4 to 5 kg/m^2 higher than those who reached AR after 7 y of age. A recent study found that the higher a child's BMI at 8 y, the lower the age of AR (Skinner et al, 2004). In addition, BMI at 2 y of age was found to have a positive correlation with BMI at 8 y.

Extensive research by Cole et al (2004) found that both the BMI centile and BMI centile crossing during early childhood are separately related to early AR. While AR is a marker for later obesity, parental and early-life BMIs are predictive of future obesity as well (Cameron and Demerath, 2002).

Nutrition, in combination with parental BMI, is an important driver of AR. Therefore, increased total calories, total protein and percent of calories from protein influence the pace of childhood growth. Hoppe et al, 2004A) reported that infants consume 3 to 4 times the protein needed for normal growth. In addition, a high food intake predicts dietary consumption in the future because Hoppe, et al (2004A) found that excess food intake tracks from 9 mo to 10 y. Other studies have also found that dietary habits don't change much over one's life without a concerted effort (Knoops et al, 2004).

As mentioned, excess caloric intake is also a factor in early AR. Based on findings of the USDA Nutrition Research Center at Houston, actual food needs of infants and young children are 11-18% less than officially recommended (Cannon, 2004). Skinner et al (2004) also reported that overestimation of children's energy needs sets the stage for childhood obesity. They reported that infants and children up to the age of 3 y may be receiving 10 to 20% more calories than needed. In addition, energy recommendations for children 4 to 6 y of age are 25% higher than needed. Thus, while early nutrition is very important to good health, excessive nutrition sets the stage for later excess weight and appears to play a major role in precocious AR and the etiology of the obesity epidemic.

Early Puberty has a Number of Potential Problems

Over the the last 200 to 250 y, the Western world has seen a progressive decline in the age of puberty (Cordain et al, 2003). In females, increased height and obesity appear to be correlated with early age of menarche. In the West, menarche has declined from about 16 to 13 y (Stoll, 1998). Over a shorter period (births between 1920 and 1986), a similar decrease was seen by South Korean women who experienced a decreased age of menarche from 16.8 to 12.7 y while height increased from 149.23 cm to 161.75 cm (Hwang et al, 2003).

A more recent report on developed countries indicates that puberty starts between 8 and 13 y and occurs ~ 9 months earlier in girls than in boys (Okasha et al, 2002). Stoll (1998) attributes advancing onset of puberty with a high fat, low fiber diet and lack of exercise. Another study found that larger BW was correlated with earlier menarche when compared to lower BW infants with the same growth rate (Silva et al, 2002). Other factors for early puberty are high birthlength and thinness in early childhood combined with rapid childhood growth (Adair, 2001). Early maturation, as indicated by skeletal growth (height) or age at peak height velocity, is related to obesity in early adulthood for both sexes (Okasha et al, 2002). Karlberg et al (2002) and Ahlgren et al (2004) also identified over nutrition, obesity and greater height as triggers for early puberty. In addition, early puberty and a high growth velocity are related to maternal weight before and during pregnancy as well as to postnatal nutrition.

Early puberty is correlated with earlier onset of hyperinsulinaemia (Stoll, 1998). In addition, early puberty tends to promote earlier cessation of height growth, and males that reach final height before 18 y have ~ 3 times the risk of prostate cancer as those that reach their final height ≥ 20 y (Habel et al, 2000). In the absence of caloric restriction, slower growth that lasts longer while maintaining a low BMI may have a beneficial effect even though a greater height may be attained. Breast cancer risk also increases progressively with each year of earlier menarche. These findings are consistent with animal studies which report that delayed puberty and reduced fertility correlate with greater longevity (Bartke et al, 2001B).

Early menarche also results in lower sex hormone binding globulin (SHBG) concentration which is undesirable in terms of increased insulin resistance and greater CVD, all-cause and cancer mortality (Gates et al, 1996). Conversely, the later the age of menarche, the higher the SHBG level.

Another reason for delaying puberty is that it allows children more time to mature before they are subjected to the forces of sexual hormones. Later puberty would enable our youth to learn more and develop mentally with less distraction from sexual stress so as to better adapt to our complex society.

Cellular, Hormonal, and Oxidative Problems with Excessive Nutrition

Excessive nutrition has obvious negative ramifications in terms of increased CVD, diabetes, cancer, and osteoporosis and reduced longevity. However, deterioration of functional systems and lower longevity start at microscopic levels, such as increased free radicals, DNA and mitochondria damage, cellular failure, and changes in the concentrations of various hormones and protein agents. While the exact mechanisms of the deleterious effects of changes in the levels or concentrations of these factors are subject to further research, their direct or indirect impact on health seems well established, especially when animal research is considered. A summary of some of these factors follows.

Cell Proliferation

Aviv et al (2003) reported that rapid cell proliferation due to robust growth early in life can accelerate the erosion of telomeres (ends of chromosomes) and thereby promote premature aging in later life. Stindl (2004) also observed that the difference in life expectancy between men and women was related to their smaller number of body cells. The problem with increased cell proliferation is that our cells have a limited capacity of replication; e.g., many cell types are limited to 50 replications before they die. (Some types of cells have over 100 potential replications.) Hayflick (1996) reported that we use up close to our limit (~ 40) of cell replications by the time we reach adulthood. Taller bodies also have more cells and thus have a higher probability of cancer initiation (Okasha et al, 2002). In addition, a larger body needs more cell replications simply to maintain this larger body mass. Thus, fewer cell duplications are left for replacement of defective, dying or dead cells during old age. Another problem with having large bodies with more cells is that they provide opportunity for cancer initiation due to free radicals created as a by product of metabolic processes and exposure to cosmic rays, x-rays, UV, and toxins from food or the environment.

It should be noted that Snell dwarf mice have delayed aging which is attributed to their small body size, low metabolic rate and temperature, and reduced number of mitotic divisions during growth resulting in preservation of replication potential in old age (Bartke, 2000). However, some researchers do not believe reduced metabolic rate and lower body temperature are major factors in increased longevity due to caloric restriction (Masoro, 2000). For example, Miller and Austad (in press) reported that small dogs have higher metabolic rates than big dogs yet live substantially longer. A possible explanation for this may be that the life prolonging benefits of smaller size more than offset the negative aspects of high metabolic rate.

Nutrition plays a key role in the proliferation of body cells which result in the development of the mature human body. Hoppe et al (2004A) reported that total energy intake, total protein intake and percent of calories from protein promote increased height and weight. In contrast, food restriction slows DNA synthesis and the rate of cell division and thus slows telomere reduction (Cameron and Demerath, 2002). Note that larger bodies have more cells than smaller one's because childhood and adolescent growth is primarily through increased numbers of cells rather than from increased cell size. However, after adulthood is reached, body size is increased mainly by larger muscle or fat cell size.

Major body size promotion hormones related to nutrition intake include growth hormone, insulin and insulin-like growth hormone-1 (IGF-1) (Cordain et al, 2003; Hoppe et al, 2004B; Miller and Austad, in press). In addition, larger young humans and animals within the same species usually have higher levels of IGF-1. Thus, protein and calories need to be reduced to levels that ensure good health, moderate growth, and increased longevity with good health. As mentioned, Hoppe et al (2004A) reported that infants received 3 to 4 times more protein than they need, and Cannon (2004) reported that the energy intake of children is about 20% greater than needed in relation to their energy expenditure. Recent findings for adults indicate a BMI of ~ 22 is ideal in terms of low mortality. Since most middle aged Westerners exceed this value, they are consuming more nutrition than needed and are promoting cellular growth in size if not quantity. Until definitive research provides guidance for an ideal dietary content and caloric level, some guidelines can be obtained from research in Hawaii, Okinawa, Kitava, Greece and China (Willcox et al, 2001; Lindeberg and Lundh, 1993; Chrysohoou et al, 2004; Chen et al, 1990). While each population has a unique diet, some common characteristics can be derived. such as, high plant and fruit intake, high fiber, low saturated fat and meat, moderate fish, and high complex carbohydrate intake. An important feature of this diet is its ability to neutralize free radicals which create cellular damage.

Intake of refined carbohydrates is also generally low in most healthful diets, and the Greeks consume relatively high-fat diets based on olive oil and cheese. Unfortunately, as populations improve their economic conditions, their diets tend to shift towards more Western dietary practices (more meat, fat, refined carbohydrates, sugar, and harmful trans fats) with negative health consequences. Specific recommendations for improved nutrition are given in a subsequent section of this chapter.

The age-adjusted incidence of cancer of the colon, rectum, pancreas, breast, endometrium and prostate has risen sharply in the developed world during the 20th century. A major cause for this increase is energy-rich diets that are high in saturated fats and easily digested carbohydrates. This diet is tied to increased insulin and IGF-1 concentrations, which in turn are related to cell proliferation (Kaaks and Lukanova, 2001). Hoppe et al (2004A) and McCarty (2000 and 2001) have also associated increased insulin or IGF-1 with high daily intakes of protein and high-glycemic carbohydrates. Growth hormone, sex hormone binding globulin, insulin-like growth factor-1 (IGF-1), creatinine, C-reactive protein, homocysteine and cholesterol also affect CVD and longevity and are impacted by dietary practices as discussed next.

Growth Hormone

Growth hormone (GH) is the main regulator of postnatal bone growth and acts by raising IGF-1 and insulin levels. Levels of GH are sensitive to nutrition, especially amino acids, such as arginine, histidine and lysine (Loveridge and Noble, 1994). However, a high protein diet reduces GH levels and lipolysis and increases the development and maintenance of high fat stores (Rolland-Cachera et al,1999). For example, obese children are characterized by high protein intake and high IGF-1 but low GH levels. In contrast, sustained low protein intake or fasting leads to increased GH levels and decreased IGF-1.

The antiaging benefits of increased GH levels have been reported widely (Bartke et al, 2001B). However, the value of increasing GH levels of apparently healthy people is controversial because controlled experiments with GH-deficient and GH-resistant animals found that lower GH promoted longevity. In addition, an 18-y prospective French study found men who had higher levels of GH were at increased risk of premature death from CVD, cancer and all causes (Maison et al, 1998). This study was based on 864 policemen 18 to 52 y of age at baseline who were free of CVD, diabetes and glucose intolerance. However, GH levels decline with age and raising low GH levels in the elderly may have beneficial results on health and longevity. Until more definitive data are available, exercise and deep sleep are not only good for health but can also increase GH levels naturally.

Insulin and Insulin-Like Growth Factor-1

Insulin controls energy metabolism and increases the activity of IGF-1 by increasing its synthesis and by decreasing binding proteins, such as IGFBP-1. Both insulin and IGF-1 stimulate anabolic processes based on the availability of energy and basic substrates, such as amino acids. Insulin and IGF-1 promote cancer by inhibiting apoptosis (programmed cell death) and by stimulating cell proliferation (Kaaks and Lukanova, 2001; Parr, 1999). In addition, insulin and IGF-1 promote the synthesis of sex hormone steroids and inhibit the synthesis of sex hormone binding globulin (SHBG), which regulates the bioavailability of circulating sex steroids to tissues (Kaaks and Lukanova, 2001; Cordain et al 2003). Elevated levels of insulin and IGF-1 are thus tied to increased cancer experienced in the developed world (Giovannucci et al, 2004). Bartke et al (2003) and Parr (1999) have suggested that low levels of insulin may be an effective way of increasing longevity. Their conclusions are based on calorie restricted animals which have much reduced insulin and IGF-1 concentrations compared to normally fed shorter lived animals. However, the minimum levels acceptable for good health are not yet defined and too low a level of IGF-1 is no doubt harmful (Hursting et al, 2003).

Centenarians have also been found to have low levels of insulin and glucose compared to their younger peers aged > 75 to < 100 y (Paolisso et al,1995). In addition, Sinaiko et al (1999) found that fasting insulin concentration was positively correlated with triglycerides, low density lipoproteins, and total cholesterol but good cholesterol (HDL-C) was negatively correlated with insulin. Their findings applied to male, female, black and white subjects when each group was evaluated separately.

It should be noted that although protein and high-glycemic foods are often identified as drivers of insulin and IGF-1 levels, Cordain et al (2002) reported that hunter-gather diets were very high in meat intake [in proportion to other food intake] but low in high-glycemic carbohydrates. In spite of high meat intake, they were free of CVD risk factors. Possible explanations for this paradox are the consumption of low-fat animals, high intake of plants and vegetables, and relatively low total calorie intake in comparison to a high energy expenditure. In addition, high levels of monounsaturated fatty acids (MUFA) and polyunsaturated fatty acids (PUFA) and higher omega-3/omega-6 ratio, antioxidants, fiber, vitamins and phytochemicals with low salt intake were favorable for low CVD.

McCarty (2000) suggested that it is the simultaneous intake of animal protein and high-glycemic foods that cause sharp rises in insulin that subsequently lead to Syndrome X. Syndrome X (also referred to as metabolic syndrome) refers to a cluster of problems, such as elevated blood sugar, hypertension, obesity, high triglycerides, low HDL and high insulin concentration. This syndrome is a precursor of type 2 diabetes and CVD and possibly some cancers. The metabolic syndrome is also associated with insulin resistance which leads to increased hyperinsulinemia, decreased hepatic IGFBP-1 production, and increased free serum IGF-1, which then promotes accelerated growth and cell proliferation (Hursting et al, 2003).

A Mediterranean study (Esposito et al, 2004) provides some interesting findings on insulin concentration [milligrams/deciliter (mg/dL)] and diet. A 27% decrease in insulin concentration resulted after subjects followed a dietary change for 2 y. In this diet, protein intake as a percentage of total calories was kept constant, but a number of other changes were made in comparison to the diet of the control group. These changes included reduced total calories, increased fiber, lower saturated fat, increased MUFA, increased omega 3 fatty acids, reduced cholesterol, and increased whole grains, fruits, vegetables, nuts and legumes. Thus, diet, independent of the percentage of protein intake, had a substantial impact on insulin levels. However, since total calories were reduced, the absolute intake of protein was reduced as well. Giovannuci et al (2004) also found excessive energy intake and a higher BMI increased insulin levels.

Insulin-Like Growth Factor-1

As mentioned, IGF-1 is a powerful mitogen for almost all the tissues of the body (Cordain et al, 2003). It also controls programmed cell death (Hursting et al, 2003). While IGF-1 can increase cancer risk at higher levels, a low concentration can also lower bone mineral density and muscle mass. In addition, IGF-1 is strongly correlated with weight and BMI. Height is also a strong marker of IGF-1 concentration, and height is positively correlated with absolute energy consumption, total daily protein intake and percentage of energy from protein during growth (Giovannucci et al, 2004; Hoppe et al, 2004A; Hursting et al, 2003).

Sex Hormone Binding Globulin

Sex hormone binding globulin (SHBG) is an important regulator of plasma sex steroids and a sensitive marker of insulin resistance (Gates et al, 1996). Low SHBG levels are tied to increased risk of type 2 diabetes, cancer, CVD and all-cause mortality (Gates et al, 1996).

Serum levels of SHBG are reduced by higher insulin levels (Lonning et al,1995; Gates et al,1996; Chen et al, 1990; Cordain et al, 2003) which as mentioned before are related to protein and high-glycemic carbohydrate intake. Taller people also tend to have lower SHBG levels. Levels of SHBG are also affected by food types; e.g., wheat lowers SHBG while rice, fish and green vegetables appear to raise it. However, total food intake is negatively related to

SHBG when it increases weight and BMI because weight and BMI are independently and negatively correlated with SHBG.

Insulin-Like Growth Factor Binding Proteins

The level of IGFBP-1 controls the availability of IGF-1, which promotes growth and cancer. Wolk et al (2004) found low levels of IGFBP-1 correlate with increased insulin resistance, obesity and CVD risk factors. They also found that as height increased in men 65 to 76 y of age, IGFBP-1 decreased with $r = -.23$, $p < .05$.

Wolk et al found that the only significant positive association between IGFBP-1 and diet was attained when carbohydrate intake was increased. They also reported that a study of women found that vegans (no animal products) had 20-40% higher IGFBP-1 and IGFBP-2 levels compared to women on vegetarian or meat diets. In addition, a higher insulin level was found to be the strongest individual predictor of lower IGFBP-1 concentration in men 55-64 y of age.

Another study (Giovannucci et al, 2004), based on a 57 to 92 y cohort, found insulin secretion positively correlated with greater intake of red meat, dairy fat, refined grains, saturated fat and trans fat and lower intake of whole grain products and fiber. Since IGFBP-1 and insulin are inversely related, their findings indicate that IGFBP-1 can be increased by reducing the consumption of animal products (red meat and dairy) and processed foods.

Decreased IGFBP-3 levels can also contribute to unregulated cell proliferation (Cordain et al, 2003). In addition, reduced IGFBP-3 levels correlate with raised insulin levels and high-glycemic carbohydrate ingestion. However, high levels of IGFBP-3 are also undesirable because they are positively correlated with pre-menopausal breast cancer in women (Renehan et al, 2004).

Serum Creatinine

Above normal serum creatinine concentration is a marker of renal function problems and increased CVD risk (Knight et al, 2004). Creatinine levels are moderately positively correlated with weight or height and are thus indirectly related to nutrition during childhood and adulthood.

C-Reactive Protein

C-reactive protein (CRP) is a marker of systemic inflammation and has been correlated with ischemic heart disease. Mendall et al (2000) found that there is a strong correlation between CRP and fatal ischemic heart disease and all-cause mortality. However, when the findings were adjusted for various other risk factors, the correlation between CRP and ischemic heart disease became small. Serum concentration of CRP was also found to be positively correlated with lower social class during childhood and greater height, BMI, smoking, insulin and fibrinogen. Since height and BMI are correlated with nutritional intake, CRP levels appear to be related to increased food intake during childhood and adulthood.

In a recent Italian Mediterranean diet study (Esposito et al, 2004), a substantial decrease (65%) in CRP concentration (mg/L) was related to changes in diet by the study participants (44 y of age) over a 2-y period. The key differences between the baseline diet (cohort 1) and intervention diet (cohort 2) were lower caloric intake, increased fiber, increased omega 3 fatty acids, reduced saturated fat, increased MUFA, reduced cholesterol, and increased olive oil, fruits, vegetables, nuts and legumes, and whole grains.

In view of a substantial number of studies published on CRP, its importance to the medical community has increased over the last 10 y. However, whether CRP is a cause or only a symptom of other risk factors is debatable. Regardless of this debate, exercise and avoiding excess food intake can reduce one's BMI and CRP and is a prudent policy for reducing systemic inflammation (which is gaining acceptance as an important cause of CVD).

Homocysteine

Homocysteine (Hcy) is present in many foods as well as synthesized by the body itself. High levels are undesirable because they increase the risk of heart attack and stroke. In addition, dietary deficiencies promote high levels of Hcy.

Homocysteine is created from the amino acid methionine and can be converted back to methionine with the presence of folic acid and vitamins B6 and B12. The normal serum level of Hcy is 5-15 micromoles/L although below 10 micromoles/L is the safest level. The Japanese and French have very little CVD and their Hcy levels are 7-8 micromoles/L (Grant, 1998). Milk more than any other animal protein increases Hcy concentrations because milk has lower levels of B vitamins which are needed to convert Hcy into useful biological products (Grant, 1998). In addition, the combination of lactose, calcium, and Hcy from non-fat milk may also promote calcification of arteries.

Among a population of elderly (~ 75 y) men and women, those with the highest intake of selenium had a 63% lower risk of having high Hcy concentration (> 14 micromoles/L) (Gonzalez et al, 2004). In contrast with other studies, protein intake was negatively correlated with Hcy for men, but this relationship disappeared when adjusted for selenium levels.

A Greek study (Chrysohoou et al, 2004) involved evaluation of the Mediterranean diet in relation to inflammatory markers in healthy adults (18-89 y of age). A tertile scoring system was established for comparing adherence to the Mediterranean diet with the highest (third) tertile marking the closest adherence to the diet. A 15% reduction in Hcy concentration was found in the third tertile. The highest compliance diet (third tertile) included the greatest reduction in high-fat dairy products, meat, and poultry, and the largest increase in the consumption of fish, fruits, nuts, non refined cereals, pulses and vegetables.

To maintain lower levels of Hcy, a diet rich in vegetables, fruits, and fortified grains is desirable. Supplements of folic acid, B6, and B12 are also recommended (Harvard Men's Health Watch, 2000).

Cholesterol

Cholesterol is a lipid found almost exclusively in animal foods and is synthesized continuously in the body primarily by the liver and adrenal cortex. Serum cholesterol facilitates the absorption and transport of fatty acids and is a precusor for the synthesis of vitamin D at the skin's surface, various steroid hormones, and sex hormones. High levels are strongly correlated with CVD and all-cause mortality. Unfavorable concentrations are generally viewed as ≥ 200 mg/dL (5.17 mmol/L). A large study by Stamler et al (2002) found that CVD and all-cause mortality increased substantially (> 8 times for CVD and > 2 times for all cause) in a graded fashion with increasing cholesterol levels starting with < 160 mg/dL, the most favorable level. Men with < 200 mg/dL had a greater life expectancy of 3.8 to 8.7 y compared to men with ≥ 240 mg/dL.

Cholesterol levels are lowered by reducing animal food consumption and increasing the intake of plant-based and high fiber foods. In addition, lowering weight is important because cholesterol tends to decrease linearly with decreasing BMI.

DNA Oxidative Damage

Oxidative damage is often cited as an important factor in chronic disease (Giovannelli et al, 2002). In an Italian study, researchers found height was strongly correlated with DNA oxidative damage. Taller height is of course tied to greater childhood and adolescent nutrition (Hoppe et al, 2004A). The intake of simple sugar was also strongly correlated with DNA oxidative damage. In addition, subjects in the lowest tertile of daily energy expenditure had the lowest DNA damage although physical activity per se was not tied to oxidative damage. A possible explanation may be that a larger body produces more free radicals due its larger energy consumption and may have a higher body temperature. Hrdlicka (1929) found taller people had a slightly higher body temperature (.3 degree F) at rest compared to shorter people. In addition, taller people have less surface area in proportion to their body weight which makes it harder to lose the heat generated during physical activity and is likely to raise their body temperature substantially compared to shorter persons. Thus, a higher rate of DNA damage would result because of this increased temperature. Giovannelli's study of human oxidative damage is supported by animal studies which have shown elevated free radicals correlate with reduced lifespan in large, lean mice (Rollo et al, 1996)

Lower Energy Intake Reduces Mortality and Increases Longevity

Mortality and longevity are related to energy intake. In the biology of aging, increased longevity due to caloric restriction is a very robust finding. Over 1500 papers (Bartke et al, 2001A and B; Barzilai and Gupta, 1999; Hursting et al, 2003; Masoro, 2000) dealing with protozoa, yeast, insects, nematodes, mice, rats, pigs, cows, dogs, non-human primates, and humans have shown a strong correlation between reduced caloric intake and lower mortality and increased longevity. Caloric restriction results in reduced body size, including reduced adiposity which is also an important longevity factor (Barzilai and Gupta, 1999; Barzilai and Gabriely, 2001). Studies with monkeys have been in progress for over 25 y and preliminary results support the findings from other animals (Bodkin et al, 2003; Miller and Austad, in press; Lane et al, 1995). Although caloric restriction has the greatest effect on increasing longevity, protein restriction is also a significant factor (Rollo et al,1996; Cameron and Demerath, 2002). While animal data are convincing, human studies involving caloric restriction have been sparse. However, there is some convincing human evidence supporting the benefits of caloric restriction without malnutrition (Samaras and Elrick, 1999). A few examples are summarized next.

Hawaiian Japanese

Energy restriction was found to reduce mortality in a study of 1915 healthy Japanese-American nonsmokers. Protein and carbohydrate intake as a percentage of total calories increased slightly with decreasing energy intake while fat intake decreased.

These men were 45-68 y at baseline and tracked for 36 y (range: 81 to 104 y). Men who consumed less than an average of 9254 kJ/d (2214 kcal/d) had the lowest mortality. Mortality was about 18% higher for an average intake of 13,426 kJ/d (3212 kcal/d) vs 7867 kJ/day (1882 kcal/d). However, those who consumed less than 4076 kJ/d (975 kcal/d) saw an increase in mortality. As expected, lower caloric intake was related to lower height at baseline. Based on BMI and weight data given by Willcox et al, the shortest men averaged 161.3 cm and the tallest 163.8 cm.

The lowest mortality was found to occur for three daily energy categories: 5208 kJ (1246 kcal), 6136 kJ (1468 kcal), and 8005 kJ (1915 kcal). If the two lowest energy intakes, 5208 kJ (1246 kcal) and 6136 kJ (1468 kcal), are applied to mainland males and adjusted for greater height and weight, the optimum mainland intake would be 6249 kJ (1503 kcal) to 7403 kJ (1771 kcal), which is close to the average found for the Madrid study discussed next. [The mainland caloric intake was increased to reflect the greater mainland height for white men in same age group; e.g., 162.6 cm (5'4") for Hawaii vs 172.7 cm (5'8") for US mainland and a weight of 63 kg (139 lb) vs 76 kg (167 lb) using the same Ponderal Index value for Hawaii and the mainland.]

Madrid

A nursing home in Madrid put a group of elderly patients on a low calorie diet every other day with excellent results compared to patients fed normally everyday (Vallejo,1956). Their mortality dropped by 50% over 3 y. The study was based on 180 people over 65 y of age. Ninety were put on reduced energy intake every other day while the other 90 ate regular meals daily. Those on regular meals consumed 9640 kJ (2306 kcal) per day and those on restricted meals received 3720 kJ (890 kcal) one day and 9640 kJ (2306 kcal) the other day. Thus, the low calorie cohort averaged 6680 kJ/d (1598 kcal/d).

While body size was not reported, they were no doubt small compared to US standards based on the historically low height of the Spanish (Samaras et al, 2004; Holzenberger et al, 1991). Thus, Americans in this age group would probably require a somewhat higher food intake.

Okinawa

Okinawa has been studied for many years and has produced strong evidence supporting the benefits of lower caloric consumption. It should be noted that Okinawan food volume is not small in spite of their low caloric intake. Thus, a large volume of food eaten does not necessarily promote obesity if low-energy density food is eaten and activity levels are high.

Okinawa is an island off mainland Japan where the children have received a substantially lower (~ 40%) caloric intake compared to the mainland. Adults also consumed a ~ 20% lower energy intake compared to the mainland (Kagawa, 1978; Willcox et al, 2001) Traditionally, Okinawans have been short and light and lead active lives into their 90s. They currently have the highest percentage of centenarians in the world. Unfortunately, in recent years the diet has become more Westernized and the health of young and middle aged Okinawans has declined.

WW II

Dietary restrictions were common in Europe during WW II. Sugar, fat and meat were reduced and plant and fiber foods increased substantially. Yet, dental caries, diabetes, and CHD declined sharply during the war but then rose shortly after the war ended when food patterns returned to prewar levels (Walker et al, 1994).

Masai

12-y study (Murray et al, 1993) was conducted of Masai natives in Africa. The health and survival of 403 undernourished and 386 normally nourished Masai 18-35 y of age were evaluated. It was found that those Masai on a lower calorie diet experienced substantially less illness (e.g., infections, especially malaria) and lost fewer days from work. Over a 12-y period, undernourishment had no impact on overall mortality. Since the age of these natives was 30-47 y at the end of the study, it appears that reduced food consumption reduced morbidity but not mortality in this relatively young age category.

Caloric Restriction and Reduced Body Size

Caloric restriction affects a number of biological parameters as well as reducing body size. Researchers have for years argued about whether caloric restricted animals live longer due to reduced calories or reduced body size. However, Rollo (2002) found that reduced growth is a more powerful factor than caloric restriction in extending longevity. In addition, Bartke's dwarf mice (genetically modified) eat as much as they want but still live longer than normal size mice (Bartke et al, 2001A; Bartke, 2000). However, they also found that the addition of caloric restriction extends the longevity of these small mice even further.

Research on dog longevity also indicates that small body size is the key to living longer. For example, small dogs live longer than big dogs without being put on calorie restricted diets (Miller and Austad, in press; Samaras et al, 2002C).

As mentioned, caloric restriction every other day appeared to work well in Madrid; however, Bartke et al (2003) cautioned that food restriction for the elderly could be harmful. For example, many of the elderly do not have well-balanced diets and may even be undernourished due to poor appetite, digestive problems and lack of funds to buy all the food they need. In addition, a diet that promotes loss of muscle mass is undesirable.

What should Children and Adults Eat?

Mother's milk includes only 7% of its energy from protein and 50% from fat (Rolland-Cachera et al, 1999). Because breast milk contains a lower protein level compared to formula milk, breast fed infants grow at a slower rate than formula fed infants. A large (>17,000 infants) study recently confirmed the growth accelerating impact of formula and cow's milk consumption on weight and length gain throughout infancy (Kramer et al, 2004). Researchers also found that as formula or milk intake increased, a greater increase in growth occurred with the biggest response seen at 3 to 6 months. However, the study did not provide strong support for the thesis that protein consumption is a major factor of growth and obesity in infancy as discussed previously.

Breast feeding and delayed introduction of solid foods appear to be beneficial to childhood health and reduced adult disease (Wilson et al, 1998). They also found that respiratory illness in children who received breast milk exclusively for at least 15 weeks (wk) was consistently lower than those that received partial breast feeding. Other benefits of breast feeding include: improves development of the infants's nervous system, protects against certain bacterial and viral infections and decreases incidence of allergies, ear infections, and diarrhea (AWHONN, 2004).

Many studies have found breast feeding reduces the risk of overweight and obesity in adulthood (Hoppe et al, 2004B). In addition, solid feeding before 15 wk was tied to increased risk of childhood wheezing, greater risk of higher percentage body fat, and greater weight. Systolic blood pressure was also 3.5 mm Hg higher in exclusively bottle fed children vs breast fed children.

It should be noted that WHO now recommends exclusive breast feeding for 6 mo for healthy mothers and full term infants (Dupont, 2003). In addition, it encourages breast feeding up to 2 y of age if diversified high-quality complementary foods are started at the 7th month.

It is interesting to note that breast fed infants are leaner than formula fed babies. In addition, they have lower body temperature and metabolic rate (Dupont, 2003). Since lower body temperature and metabolic rate are believed to promote greater longevity, this may be a further advantage of breast feeding.

In view of the over 26 million children who eat USDA school lunches on a daily basis, education is the foundation for any effort to improve the nutrition of our nation. Demas (1998) reported that research has clearly connected diet with chronic disease, such as CHD, some cancers and type 2 diabetes. She also pointed out that unless children learn about good nutrition early, trying to correct the problem in adulthood is exceptionally difficult. By providing classroom education on nutrition and hands-on cooking experiences at school, Demas (1998) found an effective way to teach children about low-fat diets and to expose them to a variety of healthful foods that they would rarely experience. She reported that experiential classroom learning about food was effective in winning acceptance of diverse, low fat foods in the school lunch program. Children in a 9-mo program ate up 3 to 20 times more new foods than a control cohort of children. Thus, early education in nutrition is a key factor in improving the eating habits of all school children.

It should be noted that vegan diets (no animal foods) lower serum lipid and IGF-1 concentrations, lower body weight substantially, reduce height, delay puberty, and decrease the risk of common Western diseases (McCarty, 1999).

Several studies have shown that people in countries following the Mediterranean diet have longer and healthier lives than those in other industrialized countries despite high levels of smoking, gaps in available health services, and relatively high BMIs averaging 26-27 kg/m^2 (Chrysohoou et al, 2004). The recommended diet includes consumption of the following:

1. Non refined cereals and products (whole-grain bread and pasta, brown rice, etc./d)
2. Fruits (4 to 6 servings/d)
3. Vegetables (2 to 3 servings/d)
4. Olive oil (as the primary lipid intake)
5. Nonfat or low-fat dairy products (1 to 2 servings/d)
6. Fish, poultry, potatoes, olives, pulses and nuts (4 to 6 servings/wk)
7. Eggs and sweets (1 to 3 servings/wk)
8. Red meat and meat products (4 to 5 servings/mo)
9. Wine (1 to 2 glasses/d)
10. Moderate fat intake and a high monounsaturated to saturated fat ratio

Traditional societies with very low CVD have low sodium intake (Lindeberg and Lundh, 1993). This is desirable because excessive salt intake can be harmful to health by increasing blood pressure or promoting osteoporosis. The study, Dietary Approaches to Stop Hypertension (DASH), found that sodium reductions from 3000 mg/d, to 2300 mg/d to 1200 mg/d cut blood pressure with each downward step in salt (UC Berkeley Wellness Letter, 2001). Most Americans eat far too much salt and the ramifications include increased risk of stroke and heart attack independent of blood pressure. The elimination of salt shakers and careful screening of food product labels can help achieve a low level of salt intake. More work is also needed to get restaurants to reduce the use of salt in their food.

The HALE project found that the Mediterranean diet can reduce the mortality of the elderly when combined with other lifestyle factors (Knoops et al, 2004). Men and women 70 to 90 y of age at baseline were tracked for 10 y and had 1/3 the all-cause, CVD and cancer mortality of those who did not follow the diet and other lifestyle factors (non-smoking, moderate alcohol level, and regular exercise). The diet was similar to the list of 10 items previously described.

A plant-based diet that is similar to the traditional Mediterranean diet has been shown to be the most important factor for survival based on the Food Habits in Later Life Cross-sectional Study conducted by the International Union of Nutritional Sciences (Darmadi-Blackberry et al, 2004). This diet favors fish and shellfish, legumes, olive oil and a high ratio of mono to unsaturated fat. Among the elderly, each 20 g increase in the intake of legumes reduced mortality risk by 7-8% and was the most important predictor of survival. A longitudinal study in Europe called HALE also found that the Mediterranean diet promotes longevity (Tufts University Health and Nutrition Letter, 2004B). Their findings support the health benefits of a diet high in fish, whole grains, fruit, nuts, vegetables and olive oil combined with a low intake of meat, dairy products, and PUFA. Moderate alcohol intake and daily exercise also reduced mortality. (Although not specified in the study, moderate alcohol intake is usually defined as ≤ 2 glasses of wine/d.)

The Seventh Day Adventists (SDA) have been studied for many years and a recent report found California SDA males live about 7.3 y longer than the average white California man and SDA women live 4.4 y longer than white California women (Willett, 2003). Their diet shuns red meat and emphasizes nuts and a vegetarian diet. (Exercise and avoidance of smoking and overweight are also important independent factors in their greater longevity.)

Does a high animal product diet increase the risk of chronic disease and obesity? On the one side, hunter gathers had a high intake of protein and appeared free of type 2 diabetes and CVD (Cordain et al, 2002). On the other side, early 20th century Asian and African populations ate high levels of unrefined carbohydrates and low levels of animal protein and were free of chronic diseases of the West. A study involving the increase in animal protein of low-protein vegetarians found that insulin levels were increased indicating that animal protein plays a role in the Western obesity epidemic (McCarty, 2000).

Menotti et al (1999) reported on food intake patterns and 25-y mortality from CHD in the Seven Countries Study. They found meat, butter, milk, cheese, margarine, and lard had a positive correlation with CHD. In addition, vegetables and fish were negatively correlated with CHD but sugar and pastries were positively correlated. Consumption of oils and alcohol had a negative correlation with CHD. Other beneficial foods included legumes, which had a large negative correlation with mortality, and cereals and vegetables which were also negatively correlated with mortality.

In one of the most comprehensive studies ever published on diet and mortality, Campbell and Chen (1994) reported that even small intakes of animal foods are tied to increases in cholesterol level and increased chronic degenerative disease mortality rates, such as cancer and heart disease. Other studies have found long-term vegan diets promote leanness and are associated with low fasting insulin levels (McCarty 2001). Campbell and Chen found that the more protein (especially animal) the Chinese ate, the more they suffered from chronic disease. Thus, the previous list of 10 dietary recommendations could result in even lower mortality if animal product consumption was further reduced. However, avoidance of animal products would require folic acid, B6 and B12 supplements.

Conclusion

The entire life cycle starting with the mother's diet and body weight needs to be addressed to deal with the obesity epidemic. Excess weight before conception, high pregnancy weight and emphasis on nutritional intake to increase birth and neonatal weight are most likely important initiators of later obesity.

Rapid or accelerated growth appears to promote chronic disease in adulthood, and increased height and body weight are also risk factors for cancer and CVD. Therefore, attaining maximum growth and development of children should be reduced in emphasis, and the consumption of a plant-based diet promoted more strongly, starting early in childhood.

A recent recommendation by WHO for a healthful BMI is now 21 to 23 kg/m^2, which is substantially lower than usually recommended. Thus, lower caloric intake with a nutritious and balanced diet that lowers body weight also promotes greater health and longevity.

References

Adair LS. Size at birth predicts age at menarche. *Pediatrics* 2001, 107, *http://www.pediatrics.org/cgi/content/full/107/4/e59.*

Ahlgren M; Melbye M; Wohfahrt J; Thorkild I; Sorensen A. Growth patterns and the risk of breast cancer in women. *New England Journal of Medicine* 2004, 351,1619-1626.

Alexander JK. Obesity and coronary heart disease. *The American Journal of the Medical Sciences* 2001, 321, 215-224.

Amador M; Bacallao J; Hermelo M. Adiposity and growth: relationship of stature at fourteen years with relative body weight at different ages and several measures of adiposity and body bulk. *European Journal of Clinical Nutrition* 1992, 46, 213-219.

Anderson JP; Ross JA; Folsom AR. Anthropometric variables, physical activity, and incidence of ovarian cancer. *Cancer* 2004, 100, 1515-1521.

Andersson SW; Bengtsson C; Hallberg L; Lapidus L; Niklasson A; Wallgren, Hulthen L. Cancer risk in Swedish women: the relation to size at birth. *British Journal of Cancer* 2001, 84, 1193-1198.

Aviv A; Levy D; Mangel M. Growth, telomere dynamics and successful and unsuccessful human aging. *Mechanisms of Ageing and Development* 2003, 124, 829-837.

AWHONN Lifelines. Promoting breast feeding. *AWHONN Lifelines* 2004, 8, 366-367.

Bartke A. Delayed aging in Ames Dwarf mice. Relationships to endocrine function and body size. In: Hekimi Z, editor. Results and Problems in Cell Differentiation. The Molecular Genetics of Aging Vol 29. *Berlin Heidelberg: Springer-Verlag;* 2000; pp 181-202.

Bartke A; Wright JC; Mattison JA; Ingram DK; Miller RA; Roth GS. Extending the lifespan of long-lived mice. *Nature* 2001A, 414, 412.

Bartke A; Coschigano K; Kopchick J; Chandrashekar V; Mattison J; Kinney B; Hauck S. Genes that prolong life: relationships of growth hormone and growth to aging and life span. *Journal of Gerontology: Biological Sciences* 2001B, 56A, B340-B349.

Bartke A; Chandrashekar V; Dominici F; Turyn D; Kinney B; Steger R; Kopchick JJ. Insulin-like growth factor I (IGF-I) and aging: controversies and new insights. *Biogerontology* 2003, 4, 1-8.

Barzilai N; Gupta G. Revisiting the role of fat mass in the life extension induced by caloric restriction. Journal of Gerontology: Biological Sciences 1999, 54A, B89-B96.

Barzilai N; Gabriely I. The role of fat depletion in the biological benefits of caloric restriction. *Journal of Nutrition* 2001, 131, 903S-906S.

Bellizzi MC; Dietz WH. Workshop on childhood obesity: summary of the discussion. *American Journal of Clinical Nutrition* 1999, 70, 173S-5S.

Bergmann RL; Richter R; Bergmann KE; Plagemann A; Brauer M; Dudenshausen JW. Secular trends in neonatal macrosomia in Berlin: influences of potential determinents. *Paediatric and Perinatal Epidemiology* 2003, 17, 244-249.

Bodkin NL; Alexander TM; Ortmeyer HK; Johnson E; Hansen BC. Mortality and morbidity in laboratory-maintained Rhesus monkeys and effects of long-term dietary restriction. *Journal of Gerontology: Biological Sciences* 2003, 58A, 212-219.

Bradley PJ. Fetal origins hypothesis in India. *The Lancet* 2001, 358, 927.

Brown LR; Renner M; Flavin C. Vital signs (Worldwatch Institute). *New York: W. W. Norton and Co.;* 1997; p 124.

Calle EE; Rodriguez C; Walker-Thurmond K; Thun MJ. Overweight, obesity, and mortality from cancer in a prospectively studied cohort of U.S. adults. *New England Journal of Medicine* 2003, 348, 1625-1638.

Campbell TC; Chen J. Diet and chronic degenerative diseases: perspectives from China. *American Journal of Clinical Nutrition.* 1994, 59 (suppl), 1152S-1161S.

Campbell TC; Parpia B; Chen J. Diet, lifestyle, and the etiology of coronary artery disease: the Cornell China Study. *American Journal of Cardiology* 1998, 82, 18T-21T.

Cameron N; Demerath EW. Critical periods in human growth and their relationship to diseases of aging. *Yearbook of Physical Anthropology* 2002, 45, 159-184.

Cannon G. Making history, and other stories. Out of the Box. *Public Health and Nutrition* 2004, 7, 3-6.

Censi L; Toti E; Pastore G; Ferro-Luzzi A. The basal metabolic rate and energy cost of standardised walking of short and tall men. *European Journal of Clinical Nutrition* 1998, 52, 441-446.

Chen J; Campbell TC; Li J; Peto R. Diet, lifestyle, and mortality in China: a study of the characteristics of 65 counties. Oxford, United Kingdom: *Oxford University press;* 1990.

Chrysohoou C; Panagiotakos DB; Pitsavos C; Das UN; Stefanadis C; Adherence to the Mediterranean diet attenuates inflammation and coagulation process in healthy adults. *Journal of the American College of Cardiology* 2004, 44, 152-158.

Cole TJ. Children grow and horses race: is the adiposity rebound a critical period for later obesity? *BMC Pediatrics* 2004, 1-7. http://www.biomedcentral.com/1471-2431/4/6.

Cordain L; Eaton SB; Brand Miller J; Mann N; Hill K. The paradoxical nature of hunter-gather diets: meat-based, yet non-atherogenic. *European Journal of Clinical Nutrition* 2002, 56 Suppl 1, S42-S52.

Cordain L; Eades MR; Eades MD. Hyperinsulinemic diseases of civilization: more than just Syndrome X. *Comparative Biochemistry and Physiology Part A* 2003, 136, 95-112.

Damon A. Secular trend in height and weight within old American families at Harvard, 1870-1965. *Amercan Journal of Physical Anthropology* 1968, 29, 45-50.

Darmadi-Blackberry I; Wahlqvist AO; Kouris-Blazos A; Steen B; Lukito W; Horie Y; Horie K. Legumes: the most important dietary predictor of survival in older people of different ethnicities. *Asia Pacific Journal of Clinical Nutrition* 2004, 13, 217-220.

Davey Smith G; Hart C; Upton M; Hole D; Gillis C; Watt G; Hawthorne V. Height and risk of death among men and women: aetiological implications of associations with cardiorespiratory disease and cancer mortality. *Journal of Epidemiology and Community Health* 2000, 54, 97-103.

de Simone G; Devereux RB; Daniels SR; Mureddu GF; Roman MJ; Kimball TR; Greco R; Witt S; Contaldo F. Stroke volume and cardiac output in normotensive children and adults. *Circulation* 1997, 95, 1837-1843.

Demas A. Low-fat school lunch programs: achieving acceptance. *American Journal of Cardiology* 1998, 82, 80T-82T.

Drury RAB. The mortality of elderly Ugandan Africans. *Tropical and Geographical Medicine* 1972, 24, 385-392.

Dupont C. Protein requirements during the first year of life. *American Journal of Clinical Nutrition* 2003, 77(suppl), 1544S-1549S.

Eaton SB; Konner M; Shostak M. Stone age in the fast lane: Chronic degenerative diseases in evolutionary perspective. *The American Journal of Medicine* 1988, 84, 739-749.

Elrick H; Samaras TT; Demas A. Missing links in the obesity epidemic. *Nutrition Research* 2002, 22, 1101-1123.

Engeland A; Tretli S; Bjorge T. Height, body mass index, and prostate cancer: a follow-up of 950,000 Norwegian men. *British Journal of Cancer* 2003A, 89, 1237-1242.

Engeland A; Tretli S; Bjorge T. Height, body mass index, and ovarian cancer: a follow -up of 1.1 million Norwegian women. *Journal of the National Cancer Institute* 2003B, 95, 1244-1248.

Eriksson JG; Forsen T; Tuomilehto J; Winter PD; Osmond C; Barker DJP. Catch-up growth in childhood and death from coronary heart disease: longitudinal study. *BMJ* 1999, 318, 427-31.

Eriksson M; Wallander M-A; Krakau I; Wedel H; Svardsudd K. The impact of birth weight on coronary heart disease morbidity and mortality in a birth cohort followed up for 85 years: a population-based study of men born in 1913. *Journal of Internal Medicine* 2004, 256, 472-281.

Esposito K; Marfella R; Ciotola M; Di Palo C; Giugliano F; D'Armiento M; D'Andrea F; Giugliano D. Effect of a Mediterranean-style diet on endothelial dysfunction and markers of vascular inflammation in the metabolic syndrome. *JAMA* 2004, 292, 1440-1446.

Falkner B; Hulman S; Kushner H. Birth weight versus childhood growth as determinents of adult blood pressure. *Hypertension* 1998, 31[part 1],145-150.

Forsen T; Eriksson JG; Tuomilehto J; Teramo K; Osmond C; Barker DJP. Mother's weight in pregnancy and coronary heart disease in a cohort of Finnish men: follow up study. *BMJ* 1997, 315, 837-840.

Frassetto LA; Todd KM; Morris, Jr RC; Sebastian A. Worldwide incidence of hip fracture in elderly women: relation to consumption of animal and vegetable foods. *Journal of Gerontology: Medical Sciences* 2000, 55A, M585-M592.

Freedman DS; Kettel Khan L; Serdula MK; Srinivasan SR; Berenson GS. BMI rebound, childhood height and obesity among adults: the Bogalusa Heart Study. *International Journal of Obesity* 2001, 25, 539-549.

Freedman DS; Kettel Khan L; Mei Z; Dietz WH; Srinivasan SR; Berenson GS. Relation of childhood height to obesity among adults: The Bogalusa Heart Study. *Pediatrics* 2002, 109, URL: http://www.pediatrics.org/cgi/content/full/109/2/e23

Friedlander Y; Paltiel O; Deutsch L; Knaanie A; Massalha S; Tiram E; Harlap S. Birthweight and relationship with infant, child and adult mortality in the Jerusalem perinatal study. *Paediatric and Perinatal Epidemiology* 2003, 17, 398-406.

Gates JR; Parpia B; Campbell TC; Junshi C. Association of dietary factors and selected plasma variables with sex hormone-binding globulin in rural Chinese women. *American Journal of Clinical Nutrition* 1996, 63, 22-31.

Giovannelli L; Saieva C; Masala G; Testa G; Salvini S; Pitozzi V; Riboli E, Dolara P; Palli D. Nutritional and lifestyle determinants of DNA oxidative damage: a study in a Mediterranean population. *Carcinogenesis* 2002, 23, 1483-1489.

Giovannucci E; Rimm EB; Liu Y; Willett WC. Height, predictors of C-peptide and cancer risk in men. *International Journal of Epidemiology* 2004, 33, 217-225.

Gonzalez S; Huerta JM; Alvarez-Uria J; Fernandez S; Patterson AM; Lasheras C. Serum selenium is associated with plasma homocysteine concentrations in elderly humans. *Journal of Nutrition* 2004, 134, 1736-1740.

Grant WB. Milk and other dietary influences on coronary heart disease. *Alternative Medical Review* 1998, 3, 281-294.

Gunnell DJ; Davey Smith G; Holly JMP; Frankel S. Leg length and risk of cancer in the Boyd Orr cohort. *BMJ* 1998, 317, 1350-1.

Gunnell DJ; Okasha M; Davey Smith G; Oliver SE; Sandhu J; Holly JMP. Height. leg length, and cancer risk: a systematic review. *Epidemiologic Reviews* 2001, 23, 313-342.

Gunnell DJ; Oliver E; Donovan JL; Peters TJ; Gillatt D; Persad R; Hamdy FC; Neal DE, Holly JMP. Do height-related variations in insulin-like growth factors underlie the associations of stature with adult chronic disease? *The Journal of Clinical Endocrinology and Metabolism* 2004, 89, 213-218.

Habel LA; Van Den Eeden SK; Freidman GD. Body size, age at shaving initiation, and prostate cancer in a large, multiracial cohort. *The Prostate* 2000, 43, 136-143.

Harvard Men's Health Watch. *Homocysteine.* December 2000, 5, 1-8.

Hayflick L. *How and Why We Age*. New York: Ballentine Books; 1996; p. 123.

Hemenway D; Azrael DR, Rimm EB, Feskanich D; Willett WC. Risk factors for hip fracture in US men aged 40-75 years. *American Journal of Public Health* 1994, 84, 1843-5.

Hemenway D; Feskanich D; Colditz GA. Body height and hip fracture: a cohort study of 90,000 women. *International Journal of Epidemiology* 1995, 24, 783-6.

Holzenberger M; Martin-Crespo RM; Vicent D; Ruiz-Torres A. Decelerated growth and longevity in men. *Archives of Gerontological Geriatrics* 1991,13, 89-101.

Hoppe C; Molgaard C; Thomsen L; Juul A; Michaelson KF. Protein intake at 9 mo of age is associated with body size but not with body fat in 10 y old Danish children. *American Journal of Clinical Nutrition* 2004A, 79, 494-501.

Hoppe C; Molgaard C; Juul A; Michaelsen KF. High intakes of skimmed milk, but not meat, increase serum IGF-1 and IGFBP-2 in eight year old boys. *European Journal of Clinical Nutrition.* 2004B: doi:10.1038/sj.ejcn.1601948: 1-6.

Hrdlicka A. *The Old Americans*. Baltimore: The Williams and Wilkins Company; 1925; pp 370 and 373.

Hursting SD; Lavigne JA; Berrigan D; Perkins SN; Barrett JC. Calorie restriction, aging, and cancer prevention: mechanisms of actions and applicability to humans. *Annual Reviews of Medicine* 2003, 54, 131-152.

Huxley R; Neil A; Collins R. Unraveling the fetal origins hypothesis: Is there really an inverse association between birthweight and subsequent blood pressure? *The Lancet* 2002, 360, 659-665.

Hypponen E; Power C; Davey Smith G. Prenatal growth, BMI, and risk of type 2 diabetes by early midlife. *Diabetes Care* 2003, 26, 2512-2517.

Hu FB; Wang B; Chen C; Jin Y; Yang J; Stampfer MJ, Xu X. Body mass index and cardiovascular risk factors in a rural Chinese population. *American Journal of Epidemiology* 2000, 151, 88-97.

Hwang JY; Shin C; Frongillo EA; Shin KR; Jo I. Secular trend in age at menarche for South Korean women born between 1920 and 1986: the Asian Study. *Annals of Human Biology* 2003, 30, 434-442.

James PT. Obesity: the worldwide epidemic. *Clinics in Dermatology,* 2004, 22, 276-280.

Jones DW; Kim JS; Andrew ME; Kim SJ; Hong YP. Body mass index and blood pressure in Korean men and women: the Korean National Blood Pressure Survey. *Journal of Hypertension* 1994, 12, 1433-1437.

Kaaks R; Lukanova A. Energy balance and cancer: the role of insulin and insulin-like growth factor-I. *Proceedings of the Nutrition Society,* 2001, 60, 91-106.

Kagawa Y. Impact of Westernization on the nutrition of Japanese: changes in physique, cancer, longevity, and centenarians. *Preventive Medicine* 1978, 7, 205-217.

Karlberg J. Secular trends in pubertal development. *Hormone Research* 2002, 57, 19-30.

Kimm SYS; Obarzanek E. Childhood obesity: a new pandemic of the new millennium. *Pediatrics* 2002, 110, 1003-1007.

Knight EL; Verhave JC; Spiegelman D; Hillege HL; De Zeeuw D; Curhan GC; De Jong PE. Factors influencing serum cystatin C levels other than renal function and the impact on renal function measurement. *Kidney International* 2004, 65, 1416-1421.

Knoops KTB; de Groot LCPGN; Kromhout D; Perrin A-E; Moreiras-Varela O; Menotti A; van Staveren WA. Mediterranean diet, lifestyle factors, and 10-year mortality in elderly European men and women. *JAMA* 2004, 292, 1433-1439.

Komlos J; Baur M. From the tallest to (one of) the fattest: the enigmatic fate of the size of the American population in the twentieth century. *Economics and Human Biology* 2004, 2, 57-74.

Krakauer J; Franklin B; Kleerekoper M; Karlsson M; Levine J. Barrel build and 10 year mortality. *Preventive Cardiology* 2004, 7, 109-115.

Kramer MS; GuoT; Platt RW; Vanilovich I; Sevkovskaya Z; Dzikovich I; Michaelsen KF; Dewey K. For the Promotion of Breastfeeding Intervention Trial Study Group. Feeding effects on growth during infancy. *Journal of Pediatrics* 2004, 145, 600-605.

Lane MA; Baer Dj; Tilmont EM; Rumpler WV; Ingram DK; Roth GS; Cutler RG. Energy balance in Rhesus monkeys (Macaca mulatta) subjected to long-term dietary restriction. *Journal of Gerontology: Biological Sciences* 1995, 50A, B295-B302.

Lapunzina P; Lopez Camelo JS; Rittler M; Castilla ED. Risks of congenital anomalies in large for gestational age infants. *Journal of Pediatrics* 2002, 140, 200-204.

Leon DA; Lithhell HO; Vagero D; Koupilova H; Mohsen R; Berglund; Lithel U-B; McKeigue PM. Reduced fetal growth rate and inceased risk of death from ischaemic heart disease: cohort study of 15,000 Swedish men and women born 1915-29. *BMJ* 1998, 317, 241-245.

Leon DA; Johansson M; Rasmussen F. Gestational age and growth rates of fetal mass are inversely associated with systolic blood pressure in young adults: an epidemiologic study of 165,136 Swedish men aged 18 years. *American Journal of Epidemiology* 2000, 152, 597-604.

Leong NM; Mignone LI; Newcomb A; Titus-Ernstoff L; Baron JA; Trentham-Dietz A; Stampfer MJ; Willett WC; Egan KM. Early life risk factors in cancer: the relation of birth weight to adult obesity. *International Journal of Cancer* 2003, 103, 789-791.

Levy D; Anderson KM; Savage DD; Kannel WB; Christiansen JC; Castelli WP. Echocardiographically detected left ventricular hypertrophy: prevalence and risk factors. *Annals of Internal Medicine* 1988, 108, 7-13.

Lindeberg S; Lundh B. Apparent absence of stroke and ischaemic heart disease in a traditional Melanesian island: a clinical study in Kitava. *Journal of Internal Medicine* 1993, 233, 269-275.

Lonning PE; Helle SI; Johannessen DC; Adlercreutz H; Lien EA; Tally M; Ekse D; Fotsis T; Anker GB; Hall K. Relations between sex hormones, sex hormone binding globulin, insulin-like growth factor-I and insulin-like growth factor binding protein-I in post-menopausal breast cancer patients. *Clinical Endocrinology* 1995, 42, 2-30.

Loveridge N; Noble BS. Control of longitudinal growth: the role of nutrition. *European Journal of Clinical Nutrition* 1994, 48, 75-84.

Luo ZC; Low LC; Karlberg J. Fetal size to final height in Hong Kong Chinese children. *Journal of Pediatric Endocrinology and Metabolism* 2000, 13, 269-79.

Mayo Clinic. *Beyond menopause. Life after estrogen.* Supplement to Mayo Clinic Health Letter, February 2005, 1-8.

Maison P; Balkau B; Simon D; Chanson P; Rosselin G; Eschwege E. Growth hormone as a risk for premature mortality in healthy subjects: data from the Paris prospective study. *BMJ* 1998, 316, 1132-1133.

Masoro EJ. Caloric restriction and aging: an update. *Experimental Gerontology* 2000, 35, 299-305.

McCarty MF. The origins of western obesity: a role for animal protein? *Medical Hypotheses* 2000, 54, 488-494.

McCarty MF. Mortality from Western cancers rose dramatically among African-Americans during the 20th century: are dietary animal products to blame? *Medical Hypotheses* 2001, 57, 169-174.

Melve KK; Guessing HK; Skjaerven R; Oyen N. Infant's length at birth: an independent effect on perinatal mortality. *Acta Obstetricia and Gynecologica Scandinavica* 2000, 79, 459-464.

Mendall MA; Strachan DP; Butland BK; Ballam L; Morris J; Sweetnam PM; Elwood PC. C-reactive protein: relation to total mortality, cardiovascular mortality and cardiovascular risk factors in men. *European Heart Journal* 2000, 21, 1584-1590.

Menotti A; Kromhout D; Blackurn H; Fidanza F; Buzina R; Nissinen A. Food intake patterns and 25-year mortality from coronary heart disease: Cross-cultural correlations in the Seven Countries Study. *European Journal of Epidemiology* 1999, 15, 507-515.

Metges C. Does dietary protein in early life affect the development of adiposity in mammals? *Journal of Nutrition* 2001, 131, 3062-2066.

Miller DD. The economies of scale. *Challenge* 1990, 33, 58-61.

Miller RA; Austad SN. Growth and Aging: Why do big dogs die young? In Handbook of the Biology of Aging 6th Edition. *Masoro* E (Ed). In press

Monteiro PO; Victora CG; Barros FC; Monteiro LM. Birth size, early childhood growth, and adolescent obesity in a Brazilian birth cohort. *International Journal of Obesity and Related Metabolic Disorders* 2003, 27,1274-82.

Murtaugh MA; Jacobs DR; Moran A; Steinberger J; Sinaiko AR. Relation of birth weight to fasting insulin, insulin resistance, and body size in adolescence. *Diabetes Care* 2003, 26, 187-192.

Murray MJ; Murray AB; Murray NJ. Does chronic undernutriion in adult Masai influence morbidity and survival? *Proceedings of the Nutrition Society* 1993, 52, 107A.

Must A; Strauss RS. Risks and consequences of childhood and adolescent obesity. *International Journal of Obesity* 1999, 23 Suppl 2, S2-S11.

Oken E; Gillman MW. Fetal origins of obesity. *Obesity Research* 2003, 11, 496-506.

Okasha M; Holly J; Davey Smith G. Childhood growth and adult cancer. *Best Practice and Research Clinical Endocrinology and Metabolism* 2002, 16, 225-241.

Ong KKL; Ahmed ML; Emmett PM; Preece MA; Dunger DB. Association between postnatal catch-up growth and obesity in childhood: prospective cohort study. *BMJ* 2000, 320, 967-71.

Ong KK; Dunger DB. Birth weight, infant growth and insulin resistance. *European Journal of Endocrinology* 2004, 151, U131-U139.

Paolisso G; Gambardella A; Balbi V; Ammendola S; D'Amore A; Varricchio M. Body compostion, body fat distribution, and resting metabolic rate in healthy centenarians. *American Journal of Clinical Nutrition* 1995, 62, 746-750.

Parr T. Insulin exposure and unifying aging. *Gerontology* 1999, 45, 121-135.

Parsons TJ; Power C; Manor O. Fetal and early life growth and body mass index from birth to early adulthood in 1958 British cohort: longitudinal study. *BMJ* 2001, 323, 1331-5.

Pietilainnen KH; Kaprio J; Rasanen M; Winter T; Rissanen A; Rose RJ. Tracking body size from birth to late adolescence: contribution of birth length, birth weight, duration of gestation, parents body size, and twinship. *American Journal of Epidemiology* 2001, 154, 21-29.

Renehan AG; Zwahlen M; Minder C; O'Dwyer ST; Shale SM; Egger M. Insulin-like growth factor (IGF-1), IGF binding protein-3, and cancer risk: systematic review and meta-regression analysis. *The Lancet* 2004, 363, 1346-53.

Rolland-Cachera MF; Deheeger M; Akrout M; Bellisle F. Influence of macronutrients on adiposity development: a follow-up study of nutrition and growth from 10 months to 8 years of age. *International Journal of Obesity* 1995, 19, 573-578.

Rolland-Cachera MF; Deheeger M; Bellisle F. Increasing prevalence of obesity among 18-year old males in Sweden: evidence for early determinents. *Acta Paediatrica* 1999, 88, 365-367.

Rollo CD; Carlson J; Sawada M. Accelerated aging of giant transgenic mice is associated with elevated free radical processes. *Canadian Journal of Zoology* 1996, 74, 606-620.

Rollo CD. Growth negatively impacts the life span of mammals. *Evolution and Development* 2002, 4, 55-61.

Roseboom TJ; van der Meulen JHP; Osmond C; Barker DJP; Ravelli ACJ; Bleker OP. Adult survival after prenatal exposure to the Dutch famine 1944-45. *Paediatric and Perinatal Epidemiology* 2001, 15, 220-225.

Rush D. Maternal nutrition and perinatal survival. *Nutrition Reviews* 2001, 59, 315-326.

Samaras TT; Storms LH. Impact of height and weight on life span. *Bulletin of the World Health Organization* 1992, 70, 259-267.

Samaras TT. *The Truth About Your Height. Exploring the myths and realities of human size and its effects on performance, health, pollution, and survival.* San Diego: Tecolote Publications;1994; p 34. (Amazon.com)

Samaras TT. How body height and weight affect our performance, longevity and survival. *Journal of the Washington Academy of Sciences* 1996, 84, 131-156.

Samaras TT; Elrick H. Height, body size and longevity. *Acta Medica Okayama* 1999, 53, 149-169.

Samaras TT; Storms LH; Elrick H. Longevity, mortality and body weight. *Ageing Research Reviews* 2002A, 1, 673-691.

Samaras TT; Storms LH. Secular growth and its harmful ramifications. *Medical Hypotheses* 2002B, 58, 93-112.

Samaras TT; Elrick H. Height, body size, and longevity: is smaller better for the human body? *Western Journal of Medicine* 2002C, 176, 206-208.

Samaras TT; Elrick H; Storms LH. Birthweight, rapid growth, cancer, and longevity: a review. *Journal of the National Medical Association* 2003A, 95, 1170-1183.

Samaras TT; Elrick H; Storms LH. Is height related to longevity? *Life Sciences* 2003B, 72, 1781-1802.

Samaras TT; Elrick H; Storms LH. Is short height really a risk factor for coronary heart disease and stroke mortality? a review. *Medical Science Monitor* 2004, 10, RA63-76.

Sawaguchi A; Sawaguchi T. Japanese national SIDS project: 1998-2000 research for the improvement of infant mortality. *Forensic Science International,* 2002, 130S, S1-S7.

Saweri W. The rocky road from roots to rice: a review of the changing food and nutrition situation in Papua New Guinea. *PNG Medical Journal* 2001, 44, 151-163.

Schouten LJ; Goldbohm A; van den Brandt PA. Height, weight, weight change, and ovarian cancer risk in the Netherlands cohort study on diet and cancer. *American Journal of Epidemiology* 2003, 157, 424-433.

Seidman DS; Gale R; Stevenson DK; Laor A; Bettane PA; Danon YL. Is the association between birthweight and height attainment independent of the confounding effect of ethnic and socioeconomic factors. *Israeli Journal of Medical Science* 1993, 29, 772-776.

Silva IDS; Stavola BLD; Mann V; Kuh D; Hardy R; Wadsworth MEJ. Prenatal factors, childhood growth trajectories and age at menarche. *International Journal of Epidemiology* 2002, 31, 405-412.

Singh PN; Sabate J; Fraser GE. Does low meat consumption increase life expectancy in humans? *American Journal of Clinical Nutrition* 2003, 78 (suppl), 526S-532S.

Singhal A; Fewtrell M; Cole TJ; Lucas A. Low nutrient intake and early growth for later insulin resistance in adolescents born preterm. *The Lancet* 2003, 361, 1089-1097.

Singhal A; Cole TJ; Fewtrell M; Deanfield J; Lucas A. Is slower early growth beneficial for long-term cardiovascular health? *Circulation* 2004, 109, 1108-1113.

Silventoinen K; Kaprio J; Koskenvuo M; Lahelma E. The association between body height and coronary heart disease among Finnish twins and singletons. *International Journal of Epidemiology* 2003, 32, 78-82.

Sinaiko AR; Donahue RP; Jacobs DR; Prineas RJ. Relation of weight and rate of increase in weight during childhood and adolescence to body size, blood pressure, fasting insulin, and lipids in young adults. The Minneapolis Children's Blood Pressure Study. *Circulation* 1999, 99, 1471-1476.

Skinner JD; Bounds W; Carruth BR; Morris M; Ziegler P. Predictors of children's body mass index: a longitudinal study of diet and growth in children aged 2-8 y. *International Journal of Obesity* 2004, 28, 476-482.

Sorensen HT; Sabroe S; Rothman KJ; Gillman M; Fischer P; Sorensen TIA. Relation between weight and length at birth and body mass index in young adulthood: cohort study. *BMJ* 1997, 315, 1137-1141.

Sorensen HT; Sabroe S; Rothman KJ; Gillman M; Steffensen FH; Fischer P; Sorensen TIA. Birth weight and length as predictors for adult height. *American Journal of Epidemiology* 1999, 149, 726-729.

Stanner SA; Bulmer K; Andres C; Lantseva OE; Borodina V; Poteen VV; Yudkin JS. Does malnutrition in utero determine diabetes and coronary heart disease in adulthood: Results form the Leningrad siege study, a cross sectional study. *BMJ* 1997, 315, 1342-1347.

Stamler J; Daviglus ML; Garside DB; Dyer AR; Greenland P; Neaton JD. Relationship of baseline serum cholesterol levels in 3 large cohorts of younger men to long-term coronary, cardiovascular, and all-cause mortality and to longevity. *JAMA* 2000, 284, 311-318.

Statistical Bulletin. Increase in survivorship since 1840. New York: Metropolitan Life Insurance Company; August 1964; Vol 45, p 3.

Stettler N; Zemel BS; Kumanyika S; Stallings VA. Infant weight gain and childhood overweight status in a multicenter cohort study. *Pediatrics* 2002, 109, 194-199.

Stindl R. Tying it all together: telomeres, sexual size dimorphism and the gender gap in life expectancy. *Medical Hypotheses* 2004, 62, 151-154.

Stoll BA. Western diet, early puberty and breast cancer risk. *Breast Cancer Research and Treatment* 1998, 49, 187-193.

Tretli S. Height and weight in relation to breast cancer morbidity and mortality. A prospective study of 570,000 women in Norway. *International Journal of Cancer* 1989, 44, 23-30.

Trichopoulou A; Gnardellis C; Benetou V; Lagiou P; Bamia C; Trichopoulos D. Lipid, protein and carbohydrate intake in relation to body mass index. *European Journal of Clinical Nutrition* 2002, 56, 37-43.

Tufts University Health and Nutrition Letter. Ask Tufts Experts. November 2004A, 22, 1-8.

Tufts University Health and Nurition Letter. Eat like a Mediterranean and live longer. November 2004B, 22, 1-8.

UC Berkeley Wellness Letter, The salt story so far. November 2001, 18, 1-8.

Ulizzi L. Absence of selection on birth weight in modern Italy. *Human Biology* 2002, 74, 525-532.

Vallejo EA. La dieta de hambre a dias alternos en la alimentacion de los viejos. *Revista Clinica Espanaola* 1956, 63, 25-27 (in Spanish).

Walker ARP; Labadarios D; Glatthaar II. Diet-related disease patterns in South African interethnic populations. In: Western Diseases. Temple JN and Burkitt DP, editors. *Western Diseases.* Totowa, New Jersey: Humana Press; 1994; p. 43.

Wen SW; Kramer MS; Platt R; Demissie K; Joseph KS; Liu S; Sauve R. Secular trends of fetal growth in Canada, 1981 to 1997. *Paediatrica Perinatal Epidemiology* 2003, 17, 347-54.

Wilcox AJ. On the importance-and the unimportance-of birthweight. *International Journal of Epidemiology* 2001, 30, 1233-1241.

Willcox BJ; Willcox DC; Suzuki M. The Okinawa Program. New York: Clarkson/Potter Publishers; 2001.

Willcox BJ; Yano K; Chen R; Willcox DC; Rodriquez BL; Masaki KH; Donlon T, Tanaka B; Curb JD. How much should we eat? The association between energy intake and mortality in a 36-year follow up study of Japanese-American men. *Journal of Gerontology: Biological Sciences* 2004, 59A, 789-795.

Wilkin TJ; Metcalf BS; Murphy MJ; Kirkby J; Jeffrey AN; Voss LD. The relative contributions of birth weight, weight change, and current weight to insulin resistance in contemporary 5 year olds. *Diabetes* 2002, 51, 3468-3472.

Willett W. Lessons from the dietary studies in Adventists and questions for the future. *American Journal of Clinical Nurition* 2003, 78 (suppl), 539S-543S.

Williams S; Dickson N. Early growth, menarche, and adiposity rebound. *The Lancet* 2002, 359, 580-581.

Williams S; Poulton R. Twins and maternal smoking: ordeals for the fetal origins hypothesis? a Cohort Study. *BMJ* 1999, 318, 897-900.

Wilson AC; Forsyth JS; Greene SA; Irvine L; Hau C; Howie PW. Relation of infant diet to childhood health: seven year follow-up of cohort of children in Dundee infant feeding study. *BMJ* 1998, 316, 21-25.

Wolk K; Larsson SC; Vessby B; Wolk A, Brismar K. Metabolic, anthropometric, and nutritional factors as predictors of circulating insulin-like growth factor binding protein-1 levels in middle-aged and elderly men. *The Journal of Endocrinology and Metabolism* 2004, 89, 1879-1884.

Zaretsky MD. Communication between identifical twins: health behavior and social factors are associated with longevity that is greater among identical than fraternal U.S. World War II veteran twins. *Journal of Gerontology: Medical Sciences* 2003, 58A, 566-572.

In: New Developments in Obesity Research
Editor: Lawrence F. Ditmier, pp. 41-74
ISBN 1-60021-296-4

Chapter 2

OBESITY, METABOLIC SYNDROME, AND WEIGHT MANAGEMENT

Jian Huang[1*] and Donna Carden[1,2*]
Department of Medicine[1] and Emergency Medicine[2] Louisiana State University Health Sciences Center-Shreveport, Louisiana, USA

ABSTRACT

The metabolic syndrome (MS) describes a cluster of cardiovascular risk factors, including central adiposity, insulin resistance or overt type 2 diabetes mellitus (DM2), atherogenic dyslipidemia, and elevated blood pressure or overt hypertension. Obesity, especially central adiposity, is associated with insulin resistance, the central feature of MS. Predictably, the incidence of MS increases with increasing body mass index (BMI) but also occurs with weight gain, even at normal weight. MS predisposes individuals to the development of DM2 and is responsible for increased all cause and coronary artery disease (CAD) mortality. MS, therefore, likely represents a latent disease process that may be viewed as preclinical diabetes or preclinical atherosclerosis. Because transitional conditions are more amenable to therapeutic intervention than late-stage disease, it is becoming increasingly recognized that early identification and treatment of MS is necessary to prevent CAD mortality. Recent clinical trials have demonstrated that weight loss through lifestyle modification with or without pharmacologic intervention can improve all components of MS. Although the exact mechanisms underlying MS remain undefined, physicians can contribute to improved long-tem clinical outcome by endorsing a healthy lifestyle that promotes weight loss or maintenance and physical activity. This

* Data for correspondence for Jian Huang, MD. Associate Professor of Medicine, Department of Medicine, Louisisana State University, Health Sciences Center. PO Box 33932 1501, Kings Highway, Shreveport, LA 71130-3932. Tel: (318)675-5621 Fax: (318)675-7176 E-mail: jhuang@lsuhsc.edu.

* Donna Carden, MD, Professor of Medicine and Emergency Medicine, Department of Emergency Medicine. Louisisana State University Health Sciences Center, 1501 Kings Highway, Shreveport, LA 71130-3932. Tel: (318)675-6659 E-mail: dcarde@lsuhsc.edu

may lead to the reduction in morbidity and mortality associated with obesity, DM2, and MS.

INTRODUCTION

The metabolic syndrome (MS) refers to a cluster of cardiovascular risk factors, including abdominal obesity, insulin resistance with or without glucose intolerance, atherogenic dyslipidemia (DLP), elevated blood pressure and a proinflammatory as well as a prothrombotic state. [1, 2] Although various organizations differ on the components and emphasis placed on risk factors associated with MS, there is agreement that central to the pathogenesis of MS is its association with insulin resistance, type 2 diabetes mellitus (DM2) and premature atherosclerosis. The clustering of risk factors associated with DM2 and premature atherosclerosis was initially identified in the 1950's and continued to be refined throughout the 1980's. [3-10] However, Raven brought special attention to this group of risk factors with the suggestion that insulin resistance was an important link to DM2 and coronary artery disease (CAD) and termed the association, "Syndrome X".[11] Subsequently, insulin resistance or hyper-insulinemia was determined to be central to the development of MS. [1, 12-24] Other components of MS have been suggested and include microalbuminuria,[19] increased uric acid, [15] lipolytic enzymes, [25] free fatty acids, [26] TNF-alpha, [27] interleukin-6, [28] C-reactive protein (CRP), [29-31] von Willebrand factor (vWF), [31] fibrinogen, [31] and plasminogen activating inhibitor 1 (PAI-1). [31, 32] The terminology and major risk factors associated with MS are summarized in Table 1.

The exact prevalence of MS is not clearly defined. One of the reasons for the lack of clarity is that different emphasis has been placed on the various components of MS by a number of organizations. (Table 2) [1, 2, 19, 22, 23] However, it is clear that the prevalence of MS varies among different geographic, ethnic, age, and gender groups. [33-45] Furthermore, the relationship of MS to DM2 and CAD has also been well established by epidemiologic and clinical studies. [11, 16, 30, 46-52]

MS, CAD RISK, AND MORBIDITY AND MORTALITY

CAD is the leading cause of death in the United States and numerous factors enhance the risk of CAD mortality. [53] For example, excess body weight is associated with shortened life expectancy, [53-55] and increased CAD mortality.[56, 57] Moreover, both pre-hypertension and overt hypertension (HTN) contribute to increased CAD risk. [58] Insulin resistance, in the presence or absence of overt DM2, is associated with atherogenic DLP characterized by elevated triglycerides (TG) and diminished high-density-lipoprotein cholesterol (HDL-C), [59, 60] and is a predictor of CAD events. [61-66] However, the overall CAD risk among individuals with MS is far greater than the simple arithmetic sum of its individual components. [67-70] Not only does MS predispose to the development of DM2 [16, 46, 47] and CAD, [30, 46, 48-50] it is also responsible for increased all-cause [49, 51, 52] as well as CAD mortality. [46, 49, 51, 68, 71, 72]

Table 1. Terminology and Components Associated with the Metabolic Syndrome

Author/Year/ (Reference)	Nomenclature	Components							
		OB	DLP	↑BP/ HTN	IR/DM	↑UA/ Gout	↑UAlb	Pro-Thromb	Sed
Vague 1956/1980 (3,9)	Android Obesity/ Diabetogenic Obesity	X			X	X			
Avogaro 1965 (4)	Three Metabolic Disorders	X	X		X				
Camus 1966 (10)	Trisyndrome		X		X	X			
Reaven 1988 (11)	Syndrome X		X	X	X				
Kaplan 1989 (13)	Deadly Quartet	X	X	X	X				
DeFronzo 1991 (14)	Insulin Resistance Syndrome	X	X	X	X				
Zimmet 1991 (15)	Syndrome X Plus	X	X	X	X	X			X
Haffner 1992 (16)	Insulin Resistance Syndrome		X	X	X				
Kesaniemi 1992 (17)	Multiple Metabolic Syndrome	X	X	X	X				
Hjermann 1992 (12)	Metabolic Cardiovascular Syndrome	X	X	X	X			X	
Despres 1993 (18)	Insulin Resistance Dyslipidemia Syndrome	X	X		X				
WHO 1998 (19)	Metabolic Syndrome	X	X	X	X			X	
Hansen 1999 (20)	Metabolic Syndrome X	X	X	X	X				
Groop 2001 (21)	Dysmetabolic Syndrome	X	X	X	X			X	
NCEP 2001 (1)	Metabolic Syndrome	X	X	X	X				
EGIR 2002 (22)	Insulin Resistance Syndrome	X	X	X	X				
AACE 2003 (23)	Insulin Resistance Syndrome	X	X	X	X				X

NCEP = National Cholesterol Education Program; WHO = World Health Organization; AACE = American Association of Clinical Endocrinologists; EGIR = European Group for the Study of Insulin Resistance; OB: Obesity, or increased waist circumference, or increased waist to hip ratio; DLP: Dyslipidemia; BP: Blood pressure; HTN: Hypertension; IR: Insulin resistance; DM: Diabetes mellitus; UA: Uric acid; UAlb: Urine albumin; Pro-Thromb: Prothrombotic state; Sed: Sedentary lifestyle.

THE ROLE OF OBESITY IN THE DEVELOPMENT OF MS

Obesity may be the most important risk factor in the development of MS. Obesity is a chronic disease associated with a number of serious health consequences. [73-75] While some causes of preventable death are declining, the death rate attributable to overweight or obesity as a result of poor diet and physical inactivity has risen markedly over the last decade and may soon exceed tobacco use as the leading cause of preventable death. [53] Obesity, particularly abdominal obesity, is a major contributor to the development of insulin resistance, DM2, HTN, and atherogenic DLP, all of which are major components of MS. [74, 76-80] Thus, abdominal obesity is critical to the development of insulin resistance which in turn, is the central feature of MS. [79-82]

The prevalence of obesity has reached epidemic proportions. [75, 83, 84] Currently, more than 65% of American adults are overweight (body mass index defined as weight in kg/ height in m^2 (BMI)>/= 25) or obese (BMI>/=30). [84] The prevalence of DM2 and MS is also near epidemic in the United States and continues to increase. [33, 34, 75, 83, 85] Seventeen million American adults have been diagnosed with DM2 and another 16 million have pre-diabetes and are thus at high risk of developing overt DM2. [86] The lifetime risk for DM2 among individuals born in 2000 in the United States is estimated at 32.8% for males and

38.5% for females. [87] It is also estimated that 27% of the adult population has MS based on recent data analysis of the Third National Health and Nutrition Examination Survey (NHANES III) (1999-2000), [33] an increase from 24% in NHANES III (1988-1994). [34] In parallel, the prevalence of MS among US adolescents also increased from 4.2% to 6.4% during the same time period. [35, 88] MS is also more prevalent in obese teens. [89] Health risks associated with MS continue from adolescence into adulthood, [90, 91] making adolescent MS a major risk factor for adult cardiovascular disease.

Table 2. Definitions of the Metabolic Syndrome

NCEP (1)	WHO (19)	AACE* (23)	EGIR* (22)
At least 3 of the following: 1. FPG>/=110mg/dl 2. Abdominal obesit: WC>40 inches or 102 cm (M); >35 inches or 88cm (F) 3. TG>/=150mg/dl 4. HDL <40mg/dl (M); <50mg/dl (F) 5. BP: >/=130/85 mmHg	IR defined as one of the following: 1. IFG 2. IGT 3. DM2 4. glucose uptake below the lowest quartile for background population under investigation under hyperinsulinemic euglycemic conditions Plus any 2 of the following: 1. BP>/=160/90mmHg or Rx for HTN 2. TG>/= 150mg/dl 3. HDL<35mg/dl (M) <39mg/dl (F) 4. BMI>30 kg/m2 and or WHR>0.9 (M) >0.85 (F) Urinary Alb excretion rate >/=20ug/min or Alb/Cr >/=20mg/g	Diagnosis depends on clinical judgment based on the following risk factors: 1. BMI>/=25 kg/m2 2. TG>/=150mg/dl 3. HDL<40mg/dl (M); <50mg/dl (F) 4. BP>/=130/85mmHg 5. FPG 110 -125mg/dl 6. 2-hour post-glucose challenge 140-200mg/dl 7. Others: personal history of CVD, HTN, GDM or IGT PCOS, NAFLD, or AN; family history of DM2, HTN or CVD; sedentary lifestyle; age>40; non-Caucasian ethnicity	Hyperinsulinemia defined as fasting insulin concentration above the upper quartile for the non-diabetic subjects Plus at least 2 of the following: 1. FPG>/=110mg/dl 2. BP>/=140/90mmHg or Rx for HTN 3. TG>180mg/dl and/or HDL<40mg/dl and/or Rx for DLP 4. WC>/=94 cm (M); >/=80 cm (F)

* Defined only for non-diabetic subjects.

NCEP = National Cholesterol Education Program; WHO = World Health Organization; AACE = American Association of Clinical Endocrinologists; EGIR = European Group for the Study of Insulin Resistance; FPG = fasting plasma glucose; WC = waist circumference; TG = triglycerides; HDL = high-density lipoprotein cholesterol; BP = blood pressure; IR = insulin resistance; IFG = impaired fasting glucose; IGT = impaired glucose tolerance; DM2 = type 2 diabetes mellitus; HTN = hypertension; M = male; F = female; BMI = body mass index; Alb = albumin; Cr = creatinine; CVD = cardiovascular disease; PCOS = polycystic ovary syndrome; GDM: gestational diabetes; NAFLD: non-alcoholic fatty liver disease; AN: acanthosis nigricans.

The total health care costs attributable to obesity are estimated at hundreds of billions of US dollars per year with at least $90 billion in direct health care costs annually in the US alone. [92] Most of the costs are associated with DM2, CAD, and HTN. [74, 93, 94] As obesity, DM2, and MS become more prevalent, these trends can be expected to translate into excess cardiovascular morbidity and mortality and higher health-care expenditures. [49, 50, 52, 56, 74, 92-96]

SUBCLINICAL ATHEROSCLEROSIS AND MS

As outlined above, the term MS has been used to describe a state of insulin resistance and associated metabolic derangements. Because it would take years to assess effects of risk factors on symptomatic cardiovascular events, surrogate markers have been used to investigate the effect of insulin resistance and MS on atherosclerosis. For example, the degree of subclinical atherosclerosis, as measured by carotid artery intima-media thickness, was found to be significantly greater in patients with MS. [97] Further, increased coronary artery calcium scores as determined by electron beam tomography, a marker of subclinical coronary atherosclerosis, was reported in the majority of asymptomatic MS patients, independent of their 10-year, Framingham risk score for CAD. [98] Compared to control subjects, patients with MS with angiographically proven normal coronary arteries also have evidence of impaired coronary blood flow suggestive of coronary microvascular endothelial dysfunction. [99] Finally, an inverse relationship between insulin sensitivity and carotid atherosclerosis was reported in the Insulin Resistance Atherosclerosis Study (IRAS). [100] After adjustment for established risk factors, the presence of MS was found to be associated with a two fold increased risk of CAD in the Atherosclerosis Risk in Community Study (ARIC) [97] Therefore, insulin resistance and MS represent important risk factors for the development and progression of atherosclerosis, even before gross pathology is recognizable using standard methods. Widespread recognition of the significant association of insulin resistance and MS with accelerated atherosclerosis and CAD progression may prompt earlier and more aggressive therapeutic intervention in insulin resistance states. Recognition of MS as a unique disease state with high CAD risk is gaining acceptance as the interventional target for primary and secondary CAD prevention. [1, 2]

OBESITY AND MS

Population-based cohort studies provide support for the hypothesis that the different components of MS are linked through obesity. [101] A key feature of MS is abdominal obesity, the deposition of excess adipose tissue in visceral compartments, which gives rise to insulin resistance. There is a progressive loss of insulin sensitivity with increased BMI [79] that is directly related to the degree of visceral adiposity. [80, 81] Waist circumference (WC) has been widely accepted as a surrogate for abdominal obesity in clinical practice. [1, 73] NHANES III data showed that, with few exceptions, within the normal weight (BMI: 18.5-24.9), overweight (BMI: 25-29.9), and class I obesity (BMI: 30-34.9) categories, those with high WC values are increasingly likely to have HTN, DM2, DLP, and MS compared with those with normal WC values. [102]

WEIGHT GAIN AND MS

Recently, the Coronary Artery Risk Development in Young Adults (CARDIA) study reported that not only a higher BMI in young adulthood but also weight gain over 15 years predicts development of MS. [103] A study of 30 nondiabetic, premenopausal women followed for 7 years showed that weight gain, especially in the form of increased visceral adipose tissue, was associated with deterioration in glucose-insulin homeostasis. [104] Just as a higher BMI is associated with an increased incidence of DM2, weight gain predicts the development of DM2, independent of baseline adiposity. [105, 106] Higher BMI and weight gain are also linked to increased disease mortality. In the Cancer Prevention Study II, a prospective study of mortality among men and women in the US, the risk of death from all causes, cardiovascular disease, as well as cancer increased with elevated BMI for both genders and all age groups. [56] In the Nurses' Health Study, a weight gain of 10 kg after age 18 in women was associated with increased mortality in middle adulthood. [107] Thus, weight gain, visceral adiposity and obesity are all associated with excess cardiovascular risk and mortality.

Weight gain and obesity result from excess dietary intake and insufficient energy expenditure. [76] Availability of calorie-dense foods and decreased activity requirements of modern society have been linked with the obesity epidemic. [76] Aside from tobacco use, poor diet and physical inactivity are recognized as the principle contributors to morbidity and mortality, being responsible for 400,000 deaths in 2000. [53]

Dietary Habits, Physical Activity and MS

Dietary habits and physical activity are modifiable risk factors for obesity and the development of multiple components associated with MS. [73, 108] Several nutrients appear to influence insulin sensitivity, although their effects appear to be modest. Total fat correlates inversely with insulin sensitivity, but only in obese individuals. [109] Increased energy intake from saturated and trans-fat is associated with insulin resistance and risk of MS or DM2, [109-115] whereas polyunsaturated fat intake decreases the incidence of DM2. [115] Low intake of dietary fiber, fruits and vegetables is also associated with increased risk of DM2 or MS. [116-118] Moreover, the type of carbohydrate in food products can be categorized by its glycemic index, defined as the ratio of 2-hour post-prandial plasma glucose to the 2-hour level following ingestion of a similar amount of glucose. For a given amount of carbohydrate, diets rich in high glycemic index foods are associated with an increased risk of DM2. [115, 116, 118] Data from NHANES III (1988-1994) indicate that a diet relatively low in carbohydrate is associated with a significantly lower risk of MS. [119]

Physical inactivity is also a risk factor for MS. [103, 119] The relationship between MS risk and lifestyle-related factors was studied using data from a representative US sample in NHANES III. According to this cross-sectional logistic model, MS risk is substantially lower in individuals who are physically active and maintain a BMI in the non-obese range. [119] In the Canada Fitness Survey observational cohort study, anthropometric and physical fitness measures were obtained in 3719 men and 3854 women 20-59 years of age. High cardiorespiratory fitness was associated with lower levels of total and abdominal obesity for any given BMI compared to those with low cardiorespiratory fitness, independent of gender. [120] In the European Youth Heart Study of 589 children, an inverse relationship was found between physical activity and MS risk even after adjusting for other potential confounders. [121] In a nondiabetic, Caucasian population from the UK, a strong inverse association between physical activity and MS was documented, especially in unfit individuals. [122]

Similarly in the Aerobics Center Longitudinal Study, an inverse relationship between MS prevalence and muscle strength and cardiorespiratory fitness in men across normal, overweight, and obese BMI categories was reported. [123] Likewise, MS prevalence was markedly lower across progressively higher levels of cardiorespiratory fitness in women of different ages. [124] Thus, there is substantial evidence that physical activity and enhanced cardiorespiratory fitness dramatically reduce the risk of MS.

Although physical activity and cardiorespiratory fitness are recognized deterrents of MS, CDC data from 2004 indicate that only 30.5% of US adults aged 18 or over engage in regular leisure-time physical activity. This estimate is down from 33.2% for the same period in 2003. [125] Decline in physical activity was also reported in US adolescent females in the National Heart, Lung, and Blood Institute Growth and Health Study. [126] Given the high and increasing prevalence of MS, poor dietary habits [53] and decline in physical activity in US adults and adolescents, [33-35, 125, 126] it is apparent that promotion of a healthy diet and regular exercise as part of a therapeutic intervention in the treatment of obesity and MS is necessary.

Metabolically Obese Normal-Weight, Metabolic Fitness and MS

Not all insulin resistant individuals are overweight or obese, nor are all overweight or obese individuals insulin resistant. BMI can underestimate or overestimate the degree of fatness depending on the individual's muscularity. [73, 74] The concept of the metabolically obese, normal-weight (MONW) individual was originally suggested over 20 years ago for those with normal or near normal BMI but with one of three conditions that improves with caloric restriction: DM2, HTN, and elevated TG. [127] Examination of phenotypic differences between MONW women and women with normal weight and metabolic profile revealed greater total as well as visceral fat and lower physical activity and energy expenditure in MONW women despite similar age, BMI and fat-free mass. [128] Recent analysis of data from 7602 adults from NHANES III confirms that MS is not limited to obese patients and, in fact, it is relatively common among normal weight or slightly overweight individuals. [129] While it is clear that MS increases with BMI, weight gain itself, even in normal weight individuals, is also associated with increased prevalence of MS. [104, 129] Thus, it has been proposed that those with MS, but a BMI of 18.5-26.9 kg/m^2, be classified as MONW. [129] Accordingly, rather than focusing primarily on body weight, many health professionals emphasize lifestyle modifications through diet and exercise to achieve the "metabolic fitness", defined as the absence of biochemical risk factors for MS, such as DLP, insulin resistance, preclinical DM2 and HTN. [130]

Abdominal Adiposity, Waist Circumference and MS

The fact that MS is common, even in normal weight individuals, lead to identification of abdominal adiposity as an independent risk factor for MS. [1] Results from NHANES III documented that abdominal adiposity and WC are more closely linked to cardiovascular disease risk factors than BMI. [131] Although WC and BMI are interrelated, WC provides an independent prediction of disease risk. [73] In fact, the Nurses' Health Study documented that WC is strongly associated with increased CAD risk in normal weight women and those with a WC > 76.2 cm (30 inches). [132] Predictably, it has been suggested that WC be included as part of the routine physical exam. [133] Given the increasing prevalence of obesity, DM2, and MS, [33, 34, 75, 83] and the increasing prevalence of abdominal obesity as measured by WC, [134, 135] preventing weight gain and overweight among persons with healthy weights and avoiding further weight gain among those already overweight are important public health goals. Appropriate attention should also be given to increase in WC even if overall weight remains stable or within the normal range. [136]

Weight Loss and MS

Since all components of MS are closely associated with obesity or the metabolically obese state, weight reduction or maintenance and physical activity are the corner stone of treatment of MS. However, treatment of obesity is challenging because body weight is determined by the integration of many biological and environmental factors. Non-pharmacological methods of weight loss are effective in the short-term but there is an unsatisfactory overall success rate with dietary treatment of obesity [137]. There is also a relative paucity of long-term studies using pharmalogic approaches to weight loss. [138] Moreover, one-third to two-thirds of weight lost, regardless of approach outside of surgery, is regained within one year and almost all is regained within five years. [139, 140]

Because of the difficulties in achieving permanent weight control, many experts emphasize adoption of life-long habits conducive to weight control and overall health rather than temporary measures for drastic weight loss. A program that encompasses regular physical activity with energy expenditure of more than 1,500 kcal/week and modest energy intake from a variety of nutritionally balanced food has the potential to meet such a treatment goal. There is mounting evidence that modest (5-10%) weight reduction through dietary modification and physical activity improves blood pressure, glucose tolerance, and lipid levels among obese patients. [73, 141-150] Many professionals recommend modest weight reduction or weight maintenance with emphasis on "metabolic fitness". For those whose health is impacted by excess body weight, the goal of weight management should be to achieve better health and well being over the long term, and not simply to lose weight.

Treatment of MS

The primary treatment goal of MS is prevention of DM2 and cardiovascular events. Currently, there are no firm guidelines on the optimal management of blood pressure, DLP, impaired glucose tolerance and insulin resistance in patients with MS because of limited data examining the efficacy of risk-modification of these MS components. Nevertheless, MS and DM2 share the common feature of insulin resistance and can be viewed as a disease continuum with DM2 representing the severe end of the metabolic derangement.

National Cholesterol Education Program (NCEP) recognizes MS as a secondary target for risk reduction therapy with specific recommendations for weight loss through lifestyle modification that affect components of MS. [1] Weight loss or prevention of weight gain improves glucose tolerance and insulin sensitivity and can delay or prevent the onset of DM2 and is therefore essential in the treatment of MS. [147, 148, 150, 151] Blood pressure control remains an important component of treatment strategies for MS, since low HDL and elevated blood pressure were identified in NHANES III as the most significant CAD predictors in MS patients. [36] Lipid modification plays a vital role in the prevention and treatment of atherosclerosis and reduction of CAD events. [1, 152] Although recent data suggest a beneficial effect of bariatric surgery for obese patients with DM2, HTN, and DLP, [153] surgical treatment for weight loss should be reserved for those in whom other treatment modalities have failed and who have clinically severe obesity and co-morbidities.

Lifestyle Intervention, Weight Loss and MS

Weight loss improves all components of MS and is central to cardiovascular risk reduction. Even modest weight loss (5-10 %) has beneficial effects on lipids, blood pressure, and blood glucose. [73, 154] A combination of diet modification and increased physical activity is recommended for long-term weight management. [1, 73, 74] Weight loss is more directly associated with restriction of caloric intake and duration of dietary adherence and not type of diet. [155, 156] Physical activity is an integral part of weight loss therapy and weight maintenance. [73]

Dietary Modification

There are countless dietary weight loss programs available. While current dietary recommendations for weight loss suggest that a healthful adult diet derive 45-65% energy from carbohydrate, 20-35% from fat with restricted saturated fat, and 10-35% from protein, [1, 73, 157-162] (Table 3.), popular diets often advocate low carbohydrate (< 10% for very low carbohydrate diet; < 45% for low carbohydrate diet) with corresponding increases in fat intake. [163-165] However, an energy-rich, high fat diet contributes to weight gain and obesity. [166]

Evidence suggests that a diet high in carbohydrate may increase triglyceride (TG), lower HDL, [167, 168] and may increase postprandial glucose and insulin levels in patients with DM2. [169] Low-fat, calorie-restricted diets, such as those recommended by the American Heart Association (AHA), National Cholesterol Education Program (NCEP) and American Diabetes Association (ADA), are relatively high in carbohydrate. [1, 158, 161] For MS patients with high TG and low HDL-C, NCEP suggests that carbohydrate intake be limited to less than 60%, perhaps as low as 50%, of total calories. Additionally, most carbohydrates should be complex such as those found in whole grains, vegetables, fruits, and low fat dairy products. [1]

Emerging data suggests that saturated fat and trans-fat are two cholesterol-increasing fatty acids, while unsaturated fat, particularly monounsaturated fat, has favorable effects on lowering LDL-C. [1] NCEP suggests that the upper limit of total fat intake be 35% with saturated fat limited to less than 7%. Saturated fat should be substituted with unsaturated, especially monounsaturated fat. [1] Dietary Guidelines for Americans 2005 were recently released and suggest similar recommendations regarding dietary fat intake. [162] Implementation of these recommendations may not only reduce LDL-C, but also lower TG and raise HDL-C in patients with MS.[1, 161]

Recently, several randomized clinical trials of low carbohydrate diets for weight loss among obese individuals have been reported. Compared to traditional caloric and fat restricted diets, low carbohydrate diets produced a significantly greater weight loss for 3 [170] to 6 months, [171-173] but not at 1-year. [172, 174] Participants on the low carbohydrate diet maintained most of their 6-month weight loss, whereas those on the caloric and fat restricted diet continued to lose weight throughout the year. [174] Significant metabolic benefits on serum TG and HDL-C were observed with low-carbohydrate diets in some, [172, 174] but not all participants and studies. [170, 171] Low carbohydrate diets were also associated with significant improvements in insulin sensitivity only at 6-months [173] and glycemic control only in a subgroup of patients with DM2 at 6-month and 12-month follow-up. [173, 174] Adherence to the low carbohydrate diet was poor [171-174] and voluntary restriction of food intake and gradual increase in carbohydrate intake was common among those on low carbohydrate diets. [171, 173]

Table 3. Dietary Guidelines for Weight Management

Recommending Agency (Reference)	Carbohydrate (%)	Fat (%)	Protein (%)	Recommendation for weight loss/maintenance
WHO (157)	55-60	20-30	15	Reduce energy intake by 500-600 kcal/d for weight loss
NIH/NAASO (73)	>/=55	</=30	15	Reduce energy intake by 500-1000 kcal/d for weight loss
ADA[1] (158)	60-70	<30	15-20	Reduce energy intake by 500-1000/d
ADA[2] (160)	Moderate fat, high carbohydrate	<30	Not specified	Reduce energy intake by 500-1000/d
AHA (161)	Not stated	</=30	15	Reduce energy intake by 500-1000/d
NCEP (1)	50-60	25-35	15	Balance energy intake and expenditure to maintain desirable weight or prevent weight gain
IOM (159)	45-65	20-35	10-35	Balance intake of essential macronutrients
USDA (162)	Not specified	20-35	Not specified	Maintain healthy weight by balancing intake of calories and exercise

WHO = World Health Organization; NIH = National Institutes of Health; NAASO = North American Association for the Study of Obesity; ADA[1] = American Diabetes Association; ADA[2] = American Dietetic Association; AHA = American Heart Association; NCEP = National Cholesterol Education Program; IOM = Institute of Medicine; USDA: United States Department of Agriculture.

Randomized clinical trials of low carbohydrate dietary interventions in obese subjects with multiple CAD risk factors or known MS have documented similar effects on weight and waist circumference reduction, decrease in TG, increase in HDL-C levels, and improvement in insulin sensitivity at 4 to 16-months follow-up. [156, 174-176] As seen in other reports, the overall adherence rates were low for all types of diets. [156] Another randomized study compared the effect of a Mediterranean-style diet with that of a control diet on CAD risk reduction among 180 overweight or obese patients with MS. The two diets were similar in total calories and percent calories from carbohydrate, fat, and protein. However, patients on the Mediterranean diet were instructed to increase intake of complex carbohydrates, fiber, mono- and polyunsaturated fat, and decrease intake of saturated fat and cholesterol. At the end of a two-year follow-up, those on the Mediterranean diet had significantly greater weight and WC reduction, decrease in blood pressure, insulin resistance, TG and increase in HDL-C. There was also a significant reduction in the number of patients who continued to meet the diagnostic criteria for MS after the two-year Mediterranean diet intervention. [177]

In a recent meta-analysis of 107 studies of low versus high carbohydrate diets, weight loss was found to be associated with reduced caloric intake and longer diet duration but not with reduced carbohydrate content. [155] Therefore, negative energy balance provides weight loss regardless of the macronutrient composition of the diet and caloric restriction is the essential component of weight loss treatment.[178] Low carbohydrate diets may facilitate short-term weight loss compared to traditional calorie and fat restricted diets. However, low carbohydrate intake appears to be unsustainable and long-term safety and efficacy results of low carbohydrate diets are lacking.

There is insufficient evidence to make recommendations for or against the use of low-carbohydrate diets, particularly for those older than age 50, for use longer than 90 days, or for diets of 20 grams of carbohydrates or less per day. [155] Although controversy exists regarding low fat versus low carbohydrate weight loss diets, [179] there is a significant body of evidence that suggests that successful weight loss can be achieved by creation of a sustained energy deficit through a combination of regular exercise and a nutritionally balanced and varied diet.

Dietary Intervention and Lipid Profile

Changes in lipid profile through weight loss diets have been reported in obese patients with insulin resistance with or without MS. A decrease in total cholesterol and TG levels was reported with a low carbohydrate, very low calorie diet among obese individuals and MS patients followed for an average of 4 months. [176] Reponses of TG and HDL-C to the low carbohydrate diet were more favorable than to the conventional low fat, calorie-restricted diet at 12-months in obese patients with DM2 and MS. [174] A study comparing Atkins, Ornish, Weight Watchers, and Zone diets for weight loss and CAD risk reduction in obese patients with known MS components documented a persistent increase in HDL in all these weight loss programs except Ornish, while decreases in TG were not significant at the end of 12-months. [156] Similar results with a low carbohydrate diet were reported in obese, hyperinsulinemic subjects at 16 months. [175] Both elevated HDL and reduced TG were observed with a low carbohydrate diet in obese, hyperlipidemic patients [180] and with the Mediterranean diet in MS patients. [177] On the other hand, a low fat, high carbohydrate diet with a simple to complex carbohydrate ratio of 1:2 and ad libitum energy intake resulted in moderate weight reduction and decrease in both total cholesterol and HDL-C after 6 months in patients with MS compared to a high fat, low carbohydrate diet. An increase in TG was observed when the ratio of simple to complex carbohydrate was reversed. [181]

Dietary Intervention and Insulin Sensitivity

In addition to weight loss, significant improvement of fasting insulin level or insulin sensitivity was observed with a low-carbohydrate or a low-fat, calorie-restricted diet in obese, hyperinsulinemic subjects or obese patients with MS components at one-year.[156, 175] However, the beneficial effects on insulin sensitivity from the low carbohydrate diet did not persist beyond 6-months in obese patients with DM2 and MS. [173, 174] On the other hand, the Mediterranean diet, rich in complex carbohydrates, fiber, and unsaturated fat, resulted in enhanced insulin sensitivity in MS patients for up to 2 years. [177] The importance of glycemic load in the development of DM2 among pre-diabetic subjects was recently demonstrated by the results of the Study to Prevent Non-Insulin-Dependent Diabetes Mellitus (STOP-NIDDM). In this large, randomized controlled trial, treatment with acarbose, an alpha-glucosidase inhibitor which slows the digestion of satrches, reduced new-onset DM2 by 25% compared to placebo after a 3-year follow-up. [182] More encouragingly, diabetes prevention studies employing therapeutic lifestyle changes (TLC) with an emphasis on low saturated fat, high-fiber diet and regular exercise have demonstrated significant improvement in glucose tolerance and up to a 58% reduction in the incidence of new DM2 along with modest weight loss. These studies were conducted in different geographic areas, among ethnically diverse populations, in the settings of intensive training program and community-based primary care and with relatively long-term follow-up. [141, 147, 148, 150, 151]

Dietary Intervention and Blood Pressure Control

Weight loss through dietary modification is associated with improved blood pressure in obese individuals.[73] Diets rich in fiber with low saturated and high unsaturated fat lower blood pressure, especially when combined with restricted sodium intake. [183] Interestingly, acarbose not only diminished the incidence of DM2, but also reduced new-onset HTN in pre-diabetic patients by 34%. [184] In one non-randomized cohort study, a very low carbohydrate, high protein diet also produced substantial reduction in weight and blood pressure in obese MS patients after 4 months. [176] However, other investigations have reported that dietary intervention did not lead to a persistent lowering of blood pressure in obese patients with manifestations of MS. [156, 174, 175] The Mediterranean diet, on the other hand, has been shown to significantly reduce blood pressure compared to a conventional, low fat diet, even after two years. [177]

Dietary Intervention and Inflammation

The effects of weight loss diets on serum inflammatory markers have also been reported in obese patients. C-reactive protein was significantly reduced in obese subjects with some or all components of MS by a low carbohydrate diet, conventional, low-fat, calorie-restricted diet, or a Mediterranean diet. [156, 175, 177] The Mediterranean diet also reduced the proinflammatory mediators interleukin-6 (IL-6), IL-7, and IL-18. [177]

Physical Activity

Exercise is an important component of weight loss therapy even though it is not associated with substantially greater weight loss than diet alone. [73] In fact, successful weight loss is best attained through combined diet and exercise to optimize energy deficit. [185] Sustained physical activity is beneficial for weight maintenance after initial weight loss is achieved through dietary modification and in the prevention of weight gain. [73, 186] Adequate physical activity is also beneficial for reducing CAD risk beyond weight loss itself. [73, 187] The effects of physical activity on cardiovascular outcomes evaluated in 2803 diabetic men in the Health Professionals' Follow-up Study demonstrated that volume and intensity of physical activity were both important determinants of reduced mortality. [188] Moderate exercise with expenditure of approximately 1000 calories on most, preferably all, days of the week is recommended for weight loss and maintenance. [73] In addition, resistance training plays an important role in achieving and maintaining healthy weight. Although studies of strength training have not documented significant changes in body weight, they have demonstrated increases in fat-free body mass and decreases in fat mass. [189, 190]

Most of the benefits of resistance training can be achieved with two 15-20-minute sessions per week. [191] Exercise should incorporate both upper ad lower body musculature and be performed in a slow and controlled manner in order to minimize injury. Current guidelines recommend a minimum of 30-minute, moderate-intensity physical activity on most, preferably all, days of the week. [73, 186, 187, 192-195] (Table 4)

Physical Activity and Lipid Profile

Physically active individuals have higher HDL and lower TG than sedentary individuals. [196] Furthermore, atherogenic DLP respond favorably to exercise training. [196] For example, aerobic exercise and resistance training increases HDL-C and decreases TG, [197-199] changes that may persist for as much as 36 months in MS patients. [200] The Health, Risk Factors, Exercise Training and Genetics (HERITAGE) Family study compared the effects of a 20-week endurance exercise program on men with favorable and unfavorable lipid profiles. Individuals with atherogenic DPL had significantly greater reductions in apolipoprotein B and greater increases in HDL-C than the control group or those with isolated low HDL-C. [201]

Table. 4 Guidelines on Exercise for CAD Risk Reduction

Recommending Agency (Reference)	Type of Prevention	Volume of Exercise	Intensity of Exercise	Duration (continuous or accumulated) of Exercise	Frequency of Exercise
HLB/NAASO (73)	Primary	1000Kcal/week	Moderate	>/=30 minutes	Most/all days of the week
NIH (187)	Primary	Not specified	Moderate	>/=30 minutes	Most/all days of the week
CDC (186)	Primary	200kcal/day	Moderate	>/=30 minutes	Most/all days of the week
ACSM (193)	Primary	Not specified	40-85%VO_2R	20-60 minutes	3-5 days/week
AHA (194)	Primary	Not specified	Not specified	30-60 minutes	4-6 days/week
ACSM (195)	Secondary	Not specified	>/= 40%VO_2R	20-60 minutes	3-5 days/week
AHA (194)	Secondary	Not specified	50-80% VO_2R	20-30 minutes	3-4 days/week

HLB/NAASO: National Heart Lung and Blood Institute/North American Association for the study of Obesity; NIH: National Institutes of Health; CDC: Centers for Disease Control; ACSM: American College of Sports Medicine; AHA: American Heart Association. VO_2R: Maximum oxygen uptake reserve.

Physical Activity and Insulin Sensitivity

Regular exercise increases insulin sensitivity in skeletal muscle [202] and exercise training even without weight loss increases insulin sensitivity in previously sedentary adults. [203] In the Finnish Diabetes Prevention Study (DPS) of 522 middle aged, pre-diabetic patients, those treated with combined diet and exercise showed significantly improved fasting glucose, 2-hour post- challenge glucose, and HbA1c at the end of one year. [141] However, the effect remained statistically significant only for HbA1c at 3 years. [148] Along with significant weight and WC reduction, DM2 incidence was reduced by 58% in the intervention group compared to the control group. [141] The US Diabetes Prevention Program (DPP) enrolled 3234 pre-diabetic patients and randomly assigned them to standard lifestyle recommendations with metformin or placebo or an intensive lifestyle modification group. Patients in the intensive lifestyle modification group also had 58% reduction in DM2 incidence, improved insulin sensitivity and significant weight loss. [150] In a randomized, controlled study of 79 normoglycemic, insulin resistant men and women, reductions in weight and WC occurred in parallel with changes in insulin sensitivity in the group treated with intensive diet and exercise after 4-months of intervention. [204]

Physical Activity and Blood Pressure Control

Physical activity is associated with reduction of blood pressure. [145] Ambulatory blood pressure measurements provide evidence that endurance training lowers blood pressure in sedentary men. [205] However, some studies have failed to show a clinically meaningful reduction in blood pressure with endurance training. [206] There are few reports on the effect of exercise on blood pressure in MS and most studies include both a diet and an exercise intervention. In the Finnish Diabetes Prevention Study, significant reductions in systolic and diastolic blood pressure in pre-diabetic patients were observed with diet plus exercise. [141] However, other investigators reported no change in blood pressure in insulin resistant patients participating in a 4-month combined diet and exercise program. [204]

Physical Activity and Inflammation

Prothrombotic and proinflammatory states are seen in individuals with MS. [2] The presence of serum inflammatory markers especially is associated with a significant risk of CAD in healthy individuals [207] and in MS patients [30] CRP in particular is related to the number of MS components present and is recognized as a risk factor for CAD. [30] Physical activity has been documented to be inversely related to levels of inflammatory markers including: fibrinogen, blood viscosity, platelet counts, coagulation factors VIII and IX, von Willebrand factor, fibrin D-dimer, tissue plasminogen activator antigen, CRP, even after adjusting for confounding factors. [208] A beneficial effect of frequent leisure-time physical activity on obesity-related inflammatory markers has also been described. [209] Finally, a significant reduction in monocyte chemoattractant protein-1 (MCP-1], IL-8 and visceral fat has bee reported in MS patients treated with exercise and pravastatin compared to those treated with pravastatin alone. [210] Thus, exercise may play a role in alleviating the prothrombotic and proinflammatory state in MS.

Pharmacological Intervention, Weight Loss and MS

Currently, there is little experimental or outcome data on cardiovascular mortality with pharmacologic weight loss agents. [138] Sibutramine, an appetite suppressant, and orlistat, a lipase inhibitor, are the only approved agents for long-term use in weight loss that have been extensively studied. Both medications have modest efficacy in promoting and maintaining weight loss, an effect that may benefit those with insulin resistance and pre-diabetes.

Sibutramine plus an AHA diet significantly increases weight loss and improves lipids when compared to placebo in obese patients with high TG and low HDL characteristic of MS. [211] Orlistat improves fasting blood glucose and fasting insulin levels in obese patients with and without DM2. [212, 213] For example, in a double-blind, prospective study, 3305 obese patients with or without IGT were randomized to TLC plus orlistat or placebo. Orlistat combined with TLC produced greater weight loss over 4 years and resulted in significant reductions in the incidence of DM2 in a subgroup with IGT. [214] In a post hoc analysis, weight loss induced by a 52-week, calorie-restricted diet plus orlistat was also associated with significant reductions in TG and increases in HDL-C in obese MS patients compared with obese subjects without MS. [215] Finally, orlistat plus a calorie-restricted diet has been reported to significantly reduce weight, WC and systolic blood pressure, and improve fasting glucose and insulin sensitivity compared to diet alone in MS patients with co-existing DM2. [216] Interestingly, orlistat-treated patients had a marked reduction in estimated 10-year risk of vascular events and were much less likely to continue to meet criteria for MS at the end of the study period. [216]

Metformin is a pharmacologic agent associated with improvement of insulin sensitivity and weight loss in obese patients without insulin resistance. [217-219] Metformin appears to decrease caloric intake by suppressing appetite, thereby potentiating weight loss. [220] Interestingly, metformin-induced weight loss appears to preferentially affect adipose tissue. [221]

The effects of metformin on weight loss and insulin sensitivity have also been documented in pre-diabetic and insulin resistant patients. For example, addition of metformin to TLC in the Diabetes Prevention Program resulted in moderate weight reduction, improved glucose tolerance, and a 31% reduction in new onset DM2. Although impressive, these effects were even less dramatic than those achieved with intensive TLC. [150] Polycystic ovary syndrome (PCOS) is a disorder strongly associated with insulin resistance. [23] Treatment of obese women with or without PCOS and obese, hyperinsulinemic adolescent girls with metformin plus a low-caloric diet resulted in weight loss comparable to sibutramine and orlistat [218] and improved insulin sensitivity. [217-219] Thus, metformin, although not an approved weight loss agent, does facilitate weight loss and improves insulin sensitivity.

Despite the positive effects of weight loss agents on insulin resistance, lipid profile and development of DM2 reported in some studies, reports examining the effect of these agents on long-term cardiovascular mortality are largely unavailable. However, in the UK Prospective Diabetes Study (UKPDS), 1704 overweight patients with newly diagnosed DM2 were randomized to treatment with diet, metformin, sulfonylurea, or insulin. Only metformin treatment, over a period of approximately 10 years, was associated with significant reductions in myocardial infarction. [222]

Other Pharmacologic Interventions

Hydroxymethylglutaryl-coenzyme A (HMG-CoA) reductase inhibitors (statin), fibrates, and thiazolidinediones (TZD), pharmacologic agents without direct weight loss effects, have been reported to improve some aspects of MS including atherogenic DLP, insulin resistance, and inflammation. Moreover, subgroup analysis of several large randomized clinical trials indicates that angiotensin converting enzyme inhibitors (ACE-I) and angiotensin receptor blockers (ARB) not only lower blood pressure but also improve insulin sensitivity [223] and reduce new-onset DM2. [224-227] Since both insulin resistance and elevated blood pressure are major components of MS, treatment of hypertensive MS patients with an ACE-I or ARB appears attractive. These reports provide direct evidence that agents not directly related to weight loss are effective in improving the metabolic profile in MS.

It appears that the beneficial effects of statins on CAD risk reduction in patients with DM2 or MS are at least partially related to lipid modification. [228-232] A recent post hoc analysis from the Scandinavian Simvastatin Survival Study (4S) documented that the absolute

CAD risk reduction with simvastatin was most evident in a subgroup of patients with MS. [232] This is in contrast to the results obtained in the Anglo-Scandinavian Cardiac Outcomes Trial-Lipid Lowering Arm (ASCOT-LLA) in which atorvastatin reduced CAD events only in subjects without MS. [233] Differences in baseline LDL may account for outcome differences in these two studies with statins being less effective if basal LDL is not high. [232]

Although unproven, it is also possible that the differences in CAD risk reduction observed between statins, may represent differences in the effect of individual drugs on glucose metabolism. For example, in the West of Scotland Coronary Prevention Study (WOSCOPS), pravastatin resulted in a 30% reduction of new-onset DM2, [234] an effect not observed with simvastatin in the Heart Protection Study [229] or atorvastatin in ASCOT-LLA [233] If this hypothesis is correct, it might be predicted that statins that prevent or delay DM2 would more effectively lower overall CAD risk than those that have less effect on DM2 prevention. Although there is indirect evidence to support this hypothesis, it remains speculative.

In contrast to statins, fibrates produce greater reductions in TG and increases in HDL-C with less LDL-C lowering effects. In a post hoc analysis of the Helsinki Heart Study (HHS), gemfibrozil reduced CAD events most in individuals with higher BMI, high TG and low HDL-C levels. [235] Similar findings were reported in the Veterans Affairs HDL Intervention Trial (VA-HIT). [236] Thus, fibrates appear to have a beneficial effect on atherogenic lipid profile as well as cardiovascular mortality among patients with insulin resistance.

TZDs increase insulin sensitivity and have favorable effects on lipid profiles in patients with insulin resistance. In these patients, TZDs are associated with significant improvements in glucose tolerance and an 89% reduction in new-onset DM2 compared to subjects not treated with any anti-diabetic agent. [237] Similar results were reported in the Troglitazone in Prevention of Diabetes (TRIPOD) study in Hispanic women with history of gestational DM and insulin resistance. [238] Despite similar salutary effects on insulin sensitivity, the TZDs have discrepant effects on lipids. [239, 240] Furthermore, concerns exist regarding weight gain from fluid retention and worsening of heart failure with the use of these agents. Therefore, although TZDs may exert beneficial effects on lipids and insulin sensitivity, their role in preventing cardiovascular events remains unclear.

As described above, MS appears to be associated with a pro-inflammatory and pro-thrombotic state [2, 28-31, 241] Atorvastatin, [242] gemfibrozil, [243] rosiglitazone, [244] and irbesartan [245] have all been reported to significantly reduce inflammatory or pro-thrombotic markers in pre-diabetic, diabetic, or MS patients. Although these interventions appear to modulate pro-inflammatory and pro-thrombotic markers, controlled trials are needed to determine whether such intervention translates into improved clinical outcomes.

Surgical Intervention, Weight Loss and MS

Weight loss surgery is an option for weight reduction in patients with severe obesity, i.e. a BMI>/= 40, or a BMI>/= 35 and co-morbid conditions. [73] Weight loss surgery provides significant, sustained weight loss in most patients for over five years. [73] Two types of operations have proven to be effective: gastric volume restriction and gastric bypass surgeries. [73] Mortality rates associated with the surgeries vary depending on co-morbidities.

A recent meta-analysis examined the impact of bariatric surgery on weight loss, operative mortality, and four obesity-associated co-morbidities, including DM2, DLP, HTN, and obstructive sleep apnea. [153] Surgery elicited substantial weight loss and operative mortality at 30 days or less was 0.1% for the restrictive procedures and 0.5% for gastric bypass procedures but increased to 1.1% for biliopancreatic diversion or duodenal switch procedures. [153] All bariatric surgeries were associated with significant improvements in glucose tolerance or even DM2 resolution with more pronounced effects observed with malabsorptive as compared to restrictive procedures. Interestingly, resolution of DM2 often occurred days following surgery, before significant weight loss was achieved. [153, 246] The temporal relationship between bariatric surgery and resolution or improvement in DM2 may therefore be related to changes in gut hormones regulating glucose metabolism rather than weight loss *per se*. [153]

Improvement in lipid profile was seen in the vast majority of patients undergoing bariatric surgery, although there are differences observed between surgical procedures. [153] Blood pressure reduction also occurred in the majority of patients undergoing bariatric surgery, but in contrast to surgical effects on DM2 and DPL, blood pressure reduction appeared to be independent of operative procedure. [153] Thus a majority of patients with DM2, HTN, and DLP, experience significant improvement or complete resolution of metabolic derangements in addition to the marked weight loss following bariatric surgery. The differences in results observed on DM2 and DLP with different forms of weight loss surgery provides indirect evidence that nutrient absorption and gut hormones play an integral role in the genesis of the metabolic abnormalities of obesity. [153]

Global Risk Reduction and MS

Recently, the INTERHEART study of 29,000 individuals from 52 countries revealed that modifiable risk factors account for more than 90% of the risk of myocardial infarction across all age, gender, and geographic groups. [247] According to a recent study using mathematical modeling, optimal control of blood pressure and lipids would result in prevention of more than 80% of CAD events in patients with MS. [248] Multifactorial intervention in patients with DM2 has been demonstrated to improve glycosylated hemoglobin values, blood pressure, and fasting lipid profile, and to produce significant reductions in complications from DM2 as well as cardiovascular events. [249] The concept of global risk reduction for health promotion and disease prevention was advocated in a recent joint scientific statement from the American Cancer Society, the American Diabetes Association, and the American Heart Association. [250] Multifactorial intervention in patients with MS may provide the mechanism for such global risk reduction.

Conclusion

Physicians face the challenge of an epidemic of obesity, DM2, and MS with severe public health consequences. Early recognition of MS and aggressive implementation of therapeutic strategies to halt the syndrome at its pre-clinical stages are keys to preventing DM2, CAD and

CAD mortality. Adequate treatment requires targeting multiple risk factors and components of MS, including obesity, insulin resistance, pre-diabetes or overt DM2, pre-hypertension or overt HTN, and atherogenic DLP. TLC remains the mainstay of treatment of obesity and MS with emphasis on adoption of a healthy diet and regular physical activity. Pharmacological intervention also appears to play an important role in the treatment of obesity, glucose intolerance, atherogenic DLP, elevated blood pressure and MS. Surgery is reserved for those with severe obesity and co-morbid conditions and those who fail to response to TLC with or without pharmacologic intervention.

Although the exact mechanisms underlying MS remain to be determined, physicians can immediately contribute to improved long-term outcome by endorsing a healthy lifestyle that promotes weight loss or at least, weight maintenance and increased physical activity. This may lead to the reduction in morbidity and mortality associated with obesity, DM2, and MS.

REFERENCES

[1] Executive Summary of The Third Report of The National Cholesterol Education Program (NCEP) Expert Panel on Detection, Evaluation, And Treatment of High Blood Cholesterol In Adults (Adult Treatment Panel III). Jama 2001;285(19):2486-97.

[2] Grundy SM, Brewer HB, Jr., Cleeman JI, Smith SC, Jr., Lenfant C. Definition of metabolic syndrome: Report of the National Heart, Lung, and Blood Institute/American Heart Association conference on scientific issues related to definition. *Circulation* 2004;109(3):433-8.

[3] Vague J. The degree of masculine differentiation of obesities: a factor determining predisposition to diabetes, atherosclerosis, gout, and uric calculous disease. *Am J Clin Nutr* 1956;4(1):20-34.

[4] Avogaro P, Creapaldi G. Essential hyperlipidemia, obesity and diabetes. *Diabetologia* 1965;1:137.

[5] Welborn TA, Breckenridge A, Rubinstein AH, Dollery CT, Fraser TR. Serum-insulin in essential hypertension and in peripheral vascular disease. *Lancet* 1966;1(7451):1336-7.

[6] Welborn TA, Wearne K. Coronary heart disease incidence and cardiovascular mortality in Busselton with reference to glucose and insulin concentrations. *Diabetes Care 1979*;2(2):154-60.

[7] Pyorala K. Relationship of glucose tolerance and plasma insulin to the incidence of coronary heart disease: results from two population studies in Finland. *Diabetes Care* 1979;2(2):131-41.

[8] Modan M, Halkin H, Almog S, et al. Hyperinsulinemia. A link between hypertension obesity and glucose intolerance. *J Clin Invest* 1985;75(3):809-17.

[9] Vague J, Vague P, Tramoni M, Vialettes B, Mercier P. Obesity and diabetes. *Acta Diabetol Lat* 1980;17(2):87-99.

[10] Camus JP. [Gout, diabetes, hyperlipemia: a metabolic trisyndrome]. *Rev Rhum Mal Osteoartic* 1966;33(1):10-4.

[11] Reaven GM. Banting lecture 1988. *Role of insulin resistance in human disease. Diabetes* 1988;37(12):1595-607.

[12] Hjermann I. The metabolic cardiovascular syndrome: syndrome X, Reaven's syndrome, insulin resistance syndrome, atherothrombogenic syndrome. *J Cardiovasc Pharmacol* 1992;20 Suppl 8:S5-10.

[13] Kaplan NM. The deadly quartet. Upper-body obesity, glucose intolerance, hypertriglyceridemia, and hypertension. *Arch Intern Med* 1989;149(7):1514-20.

[14] DeFronzo RA, Ferrannini E. Insulin resistance. A multifaceted syndrome responsible for NIDDM, obesity, hypertension, dyslipidemia, and atherosclerotic cardiovascular disease. *Diabetes Care* 1991;14(3):173-94.

[15] Zimmet PZ. Kelly West Lecture 1991. Challenges in diabetes epidemiology--from West to the rest. *Diabetes Care* 1992;15(2):232-52.

[16] Haffner SM, Valdez RA, Hazuda HP, Mitchell BD, Morales PA, Stern MP. Prospective analysis of the insulin-resistance syndrome (syndrome X). *Diabetes* 1992;41(6):715-22.

[17] Kesaniemi YA, Lilja M, Kervinen K, Rantala A. Multiple metabolic syndrome: aspects of genetic epidemiology and molecular genetics. *Ann Med* 1992;24(6):461-4.

[18] Despres JP. The insulin resistance-dyslipidemia syndrome: the most prevalent cause of coronary artery disease? *Cmaj* 1993;148(8):1339-40.

[19] Alberti KG, Zimmet PZ. Definition, diagnosis and classification of diabetes mellitus and its complications. Part 1: diagnosis and classification of diabetes mellitus provisional report of a WHO consultation. *Diabet Med* 1998;15(7):539-53.

[20] Hansen BC. The metabolic syndrome X. Ann N Y Acad Sci 1999;892:1-24.

[21] Groop L, Orho-Melander M. The dysmetabolic syndrome. *J Intern Med* 2001;250(2):105-20.

[22] Balkau B, Charles MA, Drivsholm T, et al. Frequency of the WHO metabolic syndrome in European cohorts, and an alternative definition of an insulin resistance syndrome. *Diabetes Metab* 2002;28(5):364-76.

[23] Einhorn D, Reaven GM, Cobin RH, et al. American College of Endocrinology position statement on the insulin resistance syndrome. *Endocr Pract* 2003;9(3):237-52.

[24] Foster DW. Insulin resistance--a secret killer? *N Engl J Med* 1989;320(11):733-4.

[25] Knudsen P, Eriksson J, Lahdenpera S, Kahri J, Groop L, Taskinen MR. Changes of lipolytic enzymes cluster with insulin resistance syndrome. *Botnia Study Group. Diabetologia* 1995;38(3):344-50.

[26] Ferrannini E, Baldi S, Stern M. Circulating free fatty acids are an integral part of the insulin resistance syndrome. *Diabetes* 1995;44:182A.

[27] Hotamisligil GS, Arner P, Caro JF, Atkinson RL, Spiegelman BM. Increased adipose tissue expression of tumor necrosis factor-alpha in human obesity and insulin resistance. *J Clin Invest* 1995;95(5):2409-15.

[28] Pickup JC, Mattock MB, Chusney GD, Burt D. NIDDM as a disease of the innate immune system: association of acute-phase reactants and interleukin-6 with metabolic syndrome X. *Diabetologia* 1997;40(11):1286-92.

[29] Tamakoshi K, Yatsuya H, Kondo T, et al. The metabolic syndrome is associated with elevated circulating C-reactive protein in healthy reference range, a systemic low-grade inflammatory state. *Int J Obes Relat Metab Disord* 2003;27(4):443-9.

[30] Ridker PM, Buring JE, Cook NR, Rifai N. C-reactive protein, the metabolic syndrome, and risk of incident cardiovascular events: an 8-year follow-up of 14 719 initially healthy American women. *Circulation* 2003;107(3):391-7.

[31] Yudkin JS. Abnormalities of coagulation and fibrinolysis in insulin resistance. Evidence for a common antecedent? *Diabetes Care* 1999;22 Suppl 3:C25-30.

[32] Juhan-Vague I, Thompson SG, Jespersen J. Involvement of the hemostatic system in the insulin resistance syndrome. A study of 1500 patients with angina pectoris. The ECAT Angina Pectoris Study Group. *Arterioscler Thromb* 1993;13(12):1865-73.

[33] Ford ES, Giles WH, Mokdad AH. Increasing prevalence of the metabolic syndrome among u.s. Adults. *Diabetes Care* 2004;27(10):2444-9.

[34] Ford ES, Giles WH, Dietz WH. Prevalence of the metabolic syndrome among US adults: findings from the third National Health and Nutrition Examination Survey. *Jama* 2002;287(3):356-9.

[35] Duncan GE, Li SM, Zhou XH. Prevalence and trends of a metabolic syndrome phenotype among u.s. Adolescents, 1999-2000. *Diabetes Care* 2004;27(10):2438-43.

[36] Alexander CM, Landsman PB, Teutsch SM, Haffner SM. NCEP-defined metabolic syndrome, diabetes, and prevalence of coronary heart disease among NHANES III participants age 50 years and older. *Diabetes* 2003;52(5):1210-4.

[37] Bonora E, Kiechl S, Willeit J, et al. Prevalence of insulin resistance in metabolic disorders: the Bruneck Study. *Diabetes* 1998;47(10):1643-9.

[38] Schmidt MI, Duncan BB, Watson RL, Sharrett AR, Brancati FL, Heiss G. A metabolic syndrome in whites and African-Americans. The Atherosclerosis Risk in Communities baseline study. *Diabetes Care* 1996;19(5):414-8.

[39] Karter AJ, Mayer-Davis EJ, Selby JV, et al. Insulin sensitivity and abdominal obesity in African-American, Hispanic, and non-Hispanic white men and women. The Insulin Resistance and Atherosclerosis Study. *Diabetes* 1996;45(11):1547-55.

[40] Tan CE, Ma S, Wai D, Chew SK, Tai ES. Can we apply the National Cholesterol Education Program Adult Treatment Panel definition of the metabolic syndrome to Asians? *Diabetes Care* 2004;27(5):1182-6.

[41] Lee WY, Park JS, Noh SY, Rhee EJ, Kim SW, Zimmet PZ. Prevalence of the metabolic syndrome among 40,698 Korean metropolitan subjects. *Diabetes Res Clin* Pract 2004;65(2):143-9.

[42] Ilanne-Parikka P, Eriksson JG, Lindstrom J, et al. Prevalence of the metabolic syndrome and its components: findings from a Finnish general population sample and the Diabetes Prevention Study cohort. *Diabetes Care* 2004;27(9):2135-40.

[43] Enkhmaa B SK, Anuurad E, Nogi A, Keiko K, Yamasaki M, Oyunsuren T, Yamane Y. Prevalence of the metabolic syndrome using the third report of the National Cholesterol Education Program Expert Panel on detection, evaluation, and treatment of high blood cholesterol in adults (ATP III) and the modified ATP III definitions for Japanese and Mongolians. Clinica Chimica Acta. 2005;352(1-2):105-13.

[44] Balkau B, Vernay M, Mhamdi L, et al. The incidence and persistence of the NCEP (National Cholesterol Education Program) metabolic syndrome. The French D.E.S.I.R. study. *Diabetes Metab* 2003;29(5):526-32.

[45] Park YW, Zhu S, Palaniappan L, Heshka S, Carnethon MR, Heymsfield SB. The metabolic syndrome: prevalence and associated risk factor findings in the US population from the Third National Health and Nutrition Examination Survey, 1988-1994. *Arch Intern Med* 2003;163(4):427-36.

[46] Sattar N, Gaw A, Scherbakova O, et al. Metabolic syndrome with and without C-reactive protein as a predictor of coronary heart disease and diabetes in the West of Scotland Coronary Prevention Study. *Circulation* 2003;108(4):414-9.
[47] Lorenzo C, Okoloise M, Williams K, Stern MP, Haffner SM. The metabolic syndrome as predictor of type 2 diabetes: the San Antonio heart study. *Diabetes Care* 2003;26(11):3153-9.
[48] Isomaa B, Almgren P, Tuomi T, et al. Cardiovascular morbidity and mortality associated with the metabolic syndrome. *Diabetes Care* 2001;24(4):683-9.
[49] Lakka HM, Laaksonen DE, Lakka TA, et al. The metabolic syndrome and total and cardiovascular disease mortality in middle-aged men. *Jama* 2002;288(21):2709-16.
[50] Vega GL. Results of Expert Meetings: Obesity and Cardiovascular Disease. Obesity, the metabolic syndrome, and cardiovascular disease. *Am Heart J* 2001;142(6):1108-16.
[51] Marroquin OC, Kip KE, Kelley DE, et al. Metabolic syndrome modifies the cardiovascular risk associated with angiographic coronary artery disease in women: a report from the Women's Ischemia Syndrome Evaluation. *Circulation* 2004;109(6):714-21.
[52] Malik S, Wong ND, Franklin SS, et al. Impact of the metabolic syndrome on mortality from coronary heart disease, cardiovascular disease, and all causes in United States adults. *Circulation* 2004;110(10):1245-50.
[53] Mokdad AH, Marks JS, Stroup DF, Gerberding JL. Actual causes of death in the United States, 2000. *Jama* 2004;291(10):1238-45.
[54] Allison DB, Fontaine KR, Manson JE, Stevens J, VanItallie TB. Annual deaths attributable to obesity in the United States. *Jama* 1999;282(16):1530-8.
[55] Fontaine KR, Redden DT, Wang C, Westfall AO, Allison DB. Years of life lost due to obesity. *Jama* 2003;289(2):187-93.
[56] Calle EE, Thun MJ, Petrelli JM, Rodriguez C, Heath CW, Jr. Body-mass index and mortality in a prospective cohort of U.S. adults. *N Engl J Med* 1999;341(15):1097-105.
[57] Wilson PW, D'Agostino RB, Sullivan L, Parise H, Kannel WB. Overweight and obesity as determinants of cardiovascular risk: the Framingham experience. *Arch Intern Med* 2002;162(16):1867-72.
[58] Warnock DG, Textor SC. Hypertension. *Am J Kidney Dis* 2004;44(2):369-75.
[59] Brunzell JD, Hokanson JE. Dyslipidemia of central obesity and insulin resistance. *Diabetes Care* 1999;22 Suppl 3:C10-3.
[60] Ginsberg HN, Huang LS. The insulin resistance syndrome: impact on lipoprotein metabolism and atherothrombosis. *J Cardiovasc Risk* 2000;7(5):325-31.
[61] Kuusisto J, Lempiainen P, Mykkanen L, Laakso M. Insulin resistance syndrome predicts coronary heart disease events in elderly type 2 diabetic men. *Diabetes Care* 2001;24(9):1629-33.
[62] Kannel WB, McGee DL. Diabetes and cardiovascular disease. The Framingham study. *Jama* 1979;241(19):2035-8.
[63] Pyorala M, Miettinen H, Laakso M, Pyorala K. Hyperinsulinemia predicts coronary heart disease risk in healthy middle-aged men: the 22-year follow-up results of the Helsinki Policemen Study. *Circulation* 1998;98(5):398-404.
[64] Haffner SM, Lehto S, Ronnemaa T, Pyorala K, Laakso M. Mortality from coronary heart disease in subjects with type 2 diabetes and in nondiabetic subjects with and without prior myocardial infarction. *N Engl J Med* 1998;339(4):229-34.

[65] Lempiainen P, Mykkanen L, Pyorala K, Laakso M, Kuusisto J. Insulin resistance syndrome predicts coronary heart disease events in elderly nondiabetic men. *Circulation* 1999;100(2):123-8.
[66] Fontbonne A, Charles MA, Thibult N, et al. Hyperinsulinaemia as a predictor of coronary heart disease mortality in a healthy population: the Paris Prospective Study, 15-year follow-up. *Diabetologia* 1991;34(5):356-61.
[67] Maheux P, Jeppesen J, Sheu WH, et al. Additive effects of obesity, hypertension, and type 2 diabetes on insulin resistance. *Hypertension* 1994;24(6):695-8.
[68] Reilly MP, Rader DJ. The metabolic syndrome: more than the sum of its parts? *Circulation* 2003;108(13):1546-51.
[69] Haffner SM. Impaired glucose tolerance, insulin resistance and cardiovascular disease. *Diabet Med* 1997;14 Suppl 3:S12-8.
[70] Bonora E, Targher G, Formentini G, et al. The Metabolic Syndrome is an independent predictor of cardiovascular disease in Type 2 diabetic subjects. Prospective data from the Verona Diabetes Complications Study. *Diabet Med* 2004;21(1):52-8.
[71] Ninomiya JK, L'Italien G, Criqui MH, Whyte JL, Gamst A, Chen RS. Association of the metabolic syndrome with history of myocardial infarction and stroke in the Third National Health and Nutrition Examination Survey. *Circulation* 2004;109(1):42-6.
[72] Kip KE, Marroquin OC, Kelley DE, et al. Clinical importance of obesity versus the metabolic syndrome in cardiovascular risk in women: a report from the Women's Ischemia Syndrome Evaluation (WISE) study. *Circulation* 2004;109(6):706-13.
[73] *The practical guide to the identification, evaluation, and treatment of overweight and obesity in adults.* National Institutes of Health, United States Department of Health and Human Services. Rockville, MD. NIH Publication No. 00-4080. 2000.
[74] *The Surgen General's call to action to prevent and decrease overweight and obesity.* United States Department of Health and Human Servces. Office of the Surgeon General. Rockville, MD. 2001.
[75] Mokdad AH, Ford ES, Bowman BA, et al. Prevalence of obesity, diabetes, and obesity-related health risk factors, 2001. *Jama* 2003;289(1):76-9.
[76] Kopelman PG. Obesity as a medical problem. *Nature* 2000;404(6778):635-43.
[77] Sowers JR. Obesity as a cardiovascular risk factor. *Am J Med* 2003;115 Suppl 8A:37S-41S.
[78] Bray GA. Medical consequences of obesity. *J Clin Endocrinol Metab* 2004;89(6):2583-9.
[79] Ferrannini E, Natali A, Bell P, Cavallo-Perin P, Lalic N, Mingrone G. Insulin resistance and hypersecretion in obesity. European Group for the Study of Insulin Resistance (EGIR). *J Clin Invest* 1997;100(5):1166-73.
[80] Montague CT, O'Rahilly S. The perils of portliness: causes and consequences of visceral adiposity. *Diabetes* 2000;49(6):883-8.
[81] Kelley DE, Williams KV, Price JC, McKolanis TM, Goodpaster BH, Thaete FL. Plasma fatty acids, adiposity, and variance of skeletal muscle insulin resistance in type 2 diabetes mellitus. *J Clin Endocrinol Metab* 2001;86(11):5412-9.
[82] Ferrannini E, Haffner SM, Mitchell BD, Stern MP. Hyperinsulinaemia: the key feature of a cardiovascular and metabolic syndrome. *Diabetologia* 1991;34(6):416-22.
[83] Mokdad AH, Bowman BA, Ford ES, Vinicor F, Marks JS, Koplan JP. The continuing epidemics of obesity and diabetes in the United States. *Jama* 2001;286(10):1195-200.

[84] Hedley AA, Ogden CL, Johnson CL, Carroll MD, Curtin LR, Flegal KM. Prevalence of overweight and obesity among US children, adolescents, and adults, 1999-2002. *Jama* 2004;291(23):2847-50.

[85] Kereiakes DJ, Willerson JT. Metabolic syndrome epidemic. *Circulation* 2003;108(13):1552-3.

[86] HHS, ADA warn Americans of "pre-diabetes", encorage people to take healthy steps to reduce risks. United, States Department of Health and Human Services. Available at: http://www.hhs.gov/news/press/2002press/20020327.html. 2002.

[87] Narayan KM, Boyle JP, Thompson TJ, Sorensen SW, Williamson DF. Lifetime risk for diabetes mellitus in the United States. *Jama* 2003;290(14):1884-90.

[88] Weiss R, Dziura J, Burgert TS, et al. Obesity and the metabolic syndrome in children and adolescents. *N Engl J Med* 2004;350(23):2362-74.

[89] Goodman E, Daniels SR, Morrison JA, Huang B, Dolan LM. Contrasting prevalence of and demographic disparities in the World Health Organization and National Cholesterol Education Program Adult Treatment Panel III definitions of metabolic syndrome among adolescents. *J Pediatr* 2004;145(4):445-51.

[90] Bao W, Srinivasan SR, Wattigney WA, Berenson GS. Persistence of multiple cardiovascular risk clustering related to syndrome X from childhood to young adulthood. The Bogalusa Heart Study. *Arch Intern Med* 1994;154(16):1842-7.

[91] Eisenmann JC, Welk GJ, Wickel EE, Blair SN. Stability of variables associated with the metabolic syndrome from adolescence to adulthood: the Aerobics Center Longitudinal Study. *Am J Hum Biol* 2004;16(6):690-6.

[92] Manson JE, Skerrett PJ, Greenland P, VanItallie TB. The escalating pandemics of obesity and sedentary lifestyle. A call to action for clinicians. *Arch Intern Med* 2004;164(3):249-58.

[93] Thompson D, Wolf AM. The medical-care cost burden of obesity. *Obes Rev* 2001;2(3):189-97.

[94] Hogan P, Dall T, Nikolov P. Economic costs of diabetes in the US in 2002. *Diabetes Care* 2003;26(3):917-32.

[95] Rao SV, Donahue M, Pi-Sunyer FX, Fuster V. Results of Expert Meetings: Obesity and Cardiovascular Disease. Obesity as a risk factor in coronary artery disease. *Am Heart J* 2001;142(6):1102-7.

[96] Kleinman JC, Donahue RP, Harris MI, Finucane FF, Madans JH, Brock DB. Mortality among diabetics in a national sample. *Am J Epidemiol* 1988;128(2):389-401.

[97] McNeill AM, Rosamond WD, Girman CJ, et al. Prevalence of coronary heart disease and carotid arterial thickening in patients with the metabolic syndrome (The ARIC Study). *Am J Cardiol* 2004;94(10):1249-54.

[98] Kullo IJ, Cassidy AE, Peyser PA, Turner ST, Sheedy PF, 2nd, Bielak LF. Association between metabolic syndrome and subclinical coronary atherosclerosis in asymptomatic adults. *Am J Cardiol* 2004;94(12):1554-8.

[99] Turhan H, Erbay AR, Yasar AS, Bicer A, Sasmaz H, Yetkin E. Impaired coronary blood flow in patients with metabolic syndrome: documented by Thrombolysis in Myocardial Infarction (TIMI) frame count method. *Am Heart J* 2004;148(5):789-94.

[100] Howard G, O'Leary DH, Zaccaro D, et al. Insulin sensitivity and atherosclerosis. The Insulin Resistance Atherosclerosis Study (IRAS) Investigators. *Circulation* 1996;93(10):1809-17.

[101] Maison P, Byrne CD, Hales CN, Day NE, Wareham NJ. Do different dimensions of the metabolic syndrome change together over time? Evidence supporting obesity as the central feature. *Diabetes Care* 2001;24(10):1758-63.

[102] Janssen I, Katzmarzyk PT, Ross R. Body mass index, waist circumference, and health risk: evidence in support of current National Institutes of Health guidelines. *Arch Intern Med* 2002;162(18):2074-9.

[103] Carnethon MR, Loria CM, Hill JO, Sidney S, Savage PJ, Liu K. Risk factors for the metabolic syndrome: the Coronary Artery Risk Development in Young Adults (CARDIA) study, 1985-2001. *Diabetes Care* 2004;27(11):2707-15.

[104] Lemieux S, Prud'homme D, Nadeau A, Tremblay A, Bouchard C, Despres JP. Seven-year changes in body fat and visceral adipose tissue in women. Association with indexes of plasma glucose-insulin homeostasis. *Diabetes Care* 1996;19(9):983-91.

[105] Koh-Banerjee P, Wang Y, Hu FB, Spiegelman D, Willett WC, Rimm EB. Changes in body weight and body fat distribution as risk factors for clinical diabetes in US men. *Am J Epidemiol* 2004;159(12):1150-9.

[106] Colditz GA, Willett WC, Rotnitzky A, Manson JE. Weight gain as a risk factor for clinical diabetes mellitus in women. *Ann Intern Med* 1995;122(7):481-6.

[107] Manson JE, Willett WC, Stampfer MJ, et al. Body weight and mortality among women. *N Engl J Med* 1995;333(11):677-85.

[108] Kreuter MW, Scharff DP, Brennan LK, Lukwago SN. Physician recommendations for diet and physical activity: which patients get advised to change? *Prev Med* 1997;26(6):825-33.

[109] Mayer-Davis EJ, Monaco JH, Hoen HM, et al. Dietary fat and insulin sensitivity in a triethnic population: the role of obesity. The Insulin Resistance Atherosclerosis Study (IRAS). *Am J Clin Nutr* 1997;65(1):79-87.

[110] Maron DJ, Fair JM, Haskell WL. Saturated fat intake and insulin resistance in men with coronary artery disease. The Stanford Coronary Risk Intervention Project Investigators and Staff. *Circulation* 1991;84(5):2020-7.

[111] Vessby B. Dietary fat, fatty acid composition in plasma and the metabolic syndrome. *Curr Opin Lipidol* 2003;14(1):15-9.

[112] Laaksonen DE, Lakka TA, Lakka HM, et al. Serum fatty acid composition predicts development of impaired fasting glycaemia and diabetes in middle-aged men. *Diabet Med* 2002;19(6):456-64.

[113] Lovejoy JC, Champagne CM, Smith SR, et al. Relationship of dietary fat and serum cholesterol ester and phospholipid fatty acids to markers of insulin resistance in men and women with a range of glucose tolerance. *Metabolism* 2001;50(1):86-92.

[114] Hu FB, Manson JE, Stampfer MJ, et al. Diet, lifestyle, and the risk of type 2 diabetes mellitus in women. *N Engl J Med* 2001;345(11):790-7.

[115] Salmeron J, Hu FB, Manson JE, et al. Dietary fat intake and risk of type 2 diabetes in women. *Am J Clin Nutr* 2001;73(6):1019-26.

[116] Salmeron J, Manson JE, Stampfer MJ, Colditz GA, Wing AL, Willett WC. Dietary fiber, glycemic load, and risk of non-insulin-dependent diabetes mellitus in women. *Jama* 1997;277(6):472-7.

[117] Ford ES, Mokdad AH, Giles WH, Brown DW. The metabolic syndrome and antioxidant concentrations: findings from the Third National Health and Nutrition Examination Survey. *Diabetes* 2003;52(9):2346-52.

[118] Schulze MB, Liu S, Rimm EB, Manson JE, Willett WC, Hu FB. Glycemic index, glycemic load, and dietary fiber intake and incidence of type 2 diabetes in younger and middle-aged women. *Am J Clin Nutr* 2004;80(2):348-56.

[119] Zhu S, St-Onge MP, Heshka S, Heymsfield SB. Lifestyle behaviors associated with lower risk of having the metabolic syndrome. *Metabolism* 2004;53(11):1503-11.

[120] Ross R, Katzmarzyk PT. Cardiorespiratory fitness is associated with diminished total and abdominal obesity independent of body mass index. *Int J Obes Relat Metab Disord* 2003;27(2):204-10.

[121] Brage S, Wedderkopp N, Ekelund U, et al. Features of the metabolic syndrome are associated with objectively measured physical activity and fitness in Danish children: the European Youth Heart Study (EYHS). *Diabetes Care* 2004;27(9):2141-8.

[122] Franks PW, Ekelund U, Brage S, Wong MY, Wareham NJ. Does the association of habitual physical activity with the metabolic syndrome differ by level of cardiorespiratory fitness? *Diabetes Care* 2004;27(5):1187-93.

[123] Jurca R, Lamonte MJ, Church TS, et al. Associations of muscle strength and fitness with metabolic syndrome in men. *Med Sci Sports Exerc* 2004;36(8):1301-7.

[124] Farrell SW, Cheng YJ, Blair SN. Prevalence of the metabolic syndrome across cardiorespiratory fitness levels in women. *Obes Res* 2004;12(5):824-30.

[125] Leisure-time physical activity. National Health Interview Survey. Available at http://www.cdc.gov/nchs/about/major/nhis/released200412.htm. 2004.

[126] Kimm SY, Glynn NW, Kriska AM, et al. Decline in physical activity in black girls and white girls during adolescence. *N Engl J Med* 2002;347(10):709-15.

[127] Ruderman NB, Schneider SH, Berchtold P. The "metabolically-obese," normal-weight individual. *Am J Clin Nutr* 1981;34(8):1617-21.

[128] Dvorak RV, DeNino WF, Ades PA, Poehlman ET. Phenotypic characteristics associated with insulin resistance in metabolically obese but normal-weight young women. *Diabetes* 1999;48(11):2210-4.

[129] St-Onge MP, Janssen I, Heymsfield SB. Metabolic syndrome in normal-weight Americans: new definition of the metabolically obese, normal-weight individual. *Diabetes Care* 2004;27(9):2222-8.

[130] Nisoli E, Carruba MO. Emerging aspects of pharmacotherapy for obesity and metabolic syndrome. *Pharmacol Res* 2004;50(5):453-69.

[131] Zhu S, Wang Z, Heshka S, Heo M, Faith MS, Heymsfield SB. Waist circumference and obesity-associated risk factors among whites in the third National Health and Nutrition Examination Survey: clinical action thresholds. *Am J Clin Nutr* 2002;76(4):743-9.

[132] Rexrode KM, Carey VJ, Hennekens CH, et al. Abdominal adiposity and coronary heart disease in women. *Jama* 1998;280(21):1843-8.

[133] Wang J. Waist circumference: a simple, inexpensive, and reliable tool that should be included as part of physical examinations in the doctor's office. *Am J Clin Nutr* 2003;78(5):902-3.

[134] Ford ES, Mokdad AH, Giles WH. Trends in waist circumference among U.S. adults. *Obes Res* 2003;11(10):1223-31.

[135] Okosun IS, Choi ST, Boltri JM, et al. Trends of abdominal adiposity in white, black, and Mexican-American adults, 1988 to 2000. *Obes Res* 2003;11(8):1010-7.

[136] Willett WC, Dietz WH, Colditz GA. Guidelines for healthy weight. *N Engl J Med* 1999;341(6):427-34.

[137] Ayyad C, Andersen T. Long-term efficacy of dietary treatment of obesity: a systematic review of studies published between 1931 and 1999. *Obes Rev* 2000;1(2):113-9.

[138] Padwal R, Li SK, Lau DC. Long-term pharmacotherapy for overweight and obesity: a systematic review and meta-analysis of randomized controlled trials. *Int J Obes Relat Metab Disord* 2003;27(12):1437-46.

[139] Panel. NTAC. Methods for voluntary weight loss and control. *Ann Intern Med* 1993;119:764-70.

[140] Wadden TA. Treatment of obesity by moderate and severe caloric restriction. Results of clinical research trials. *Ann Intern Med* 1993;119(7 Pt 2):688-93.

[141] Tuomilehto J, Lindstrom J, Eriksson JG, et al. Prevention of type 2 diabetes mellitus by changes in lifestyle among subjects with impaired glucose tolerance. *N Engl J Med* 2001;344(18):1343-50.

[142] Wassertheil-Smoller S, Blaufox MD, Oberman AS, Langford HG, Davis BR, Wylie-Rosett J. The Trial of Antihypertensive Interventions and Management (TAIM) study. Adequate weight loss, alone and combined with drug therapy in the treatment of mild hypertension. *Arch Intern Med* 1992;152(1):131-6.

[143] Wing RR, Koeske R, Epstein LH, Nowalk MP, Gooding W, Becker D. Long-term effects of modest weight loss in type II diabetic patients. *Arch Intern Med* 1987;147(10):1749-53.

[144] Osterman J, Lin T, Nankin HR, Brown KA, Hornung CA. Serum cholesterol profiles during treatment of obese outpatients with a very low calorie diet. Effect of initial cholesterol levels. *Int J Obes Relat Metab Disord* 1992;16(1):49-58.

[145] Halbert JA, Silagy CA, Finucane P, Withers RT, Hamdorf PA, Andrews GR. The effectiveness of exercise training in lowering blood pressure: a meta-analysis of randomised controlled trials of 4 weeks or longer. *J Hum Hypertens* 1997;11(10):641-9.

[146] Hu FB, Sigal RJ, Rich-Edwards JW, et al. Walking compared with vigorous physical activity and risk of type 2 diabetes in women: a prospective study. *Jama* 1999;282(15):1433-9.

[147] Pan XR, Li GW, Hu YH, et al. Effects of diet and exercise in preventing NIDDM in people with impaired glucose tolerance. The Da Qing IGT and Diabetes Study. *Diabetes Care* 1997;20(4):537-44.

[148] Lindstrom J, Louheranta A, Mannelin M, et al. The Finnish Diabetes Prevention Study (DPS): Lifestyle intervention and 3-year results on diet and physical activity. *Diabetes Care* 2003;26(12):3230-6.

[149] Hardman AE. Interaction of physical activity and diet: implications for lipoprotein metabolism. *Public Health Nutr* 1999;2(3A):369-76.

[150] Knowler WC, Barrett-Connor E, Fowler SE, et al. Reduction in the incidence of type 2 diabetes with lifestyle intervention or metformin. *N Engl J Med* 2002;346(6):393-403.

[151] Eriksson KF, Lindgarde F. Prevention of type 2 (non-insulin-dependent) diabetes mellitus by diet and physical exercise. The 6-year Malmo feasibility study. *Diabetologia* 1991;34(12):891-8.

[152] Yusuf S. Two decades of progress in preventing vascular disease. *Lancet* 2002;360(9326):2-3.

[153] Buchwald H, Avidor Y, Braunwald E, et al. Bariatric surgery: a systematic review and meta-analysis. *Jama* 2004;292(14):1724-37.

[154] Van Gaal LF, Wauters MA, De Leeuw IH. The beneficial effects of modest weight loss on cardiovascular risk factors. *Int J Obes Relat Metab Disord* 1997;21 Suppl 1:S5-9.

[155] Bravata DM, Sanders L, Huang J, Krumholz HM, Olkin I, Gardner CD. Efficacy and safety of low-carbohydrate diets: a systematic review. *Jama* 2003;289(14):1837-50.

[156] Dansinger ML, Gleason JA, Griffith JL, Selker HP, Schaefer EJ. Comparison of the Atkins, Ornish, Weight Watchers, and Zone diets for weight loss and heart disease risk reduction: a randomized trial. *Jama* 2005;293(1):43-53.

[157] *Obesity: preventing and managing the global epidemic.* Report of a WHO Consultation Technical Report Series 894. World Health Organization. Geneva, Switzerland. 2000.

[158] Franz MJ, Bantle JP, Beebe CA, et al. Nutrition principles and recommendations in diabetes. *Diabetes Care* 2004;27 Suppl 1:S36-46.

[159] *Dietary reference intakes for energy, carbohydrate, fiber, fat, fatty acids, cholesterol, protein, and aminoacids (macronutrients).* Institute of Medicine. Washington, DC: National Academy Press;. 2002.

[160] Cummings S, Parham ES, Strain GW. Position of the American Dietetic Association: weight management. *J Am Diet Assoc* 2002;102(8):1145-55.

[161] Krauss RM, Eckel RH, Howard B, et al. AHA Dietary Guidelines: revision 2000: A statement for healthcare professionals from the Nutrition Committee of the American Heart Association. *Circulation* 2000;102(18):2284-99.

[162] Dietary Guidelines for Americans 2005. United States Department of Health and Human Services, Department of Agriculture. Available at *http://www.healthierus.gov/ dietaryguidelines/*. 2005.

[163] Sears B LB. *The zone: a dietary road map.* New York, NY: Harper Collins. 1995.

[164] Atkins R. *Dr. Atkin's new diet revolution.* New York, NY: Avon Books. 1998.

[165] Agatston A. *The South Beach diet.* Emmaus, PA: Rodale Press. 2003.

[166] Astrup A. Healthy lifestyles in Europe: prevention of obesity and type II diabetes by diet and physical activity. *Public Health Nutr* 2001;4(2B):499-515.

[167] Knittle JL, Ahrens EH, Jr. Carbohydrate Metabolism in Two Forms of Hyperglyceridemia. *J Clin Invest* 1964;43:485-95.

[168] Katan MB. Effect of low-fat diets on plasma high-density lipoprotein concentrations. *Am J Clin Nutr* 1998;67(3 Suppl):573S-6S.

[169] Rasmussen OW, Thomsen C, Hansen KW, Vesterlund M, Winther E, Hermansen K. Effects on blood pressure, glucose, and lipid levels of a high-monounsaturated fat diet compared with a high-carbohydrate diet in NIDDM subjects. *Diabetes Care* 1993;16(12):1565-71.

[170] Aude YW, Agatston AS, Lopez-Jimenez F, et al. The national cholesterol education program diet vs a diet lower in carbohydrates and higher in protein and monounsaturated fat: a randomized trial. *Arch Intern Med* 2004;164(19):2141-6.

[171] Brehm BJ, Seeley RJ, Daniels SR, D'Alessio DA. A randomized trial comparing a very low carbohydrate diet and a calorie-restricted low fat diet on body weight and cardiovascular risk factors in healthy women. *J Clin Endocrinol Metab* 2003;88(4):1617-23.

[172] Foster GD, Wyatt HR, Hill JO, et al. A randomized trial of a low-carbohydrate diet for obesity. *N Engl J Med* 2003;348(21):2082-90.

[173] Samaha FF, Iqbal N, Seshadri P, et al. A low-carbohydrate as compared with a low-fat diet in severe obesity. *N Engl J Med* 2003;348(21):2074-81.

[174] Stern L, Iqbal N, Seshadri P, et al. The effects of low-carbohydrate versus conventional weight loss diets in severely obese adults: one-year follow-up of a randomized trial. *Ann Intern Med* 2004;140(10):778-85.

[175] Brinkworth GD, Noakes M, Keogh JB, Luscombe ND, Wittert GA, Clifton PM. Long-term effects of a high-protein, low-carbohydrate diet on weight control and cardiovascular risk markers in obese hyperinsulinemic subjects. *Int J Obes Relat Metab Disord* 2004;28(5):661-70.

[176] Case CC, Jones PH, Nelson K, O'Brian Smith E, Ballantyne CM. Impact of weight loss on the metabolic syndrome. *Diabetes Obes Metab* 2002;4(6):407-14.

[177] Esposito K, Marfella R, Ciotola M, et al. Effect of a mediterranean-style diet on endothelial dysfunction and markers of vascular inflammation in the metabolic syndrome: a randomized trial. *Jama* 2004;292(12):1440-6.

[178] Bray GA. Low-carbohydrate diets and realities of weight loss. *Jama* 2003;289(14):1853-5.

[179] Connor WE, Connor SL. Should a low-fat, high-carbohydrate diet be recommended for everyone? The case for a low-fat, high-carbohydrate diet. *N Engl J Med* 1997;337(8):562-3; discussion 6-7.

[180] Yancy WS, Jr., Olsen MK, Guyton JR, Bakst RP, Westman EC. A low-carbohydrate, ketogenic diet versus a low-fat diet to treat obesity and hyperlipidemia: a randomized, controlled trial. *Ann Intern Med* 2004;140(10):769-77.

[181] Poppitt SD, Keogh GF, Prentice AM, et al. Long-term effects of ad libitum low-fat, high-carbohydrate diets on body weight and serum lipids in overweight subjects with metabolic syndrome. *Am J Clin Nutr* 2002;75(1):11-20.

[182] Chiasson JL, Josse RG, Gomis R, Hanefeld M, Karasik A, Laakso M. Acarbose for prevention of type 2 diabetes mellitus: the STOP-NIDDM randomised trial. *Lancet* 2002;359(9323):2072-7.

[183] Sacks FM, Svetkey LP, Vollmer WM, et al. Effects on blood pressure of reduced dietary sodium and the Dietary Approaches to Stop Hypertension (DASH) diet. DASH-Sodium Collaborative Research Group. *N Engl J Med* 2001;344(1):3-10.

[184] Chiasson JL, Josse RG, Gomis R, Hanefeld M, Karasik A, Laakso M. Acarbose treatment and the risk of cardiovascular disease and hypertension in patients with impaired glucose tolerance: the STOP-NIDDM trial. *Jama* 2003;290(4):486-94.

[185] Wood PD, Stefanick ML, Williams PT, Haskell WL. The effects on plasma lipoproteins of a prudent weight-reducing diet, with or without exercise, in overweight men and women. *N Engl J Med* 1991;325(7):461-6.

[186] Pate RR, Pratt M, Blair SN, et al. Physical activity and public health. A recommendation from the Centers for Disease Control and Prevention and the American College of Sports Medicine. *Jama* 1995;273(5):402-7.

[187] Physical activity and cardiovascular health. NIH Consensus Development Panel on Physical Activity and Cardiovascular Health. *Jama* 1996;276(3):241-6.

[188] Tanasescu M, Leitzmann MF, Rimm EB, Hu FB. Physical activity in relation to cardiovascular disease and total mortality among men with type 2 diabetes. *Circulation* 2003;107(19):2435-9.

[189] Cullinen K, Caldwell M. Weight training increases fat-free mass and strength in untrained young women. *J Am Diet Assoc* 1998;98(4):414-8.

[190] Mazzetti SA, Kraemer WJ, Volek JS, et al. The influence of direct supervision of resistance training on strength performance. *Med Sci Sports Exerc* 2000;32(6):1175-84.
[191] Feigenbaum MS, Pollock ML. Prescription of resistance training for health and disease. *Med Sci Sports Exerc* 1999;31(1):38-45.
[192] Thompson PD, Buchner D, Pina IL, et al. Exercise and physical activity in the prevention and treatment of atherosclerotic cardiovascular disease: a statement from the Council on Clinical Cardiology (Subcommittee on Exercise, Rehabilitation, and Prevention) and the Council on Nutrition, Physical Activity, and Metabolism (Subcommittee on Physical Activity). *Circulation* 2003;107(24):3109-16.
[193] Pollock ML, Gaesser GA, Butcher JD, et al. ACSM position stand: the recommended quantity and quality of exercise for developing and maintaining cardiorespiratory and muscular fitness, and flexibility in healthy adults. *Med Sci Sports Exerc* 1998;30(6):975-91.
[194] Fletcher GF. How to implement physical activity in primary and secondary prevention. A statement for healthcare-professionals from the Task Force on Risk-reduction, American Heart Association. *Circulation* 1997;96(1):355-7.
[195] *American College of Sports Medicine. Guidelines for exercise testing and prescription, 6th ed.* Baltimore, MD: Lippincott Williams & Wilkins;. 2000.
[196] Durstine JL, Haskell WL. Effects of exercise training on plasma lipids and lipoproteins. *Exerc Sport Sci Rev* 1994;22:477-521.
[197] Durstine JL, Grandjean PW, Cox CA, Thompson PD. Lipids, lipoproteins, and exercise. *J Cardiopulm Rehabil* 2002;22(6):385-98.
[198] Halbert JA, Silagy CA, Finucane P, Withers RT, Hamdorf PA. Exercise training and blood lipids in hyperlipidemic and normolipidemic adults: a meta-analysis of randomized, controlled trials. *Eur J Clin Nutr* 1999;53(7):514-22.
[199] Tucker LA, Silvester LJ. Strength training and hypercholesterolemia: an epidemiologic study of 8499 employed men. *Am J Health Promot* 1996;11(1):35-41.
[200] Christ M, Iannello C, Iannello PG, Grimm W. Effects of a weight reduction program with and without aerobic exercise in the metabolic syndrome. *Int J Cardiol* 2004;97(1):115-22.
[201] Couillard C, Despres JP, Lamarche B, et al. Effects of endurance exercise training on plasma HDL cholesterol levels depend on levels of triglycerides: evidence from men of the Health, Risk Factors, Exercise Training and Genetics (HERITAGE) Family Study. *Arterioscler Thromb Vasc Biol* 2001;21(7):1226-32.
[202] Goodyear LJ, Kahn BB. Exercise, glucose transport, and insulin sensitivity. *Annu Rev Med* 1998;49:235-61.
[203] Duncan GE, Perri MG, Theriaque DW, Hutson AD, Eckel RH, Stacpoole PW. Exercise training, without weight loss, increases insulin sensitivity and postheparin plasma lipase activity in previously sedentary adults. *Diabetes Care* 2003;26(3):557-62.
[204] McAuley KA, Williams SM, Mann JI, et al. Intensive lifestyle changes are necessary to improve insulin sensitivity: a randomized controlled trial. *Diabetes Care* 2002;25(3):445-52.
[205] Van Hoof R, Hespel P, Fagard R, Lijnen P, Staessen J, Amery A. Effect of endurance training on blood pressure at rest, during exercise and during 24 hours in sedentary men. *Am J Cardiol* 1989;63(13):945-9.

[206] Gilders RM, Dudley GA. Endurance exercise training and treatment of hypertension. The controversy. *Sports Med* 1992;13(2):71-7.

[207] Pai JK, Pischon T, Ma J, et al. Inflammatory markers and the risk of coronary heart disease in men and women. *N Engl J Med* 2004;351(25):2599-610.

[208] Wannamethee SG, Lowe GD, Whincup PH, Rumley A, Walker M, Lennon L. Physical activity and hemostatic and inflammatory variables in elderly men. *Circulation* 2002;105(15):1785-90.

[209] Pischon T, Hankinson SE, Hotamisligil GS, Rifai N, Rimm EB. Leisure-time physical activity and reduced plasma levels of obesity-related inflammatory markers. *Obes Res* 2003;11(9):1055-64.

[210] Troseid M, Lappegard KT, Claudi T, et al. Exercise reduces plasma levels of the chemokines MCP-1 and IL-8 in subjects with the metabolic syndrome. *Eur Heart J* 2004;25(4):349-55.

[211] Dujovne CA, Zavoral JH, Rowe E, Mendel CM. Effects of sibutramine on body weight and serum lipids: a double-blind, randomized, placebo-controlled study in 322 overweight and obese patients with dyslipidemia. *Am Heart J* 2001;142(3):489-97.

[212] Hollander PA, Elbein SC, Hirsch IB, et al. Role of orlistat in the treatment of obese patients with type 2 diabetes. A 1-year randomized double-blind study. *Diabetes Care* 1998;21(8):1288-94.

[213] Ballinger A, Peikin SR. Orlistat: its current status as an anti-obesity drug. *Eur J Pharmacol* 2002;440(2-3):109-17.

[214] Torgerson JS, Hauptman J, Boldrin MN, Sjostrom L. XENical in the prevention of diabetes in obese subjects (XENDOS) study: a randomized study of orlistat as an adjunct to lifestyle changes for the prevention of type 2 diabetes in obese patients. *Diabetes Care* 2004;27(1):155-61.

[215] Reaven G, Segal K, Hauptman J, Boldrin M, Lucas C. Effect of orlistat-assisted weight loss in decreasing coronary heart disease risk in patients with syndrome X. *Am J Cardiol* 2001;87(7):827-31.

[216] Didangelos TP, Thanopoulou AK, Bousboulas SH, et al. The ORLIstat and CArdiovascular risk profile in patients with metabolic syndrome and type 2 DIAbetes (ORLICARDIA) Study. *Curr Med Res Opin* 2004;20(9):1393-401.

[217] Pasquali R, Gambineri A, Biscotti D, et al. Effect of long-term treatment with metformin added to hypocaloric diet on body composition, fat distribution, and androgen and insulin levels in abdominally obese women with and without the polycystic ovary syndrome. *J Clin Endocrinol Metab* 2000;85(8):2767-74.

[218] Gokcel A, Gumurdulu Y, Karakose H, et al. Evaluation of the safety and efficacy of sibutramine, orlistat and metformin in the treatment of obesity. *Diabetes Obes Metab* 2002;4(1):49-55.

[219] Freemark M, Bursey D. The effects of metformin on body mass index and glucose tolerance in obese adolescents with fasting hyperinsulinemia and a family history of type 2 diabetes. *Pediatrics* 2001;107(4):E55.

[220] Yki-Jarvinen H, Nikkila K, Makimattila S. Metformin prevents weight gain by reducing dietary intake during insulin therapy in patients with type 2 diabetes mellitus. *Drugs* 1999;58 Suppl 1:53-4; discussion 75-82.

[221] Stumvoll M, Nurjhan N, Perriello G, Dailey G, Gerich JE. Metabolic effects of metformin in non-insulin-dependent diabetes mellitus. *N Engl J Med* 1995;333(9):550-4.

[222] Effect of intensive blood-glucose control with metformin on complications in overweight patients with type 2 diabetes (UKPDS 34). UK Prospective Diabetes Study (UKPDS) Group. *Lancet* 1998;352(9131):854-65.

[223] Pollare T, Lithell H, Berne C. A comparison of the effects of hydrochlorothiazide and captopril on glucose and lipid metabolism in patients with hypertension. *N Engl J Med* 1989;321(13):868-73.

[224] Hansson L, Lindholm LH, Niskanen L, et al. Effect of angiotensin-converting-enzyme inhibition compared with conventional therapy on cardiovascular morbidity and mortality in hypertension: the Captopril Prevention Project (CAPPP) randomised trial. *Lancet* 1999;353(9153):611-6.

[225] Yusuf S, Sleight P, Pogue J, Bosch J, Davies R, Dagenais G. Effects of an angiotensin-converting-enzyme inhibitor, ramipril, on cardiovascular events in high-risk patients. The Heart Outcomes Prevention Evaluation Study Investigators. *N Engl J Med* 2000;342(3):145-53.

[226] Lindholm LH, Ibsen H, Dahlof B, et al. Cardiovascular morbidity and mortality in patients with diabetes in the Losartan Intervention For Endpoint reduction in hypertension study (LIFE): a randomised trial against atenolol. *Lancet* 2002;359(9311):1004-10.

[227] Julius S, Kjeldsen SE, Weber M, et al. Outcomes in hypertensive patients at high cardiovascular risk treated with regimens based on valsartan or amlodipine: the VALUE randomised trial. *Lancet* 2004;363(9426):2022-31.

[228] Wilhelmsen L, Pyorala K, Wedel H, Cook T, Pedersen T, Kjekshus J. Risk factors for a major coronary event after myocardial infarction in the Scandinavian Simvastatin Survival Study (4S). Impact of predicted risk on the benefit of cholesterol-lowering treatment. *Eur Heart* J 2001;22(13):1119-27.

[229] Collins R, Armitage J, Parish S, Sleigh P, Peto R. MRC/BHF Heart Protection Study of cholesterol-lowering with simvastatin in 5963 people with diabetes: a randomised placebo-controlled trial. *Lancet* 2003;361(9374):2005-16.

[230] Sacks FM, Tonkin AM, Craven T, et al. Coronary heart disease in patients with low LDL-cholesterol: benefit of pravastatin in diabetics and enhanced role for HDL-cholesterol and triglycerides as risk factors. *Circulation* 2002;105(12):1424-8.

[231] Colhoun HM, Betteridge DJ, Durrington PN, et al. Primary prevention of cardiovascular disease with atorvastatin in type 2 diabetes in the Collaborative Atorvastatin Diabetes Study (CARDS): multicentre randomised placebo-controlled trial. *Lancet* 2004;364(9435):685-96.

[232] Pyorala K, Ballantyne CM, Gumbiner B, et al. Reduction of cardiovascular events by simvastatin in nondiabetic coronary heart disease patients with and without the metabolic syndrome: subgroup analyses of the Scandinavian Simvastatin Survival Study (4S). *Diabetes Care* 2004;27(7):1735-40.

[233] Sever PS, Dahlof B, Poulter NR, et al. Prevention of coronary and stroke events with atorvastatin in hypertensive patients who have average or lower-than-average cholesterol concentrations, in the Anglo-Scandinavian Cardiac Outcomes Trial--Lipid

Lowering Arm (ASCOT-LLA): a multicentre randomised controlled trial. *Lancet* 2003;361(9364):1149-58.

[234] Freeman DJ, Norrie J, Sattar N, et al. Pravastatin and the development of diabetes mellitus: evidence for a protective treatment effect in the West of Scotland Coronary Prevention Study. *Circulation* 2001;103(3):357-62.

[235] Tenkanen L, Manttari M, Manninen V. Some coronary risk factors related to the insulin resistance syndrome and treatment with gemfibrozil. Experience from the Helsinki Heart Study. *Circulation* 1995;92(7):1779-85.

[236] Robins SJ, Rubins HB, Faas FH, et al. Insulin resistance and cardiovascular events with low HDL cholesterol: the Veterans Affairs HDL Intervention Trial (VA-HIT). *Diabetes Care* 2003;26(5):1513-7.

[237] Durbin RJ. Thiazolidinedione therapy in the prevention/delay of type 2 diabetes in patients with impaired glucose tolerance and insulin resistance. *Diabetes Obes Metab* 2004;6(4):280-5.

[238] Buchanan TA, Xiang AH, Peters RK, et al. Preservation of pancreatic beta-cell function and prevention of type 2 diabetes by pharmacological treatment of insulin resistance in high-risk hispanic women. *Diabetes* 2002;51(9):2796-803.

[239] Chilcott J, Tappenden P, Jones ML, Wight JP. A systematic review of the clinical effectiveness of pioglitazone in the treatment of type 2 diabetes mellitus. *Clin Ther* 2001;23(11):1792-823; discussion 1.

[240] Freed MI, Ratner R, Marcovina SM, et al. Effects of rosiglitazone alone and in combination with atorvastatin on the metabolic abnormalities in type 2 diabetes mellitus. *Am J Cardiol* 2002;90(9):947-52.

[241] Rutter MK, Meigs JB, Sullivan LM, D'Agostino RB, Sr., Wilson PW. C-reactive protein, the metabolic syndrome, and prediction of cardiovascular events in the Framingham Offspring Study. *Circulation* 2004;110(4):380-5.

[242] Costa A, Casamitjana R, Casals E, et al. Effects of atorvastatin on glucose homeostasis, postprandial triglyceride response and C-reactive protein in subjects with impaired fasting glucose. *Diabet Med* 2003;20(9):743-5.

[243] Despres JP, Lemieux I, Pascot A, et al. Gemfibrozil reduces plasma C-reactive protein levels in abdominally obese men with the atherogenic dyslipidemia of the metabolic syndrome. *Arterioscler Thromb Vasc Biol* 2003;23(4):702-3.

[244] Haffner SM, Greenberg AS, Weston WM, Chen H, Williams K, Freed MI. Effect of rosiglitazone treatment on nontraditional markers of cardiovascular disease in patients with type 2 diabetes mellitus. *Circulation* 2002;106(6):679-84.

[245] Sola S, Mir MQ, Cheema FA, et al. Irbesartan and lipoic acid improve endothelial function and reduce markers of inflammation in the metabolic syndrome: results of the Irbesartan and Lipoic Acid in Endothelial Dysfunction (ISLAND) study. *Circulation* 2005;111(3):343-8.

[246] Pories WJ, Swanson MS, MacDonald KG, et al. Who would have thought it? An operation proves to be the most effective therapy for adult-onset diabetes mellitus. *Ann Surg* 1995;222(3):339-50; discussion 50-2.

[247] Yusuf S, Hawken S, Ounpuu S, et al. Effect of potentially modifiable risk factors associated with myocardial infarction in 52 countries (the INTERHEART study): case-control study. *Lancet* 2004;364(9438):937-52.

[248] Wong ND, Pio JR, Franklin SS, L'Italien GJ, Kamath TV, Williams GR. Preventing coronary events by optimal control of blood pressure and lipids in patients with the metabolic syndrome. *Am J Cardiol* 2003;91(12):1421-6.

[249] Gaede P, Vedel P, Larsen N, Jensen GV, Parving HH, Pedersen O. Multifactorial intervention and cardiovascular disease in patients with type 2 diabetes. *N Engl J Med* 2003;348(5):383-93.

[250] Eyre H, Kahn R, Robertson RM, et al. Preventing cancer, cardiovascular disease, and diabetes: a common agenda for the American Cancer Society, the American Diabetes Association, and the American Heart Association. *Circulation* 2004;109(25):3244-55.

In: New Developments in Obesity Research
Editor: Lawrence F. Ditmier, pp. 75-83
ISBN 1-60021-296-4

Chapter 3

IMPLICATIONS OF RESISTIVE FORCE SELECTION DURING BRIEF HIGH INTENSITY EXERCISE IN OVERWEIGHT SUBJECTS

Julien S. Baker * ***and Bruce Davies***
Health and Exercise Science Research Laboratory, School of Applied Science, University of Glamorgan, Pontypridd, Wales

ABSTRACT

Objective: The purpose of this study was to compare the maximal exercise performance of overweight subjects during friction braked cycle ergometry of 10 s duration when resistive forces reflected total-body mass (TBM) or fat-free mass (FFM).

Research Methods and Procedures: Data collection was administered using a single blind, cross - over design. Subjects were assigned in a random fashion to either the TBM or FFM protocol. 11 healthy, overweight, male university students (age 22.3 ± 2 yrs, body fat 27.1 ± 2 %) participated in the study.

Body composition was assessed using underwater weighing procedures. Fat mass was calculated from body density values. Power outputs were calculated using high intensity cycle ergometry.

Results: Increases ($P < 0.01$) in peak power output (PPO) were found between TBM and FFM (1029 ± 98 W TBM vs. 1397 ± 146 W FFM). Significant decreases ($P < 0.05$) were observed for the time taken to reach PPO (4 ± 1.4 s TBM vs 2 ± 1 s FFM). Pedal velocity significantly increased ($P < 0.01$) during the FFM protocol (98.2 ± 9.7 rpm TBM vs. 139.6 ± 6.6 rpm FFM) and resistive forces significantly decreased (7.5 ± 0.4 kg TBM vs. 5.5 ± 0.4 kg FFM).

Discussion: The findings of this study suggest that greater peak power outputs are obtainable when resistive forces reflect FFM as opposed to TBM. The results may have implications that relate to both the pathology of disease, related clinical examinations and exercise prescription in overweight populations.

* Address for correspondence: Dr Julien Baker. E.Mail: jsbaker@glam.ac.uk; Tel 01443 482972; Fax 01443 482285

Key Words: Peak power output, Total Body Mass (TBM), Fat Free Mass (FFM).

Introduction

High intensity cycle ergometry has been widely employed to assess indices of muscle performance during maximal exercise. Traditionally, the resistive force established for such a test is determined from total-body mass (TBM) multiplied by a ratio standard e.g.; 75g.kg x TBM [1]. More recent studies [2, 3, 4, 5] have shown that optimisation protocols that utilise manipulations in both resistive force and velocity can produce higher peak power outputs (PPO). Attempts have also been made to use the volume of the leg in the selection of optimal forces during cycle ergometry exercise [6, 7, 8] and as an indicator of maximal exercise performance [9]. It is also common for high intensity performance to be expressed relative to body mass [4, 8, 10, 11, 12] or cross sectional area of the thigh [13, 14, 15]. McInnis & Balady, [16] observed that individuals who weigh the same may have very different body compositions. Van mil et al. [17] recommended that optimal performance during high intensity cycle ergometry is highly related to an individual's lean-body mass, or to the mass of the muscles that contribute to the test. Inbar et al. [4] have suggested that fat-free mass or active muscle mass may provide more realistic assessments of anaerobic ability during high intensity exercise performance. These statements agree with the earlier suggestions of Wilkie, [18] who stated that resistive forces used during cycle ergometry should be matched as closely as possible to active muscle output. Body mass and not composition is the most commonly used index to determine resistive force selection during high intensity cycle ergometer exercise, and because power is the product of both resistive force and velocity over or underestimations in power performance and assessment may occur. This may be particularly pronounced in overweight populations. Recent research in our laboratory [19, 20] has clearly identified an increase in peak power output in athletic populations when resistive forces used during high intensity cycle ergometry are calculated from fat-free mass as opposed to total-body mass. The lighter resistive forces used resulted in a significant increase in pedal revolutions that contributed to the measured increases in peak power output. The main purpose of this study was to evaluate the friction loaded cycle ergometer performance of a group of overweight university students using two resistive force selection protocols namely, total-body mass (TBM) and fat-free mass (FFM). The TBM resistive force protocol would be inclusive of the fat component of body composition, whereas the FFM protocol would not. The null hypothesis was that statistically similar peak power outputs and pedal revolutions would be obtained for the two resistive force selection procedures. Rejection of the null hypotheses would suggest that resistive force selection based on total body mass may have a negative effect on the power profiles obtained.

Methods

Eleven male university students volunteered to participate in the experiment. Subjects were considered eligible for the study if they were healthy, free from injury and disease and had body compositions consisting of at least 25 % body fat. These details were derived from a pre activity proforma that assessed body fat in the first instance with skin-fold thicknesses. The values obtained for individual subjects were then confirmed using underwater weighing techniques. A body fat content of greater than twenty five percent was used as the criterion value defining overweight in white male subjects. Zhu et al. [21] used greater than 21 % body fat to define overweight and greater than 29 % to define obesity. All subjects participating in this study were at least overweight and 18% were classified as obese. Experimental procedures were approved by the University ethics committee and all subjects provided written informed consent. A 10 s cycle ergometer protocol was selected to ensure that enough time was allocated to collect a true peak power output value for all subjects during data collection for both protocols, and to ensure that the experimental test duration was not too demanding. This also helped with subject compliance. An optimisation procedure was avoided for the reasons stated previously. During this study resistive forces used were calculated using 75g.kg in conjunction with a TBM or FFM protocol. Prior to testing all subjects were fully habituated to experimental procedures. Habituation was performed at the same time of day as testing sessions (morning). Subjects were randomly assigned to either the TBM or FFM protocol. Each subject returned to perform the remaining test protocol with one week intervening between tests. Data collection was administered using a single blind, cross - over design. All subjects attended the laboratory after completing an overnight fast in an attempt to attenuate the influence of diet on performance.

Anthropometric Measures

Body mass, stature and body composition was determined prior to exercise using a calibrated balanced weighing scale (Seca, UK), stadiometer (Seca, UK) and underwater weighing procedures respectively [22]. Nude body mass was measured to the nearest 0.1 kg and stature to 0.1 cm. Relative body fat was estimated from body density [23]. Residual lung volume was measured using a modified oxygen re-breathing method [24]. FFM mass was determined by subtracting fat mass from TBM.

Cycle Ergometer Test Protocol

A cycle ergometer (Monark 864) was calibrated prior to warm up and data collection. The calibration procedure followed the guidelines for friction loaded ergometers outlined by Coleman [25]. Saddle heights were adjusted individually to accommodate partial flexion of the knee between 170° to 175° (with 180° denoting a straight leg position) in middle dead centre during the down stroke. Feet were firmly supported by toe clips and straps, and subjects were instructed to remain seated during the test. All subjects were verbally encouraged to perform maximally during testing and performed a standardised 5 minute warm up following procedures outlined by Jaskolska et al. [5]. Subjects were given a rolling start at approximately 60 rpm followed by a three second count down. On the command 'go!' subjects pedalled maximally, individual resistive forces were applied simultaneously and data capture initiated. Values for PPO (peak power output; the highest value recorded for power in any one second of the total exercise period) and PR (pedal revolutions; the greatest pedal velocity recorded over the exercise period) were determined using an inertia corrected computer programme [25]. Resistive forces used for the TBM or FFM cycle ergometer protocol were calculated by multiplying individual subject total-body mass or fat-free mass by 75g.kg. Data transfer was made possible using a suitably mounted sensor unit and power supply attached to the fork of the ergometer interfaced to a computer. The sampling frequency of the sensor was 18.2 Hz. Reliability of the cycle ergometer test as a measure of anaerobic power has been reported as $r = 0.92$, $P < 0.01$; $R^2 = 0.86$ [7] for a 30 s test. Test re-test correlations obtained in our laboratory for a 10 s test using overweight subjects was, $r = 0.89$, $P < 0.01$, $R^2 = 0.79$ and $r = 0.94$, $P < 0.01$; $R^2 = 0.88$ for TBM and FFM respectively. Heart rate (HR) values were recorded using short range telemetry at rest and immediately post exercise, for both TBM and FFM protocols (Sport Tester PE 3000, Polar Electro Finland).

Statistical Procedures

Following conformation of a normal distribution of data (SPSS statistical computer program), Student's paired t-test was used to examine the significance of differences between power values generated using the TBM or FFM protocols. Pearson's product correlation was used to investigate the degree of linear relationship between tests. Significance was accepted at P < 0.0.5 and all data are presented as Mean ± SDs.

Results

Physiological and anthropometric characteristics of subjects are given in table 1. Performance variables are presented in table 2. Increases ($P < 0.01$) in peak power output (PPO) were observed between TBM and FFM (1029 ± 98 W TBM vs 1397 ± 146 W FFM). Significant decreases ($P < 0.01$) were noted for the time taken to reach PPO (4 ± 1.4 s TBM vs 2 ± 1 s FFM). Pedal velocity significantly increased ($P < 0.01$) during the FFM protocol (98.2 ± 9.7 rpm TBM vs. 139.6 ± 6.6 rpm FFM) and resistive forces significantly decreased (7.5 ± 0.4 kg TBM vs 5.5 ± 0.4 kg FFM). Heart rate values increased significantly from rest to immediately post exercise for both the TBM and FFM protocols ($P < 0.01$). However, there were no differences recorded between protocols for heart rate (table 2) pre or post exercise ($P > 0.05$).

Table 1. Subject Means ± (standard deviations) for age and anthropometric characteristics (n = 11)

Variable	
Age (yrs)	22.3 ± 2
Stature (cms)	188.5 ± 7
TBM (kg)	100 ± 5*
% Fat	27.1 ± 2
FFM (kg)	73.2 ± 5*

* Sig P < 0.05

Values are Means ± SD

Table 2. Means ± (standard deviations) for cycle ergometer test results (n=11)

Variable	TBM	FFM	Sig
PPO (W)	1029 ± 98	1379 ± 146	P < 0.01
T to PPO (s)	4 ± 1.4	2 ± 1	P < 0.01
R/f (kg)	7.5 ± 0.4	5.5 ± 0.4	P < 0.01
P/revs (rpm)	98 ± 8	139 ± 6	P < 0.01
HR (bpm) pre	68 ± 10	66 ± 11	P > 0.05
HR (bpm) post	167 ± 8	165 ± 8	P > 0.05

Values are Means ± SD.
PPO Peak power output.
T to PPO Time to peak power output.
R/f Resistive force.
P/revs Pedal revolutions.
HR Heart Rate.

DISCUSSION

Results from this study indicate that PPO values increased when resistive forces were determined using the FFM resistive force selection protocol. The lighter forces enabled subjects to reach a higher pedalling velocity in the early stages of the test. This can be demonstrated by the significantly faster times taken to reach PPO, and can be further verified by the significant increase in pedal velocity when TBM and FFM were compared. These findings are in agreement with others [18, 26] who suggested that greater PPO values would be generated by lighter forces and greater velocities. The lighter resistive forces used appear to facilitate a 'maximisation' of pedal velocity that contributes to the increases in the power outputs recorded during this type of activity. Critics of friction loaded cycle ergometry point to the continual acceleration and deceleration that occurs during performance. These factors may impose limits on the total time that muscle is able to maintain an optimal velocity of contraction [27, 28]. Force-velocity relationships have been obtained under isokinetic conditions [29] which mean that maximal force has been measured at constant velocities. Although valid in a laboratory setting, the external validity of this approach is questionable. Natural movements are not isokinetic but consist of acceleration and decelerations which is the product of a variety of isotonic, isometric and isokinetic contractions. Therefore, generalisations relating to high intensity performance may be difficult [30, 31]. Higher PPO obtained for FFM may reflect greater contributions to power calculations by the moment of inertia and frictional torque characteristics inherent in cycle ergometer flywheel dynamics [26]. However, this study utilised an inertia corrected computer programme, therefore any mechanical influence on the power values obtained would have been considered for both experimental conditions. The results obtained suggest that the FFM protocol closely approximates optimal muscular force and velocity mechanics. This observation agrees with the suggestions of Wilkie, [18] who stated that force should be matched to the capacity of muscle to exploit the full force velocity relationship. Cycle ergometer tests utilising forces that are too heavy may compromise muscular integrity, resulting in premature failure. In addition, they can result in myofibril disruption and associated muscle damage, and may not represent fully the maximal attainable power profiles for individual subjects. During the FFM protocol, the heaviest resistive force (6.3 kg) produced the greatest PPO (1592 W) at a velocity of 143 rpm. This force represented a decrease of 2 kg compared to the resistive force selected based on TBM computations. Also, the power value recorded using the FFM resistive force selection criteria increased by 410 Watts. This individuals total body mass was 110 kg with a body fat value of 26%. The subject with the greatest body fat (30%) increased his power profile by 211W using the FFM method. However, it is possible that other factors influenced force development that is independent of body composition indices. These may include muscle fiber type, enzyme activity and muscle cross sectional area [14]. It is unlikely that physical activity was influential in power production because all subjects were fairly inactive. The higher PPO obtained using the FFM protocol underline the inconsistent relationship between muscle mass and total-body mass. This can be emphasised by the range of fat values observed for individual subjects. Mathematics relating to the force velocity relationship during muscular contraction indicates that the greater the resistive force on the flywheel, the slower the maximal pedalling velocity. Vandewalle et al. [11] stated that maximal muscular power is obtained at optimal values of both force and velocity. Our data suggest that resistive forces calculated from TBM have overestimated the force required to elicit optimal PPO, and therefore attenuate pedal velocity.

Previously, attempts have been made to use the volume of the leg in the selection of optimal resistive forces [6, 7]. These investigators attempted to express the values generated in units of fat-free mass, or recorded the linear relationship observed between leg volume and power output. The studies expressed their results relative to the resistive forces used, and PPO values were not generated by directly calculating individual resistive forces from the fat-free mass component of body composition. This may help to explain the lack of statistical relationship observed. Duche et al. [31] suggested that cycling peak power was significantly lower in obese than non-obese adolescents and that the differences observed disappeared when they were related to FFM. They also demonstrated that FFM was the most important explanatory variable in the variance of cycling peak power indicating that cycling peak power was dependent on resistive force in obese adolescents but independent in non-obese. The study also stated that the resistive force used for the obese population appear to have been too heavy for the subjects to reach the optimal velocity required to obtain a true PPO. These suggestions agree with the findings of this study. During an optimisation procedure power profiles may improve as a result of increasing or decreasing both force and pedalling velocity [4]. Our results demonstrate that increases in PPO during the FFM protocol occurred because of substantial increases in pedal velocity and a decrease in resistive force. Friction loaded cycle ergometry has been regarded as a useful device in the assessment of high intensity performance and for its ability to set optimal and realistic resistive forces [10]. The present findings clearly suggest that resistive forces used during high intensity cycle ergometry should be calculated using the FFM protocol if maximal power outputs and associated biochemical and physiological measurements are desirable. Power values should provide comparative and realistic assessments of between and within subject analysis that are generated by active muscle tissue and exclude the non productive fat mass. These findings are in contrast with previous authors [9, 7] who reported that body mass and leg volume were of little predictive importance during the early portion (first 30 s) of a high intensity cycle ergometer test (2 mins duration). However, other researchers [4, 17, 30, 33] are in agreement with the present findings. Our findings indicate that the PPO's generated during cycle ergometry are more related to FFM than TBM in a group of overweight subjects. Resistive forces used for brief high intensity cycle ergometry of 10 s duration that are calculated using TBM, significantly underestimate the attainable PPO when compared to FFM. These findings have far reaching clinical and physiological implications during the assessment of high intensity performance and may lead to misdiagnosis in both the pathology of disease and subsequent clinical examination. Misguided exercise prescription based on TBM resistive force selection will compromise overweight and obese subjects to a greater extent than populations with less body fat. In addition to an under assessment of power production we have observed hematological increases in oxidative stress and muscle damage when the two protocols were compared [34]. As a result, the risk of muscle damage and related disorders during high intensity exercise may be more pronounced in overweight and obese populations when a TBM protocol is utilised in preference to FFM.

REFERENCES

[1] Aylon A, Inbar O, Bar - Or O. Relationships among measurements of explosive strength and anaerobic power. In RC Nelson and CA Morehouse (Eds) *International series on sports sciences,* Vol 1, Biomechanics IV, 1974; 572 - 577.

[2] Dotan R, Bar- Or O. Load optimization for the Wingate anaerobic test. *Eur J of Appl Physiol.* 1984; 51: 409 -17.

[3] Winter EM, Brookes FBC, Hamley EJ. Optimized loads for external power output during brief, maximal cycling. *J of Sport Sci.* 1989; 7: 69 - 70.

[4] Inbar O, Bar – Or O, Skinner S. *The Wingate Anaerobic Test. Human Kinetics.* Leeds 1996.

[5] Jaskolska A, Goossens P, Veenstra B, Jakolski A, Skinner JS. Comparison of treadmill and cycle ergometer measurements of force - velocity relationships and power outputs. *Inter J of Sports Med.* 1999; 20: 192 - 197.

[6] Evans JA, Quinney HA. Determination of resistance settings for anaerobic power testing. *Can J of Appl Sport Sci.* 1981; 6: 53 - 6.

[7] Patton J, Murphy M, Frederick F. Maximal power outputs during the Wingate anaerobic test. *Int J of Sports Med.* 1985; 6: 82 - 5.

[8] Mannion AF, Jakeman PM. *Comparison of velocity dependent and time dependent measures of anaerobic work capacity.* In Kinanthropometry III. (edited by T. Reilly, J. Watkins, and J. Borms) 1986; 301 - 307. London. E and F N. Spon.

[9] Katch V. Body weight, leg volume, leg weight and leg density as determiners of short duration work performance on the bicycle ergometer. *Med Sci Sports.* 1974; 6: 267 - 270.

[10] Nakamura Y, Mutoh Y, Myashita, M. Determination of the peak power output during maximal brief pedalling bouts. *J of Sports Sci.* 1985; 3: 181 - 187.

[11] Vandewalle H, Peres G Heller, J, Monad H. All out anaerobic capacity tests on cycle ergometers. A comparative study on men and women. *Euro J of Appl Physiol.* 1985; 54: 222 - 9.

[12] Sargeant AJ, Dolan P. *Optimal velocity of muscle contraction for short term (anaerobic) power output in children and adults.* In Children and Exercise XII. (edited

by J. Rutenfraantz, R. Mocellin and F. Klimdt) 1986; pp 39-42 Champaign, Ill. Human kinetics.

[13] Ikai M, Fukunaga T. Calculation of muscle strength per unit cross-sectional area of human muscle by means of ultrasonic measurement. *Internationale Zeitschrift fur Angewandte Physiologie Einschlesslich Arbeitsphysiologie.* 1968; 26: 26-32

[14] Maughan RJ, Watson JS, Weir J. Strength and cross sectional area of human skeletal muscle. *J of Physiol.* 1983; 338: 37- 49.

[15] Davies CTM. Strength and mechanical properties of muscle in children and young adults. *Scandinavian J of Sports Sci.* 1985; 7: 11-15.

[16] McInnis KJ, Balady GJ. Effects of body composition on oxygen uptake during treadmill exercise: Body builder's verses weight matched men. *Res Quar for Exer and Sport.* 1999; 70: 150-156.

[17] Van mil E, Schoeber N, Calvert R, Bar Or O. Optimization of force in the Wingate Test for children with a neuromuscular disease. *Med Sci in Sport and Exer.* 1996; 28: 1087 - 1092.

[18] Wilkie D R. Man as a source of mechanical power. *Ergonomics.* 1960; 3: 1 - 8.

[19] Baker JS, Bailey D, Davies B. The relationship between total-body mass, fat-free mass and cycle ergometer power components during 20 seconds of maximal exercise. *J of Sci and Med in Sport.* 2001; 4: 1 - 9.

[20] Baker JS, Bailey D, Davies B. Catecholamine responses to high intensity cycle ergometer exercise: Body mass or body composition. *J of Physiol and Biochem.* 2003; 59: 77-84.

[21] Zhu S, Wang Z, Shen W, Heymsfield SB, Heshka S. Percentage body fat ranges associated with metabolic syndrome risk: Results based on the third National Health and Nutrition Examination Survey (1988-1994). *Amer J of Clin Nut.* 2003; 78: 228-35.

[22] Behnke AR, Wilmore JH. *Evaluation of body build and composition.* (Englewood Cliffs, N. J) Prentice Hall Inc publications Hertfordshire, 1974; 20-24.

[23] Siri WE. Gross composition of the body. In : Lawrence, J.H. and Tobias, C.A. *Advances in Biological and Medical physics.* IV. New York Academic Press, 1956; 239-280.

[24] Wilmore JH, Vodak PA, Parr RB, Girandola RN, Billing JE. Further simplification of a method for determination of residual lung volume. *Med Sci Sport and Exer.* 1980; 12: 216-218.

[25] Coleman S. Corrected Wingate Anaerobic Test. *Cranlea users handbook.* Cranlea and Co. Sandpits Lane, Birmingham, England 1996.

[26] Lakomy H. Measurement of work and power output using friction-loaded cycle ergometers. *Ergonomics.* 1986; 29: 509 -17.

[27] Sargeant A J, Honiville E, Young A. Maximal leg force and power output during short term dynamic exercise. *J of Appl Physiol.* 1981; 51: 1175 - 1182.

[28] McCartney N, Heigenhauser GJF, Jones NL. Power output and fatigue in human muscle in maximal cycling exercise. *J of Appl Physiol.* 1983; 55: 218 - 224.

[29] Sargeant AJ, Dolan P, Young A. Optimal velocity for maximal short term anaerobic power output in cycling. *Inter J of Sports Med.* 1984; 5: 124 - 125.

[30] Duche P, Ducher G, Lazzer S, Dore E, Tailhardat M, Bedu M. Peak power in obese and nonobese adolescents: effects of gender and braking force. *Med Sci Sports Exerc.* 2002; 34: 2072- 2078.

[31] Komi PV. Training of muscle strength and power: Interaction of neuromotor, hypertrophic and mechanical factors. *Inter Jour of Sports Med.* 1986; 7: 10 – 15.

[32] Winter EM, Brookes FBC, Hamley EJ. Maximal exercise performance and lean leg volume in men and women. *J of Sport Sci.* 1991; 9: 3 - 13.

[33] Blimkie CJ R, Roche JT, Hay, T, Bar – Or O. Anaerobic power of arms in teenage boys and girls: relationship to lean body tissue. *Euro J of Appl Physiol.* 1988; 57: 677-683.

[34] Baker JS, Bailey DM, Hullin D, Young I, Davies B. Metabolic implications of resistive force selection for oxidative stress and markers of muscle damage during 30 s of high intensity exercise. *Euro J of Appl Physiol.* 2004; 92: 321 - 327.

In: New Developments in Obesity Research
Editor: Lawrence F. Ditmier, pp. 85-119
ISBN 1-60021-296-4

Chapter 4

ETHNIC DISPARITIES IN OVERWEIGHT AND OBESITY IN THE U.S.: A REVIEW OF SOCIO-CULTURAL CONTRIBUTIONS

Abigail M. Judge
Department of Psychology,
University of North Carolina at Chapel Hill
Jennifer J. Thomas
Department of Psychology, Yale University
Anne E. Becker
Eating Disorders Clinical and Research Program, Department of Psychiatry,
Massachusetts General Hospital,
Department of Social Medicine, Harvard Medical School

ABSTRACT

The prevalence of overweight and obesity is disproportionately high among Black, Latina, and Native American women compared with non-Hispanic White women in the United States. The basis for this disparity is likely multifactorial with genetic, environmental, and behavioral contributions, yet remains incompletely understood. This chapter examines the evidence for culture-specific attitudes and behaviors that may enhance risk for obesity with a special focus on differences in body image and disordered eating across ethnic minority groups in the United States. In addition, differences in patterns of physical activity, culturally informed explanatory models of health, and dietary practices across ethnic groups are reviewed. Clarification of the socio-cultural contributions to differential risk for disordered eating and obesity will be essential to the development of effective and culturally sensitive interventions across ethnically diverse populations.

INTRODUCTION

The prevalence of overweight and obesity has markedly increased over the past several decades in the United States (U.S.). It is estimated that 65.1% of American adults are overweight, including 30.4% who are obese (Hedley et al., 2004). Epidemiologic studies have demonstrated the profound health, social, and economic consequences of overweight and obesity (Colditz, 1999; Gortmaker, Must and Perrin, 1993). Obesity is believed to be the second leading cause of death in the U.S., and is associated with numerous comorbid illnesses (Bray, 2004; Field et al., 2001). Overweight and obese individuals also suffer social disadvantages such as stigma and discrimination related to employment, education, and health care (Gortmaker et al., 1993; Seidell, 1998; Puhl and Brownell, 2001).

In addition to individual health and social consequences, health care costs associated with obesity are extremely high. A recent study estimated annual medical spending due to overweight and obesity to be as much as $92.6 billion in 2002, or 9.1% of U.S. health expenditures (Finkelstein, Fiebelkorn, and Wang, 2003). According to the Wolf and Colditz (1998), the indirect costs related to obesity and overweight are $56 billion dollars annually, comparable to the economic costs of cigarette smoking.

Substantial variation in the prevalence of overweight and obesity across socio-economically and ethnically diverse populations in the U.S. suggests socio-cultural contributions to obesity risk. Specifically, the age-adjusted prevalence of overweight and obesity is higher among ethnic minority women compared with non-Hispanic White women in the U.S., with the highest prevalence observed in Mexican American and non-Hispanic Black women. Approximately 71.7% of adult Mexican American and 77.2% of non-Hispanic Black women are overweight (body mass index [BMI] $\geq$ 25 kg/m^2), including 38.4% of Mexican American and 49.0% of non-Hispanic Black women who are obese (BMI $\geq$ 30 kg/m^2; Hedley et al., 2004). Notwithstanding the well-established genetic underpinnings of obesity, with an estimated 25-40% of the variance in weight explicable by genetic factors (Bouchard, 1994; Price, 2002), the disproportionately high prevalence of obesity among ethnic minority women suggests important environmental contributions to risk.

Clarification of the socio-cultural contributions to differential risk for disordered eating and obesity is likely to enhance the efficacy and cultural sensitivity of weight-related interventions across ethnically diverse populations. Thus the purpose of this review is to summarize relevant literature describing the multiple dimensions of socio-cultural contributions to risk for obesity, including physical activity patterns, dietary and feeding practices, body image, disordered eating, and help-seeking behaviors for weight related disorders across ethnically diverse populations. Whereas our focus is limited here to U.S. ethnic minority populations for whom rates of overweight and obesity are elevated (i.e., African Americans, Latinos, and Native Americans), we anticipate that some of the conclusions can potentially be more broadly applied across populations. Specifically, some of this literature suggests potentially modifiable risk factors for overweight that may be addressed in preventive and interventional programs.

We also begin with a few important caveats. Because the literature pertaining to possible socio-cultural factors that contribute to ethnic disparities in obesity is vast, this is not an exhaustive review, but rather a starting point for further discussion. Second, we largely review findings based on quantitative data. As a result, we do not offer a cultural analysis of why certain behaviors and attitudes possibly promoting obesity are found more commonly among some ethnic populations than others. Third, although we are primarily interested in describing the impact of ethnicity-specific socio-cultural factors on obesity in the U.S., most of the studies we review present data based on an oversimplified categorization of populations that both conflates race and ethnicity and lumps diverse ethnic groups together for the purposes of analysis. Given the substantial ethnic heterogeneity within these large categories, attributing a monolithic "culture" to them would neither be possible nor meaningful. Finally, ethnicity may imply aspects of shared experience, traditions, and values that shape environments, behaviors, and attitudes that, in turn, influence health outcomes. However, we recognize that ethnic identification—including identification with multiple ethnic groups—and related attitudes and practices vary considerably among individuals within ethnic populations. The recognition that additional contextual influences such as time and place introduce further variation invites caution against overgeneralization about the association of ethnicity with certain practices and patterns from these findings. Rather, taken together, these findings suggest promising avenues for understanding and addressing ethnic disparities in obesity in the U.S. We emphasize that further research is essential to characterize and clarify the contributions of socio-cultural differences to these ethnic disparities in obesity.

Socioeconomic Factors, Ethnicity, and Obesity

In contrast to the directly proportional relationship between socioeconomic status (SES) and obesity in developing nations, SES is inversely proportional to obesity among women in developed countries (Dykes, Brunner, Martikainen, and Wardle, 2004; Ettner and Grzywacz, 2003; Sobal and Stunkard, 1989). Because the U.S. ethnic minority populations discussed here earn a lower median income and are more likely to live below the poverty line than Caucasians (DeNavas-Walt, Proctor, and Mills, 2004), SES may confound or modify the relation between ethnicity and obesity.

Overall, data from U.S. populations suggest that higher SES individuals consume a more nutritious diet. Greater educational attainment, for example, has been linked to the consumption of more fruit and fewer high-fat meat products (Devine, Wolfe, Frongillo, and Bisogni, 1999; Tillotson, Bartsch, Gorder, Grandits, and Stamler, 1997), and low income has been associated with intake of less fiber and more fat (Ballew and Sugerman, 1995). This association may be due in part to the lower accessibility of healthful food options in low income neighborhoods. One study found that, compared to wealthier areas, lowest wealth census tracts contained only one third as many supermarkets but significantly more convenience stores (Morland, Wing, Diez Roux, and Poole, 2002). Furthermore, the fewer the number of supermarkets in the vicinity, the less likely Black Americans were to meet USDA recommendations for fruit and vegetable intake (Morland, Wing, and Roux, 2002). These investigations from the U.S. and another from the United Kingdom (Wrigley, Warm, Margetts, and Whelan, 2002) suggest that "food deserts" — low income districts in which the majority of residents do not live within walking distance of fruit and vegetable retailers (Beaumont, Lang, Leather, and Mucklow, 1995) — are a contributing factor to the poor diet among low SES populations.

In addition to consuming poorer diets, low SES individuals engage in less physical activity than their higher SES counterparts (Lee and Cubbin, 2002; Parks, Housemann, and Brownson, 2003; Wilson, Kirtland, Ainsworth, and Addy, 2004). This association may be the result of fewer appropriate venues for physical activity as well as safety concerns in less affluent neighborhoods (Molnar, Gortmaker, Bull, and Buka, 2004; Wilson et al., 2004). However, the extent to which socioeconomic variation accounts for ethnic differences in activity remains unclear. Lindquist, Reynolds, and Goran (1999) found few differences in physical activity levels between Caucasian and African American children after controlling for SES, but Crespo, Smit, Andersen, Carter-Pokras, and Ainsworth (2000) found that significant cross-ethnic differences remained among Mexican-Americans, Caucasians, and African Americans even after adjusting for multiple indicators of SES.

Despite SES differences in diet and activity, the extent to which socioeconomic factors contribute to ethnic disparities in obesity prevalence remains unknown. For example, there is evidence that the inverse relationship between SES and obesity is stronger for Caucasians than African-Americans (Averett and Korenman, 1999; Kimm, et al., 1996). In addition, data from the National Health and Nutrition Examination Surveys demonstrate that the relationship between SES and obesity has weakened significantly in the past three decades as obesity has increased across all SES categories (Zhang and Wang, 2004). It is also important to note that two studies based on large community-based samples reported that ethnic differences in weight persisted even after adjusting for SES (Shea et al., 1991; Winkleby, Gardner, and Barr-Taylor, 1996). Thus, available data suggest that ethnicity contributes independently to risk for obesity (although socioeconomic factors may modify and partially mediate this effect). Given the importance of ethnic differences that are independent of SES, we begin our review of ethnicity and relevant sociocultural factors.

Ethnicity and Physical Activity

Physical activity level is inversely correlated with adiposity (Hunter et al., 1997) and BMI (Wolf et al., 1993). Recommendations for optimal physical activity for adults range from 30 to 60 minutes of moderate intensity activity per day (Brooks, Butte, Rand, Flatt, and Caballero, 2004).

Data from several independent research teams indicate that Caucasians engage in significantly more physical activity than either African Americans (Gordon-Larsen, McMurray, and Popkin, 1999) or Hispanics (Perry, Rosenblatt, and Wang, 2004; Wolf et al., 1993). A recent large-scale population study found that 41.2% of African American women and 45.7% of Mexican American women compared to only 23.5% of Caucasian women were completely inactive during leisure time; these differences were statistically significant, even after controlling for age and multiple indices of SES (Crespo et al., 2000). There is some evidence that American Indian women are even less physically active than Caucasian, African American, and Hispanic women (King et al., 2000). Ethnic discrepancies in physical activity are consistently observed across age groups, including children (Bradley, McMurray, Harrell, and Deng, 2000), young adults (Dowda, Ainsworth, Addy, Saunders, and Riner, 2003), and older adults (Brownson et al., 2001; Wilcox, Castro, King, Housemann, and Brownson, 2000).

Ethnic minority women may be at especially high risk for physical inactivity. Available data indicate that women are less physically active than men across a variety of ethnic groups, including Caucasians, African Americans, and Hispanics (e.g., Ainsworth, Berry, Schnyder, and Vickers, 1992; Crespo et al., 2000). Gender discrepancies in activity levels may emerge during youth and increase with age. In one investigation, Caucasian and African American schoolgirls were significantly more likely than their male counterparts to engage in sedentary activities during leisure time (Bradley et al., 2000).

The antecedents of cross-ethnic differences in physical activity are poorly understood, although some unique barriers may impede activity among ethnic minority women. For example, urban American Indian women cited lack of child care and health problems as barriers to regular exercise (Harnack, Sherwood, and Story, 1999). Similarly, focus groups of Black, Hispanic, and American Indian women identified availability, cost, lack of time, and health concerns as perceived barriers to increasing physical activity (Eyler et al., 1998), although these same factors were not correlated with activity levels in a quantitative follow-up study (Eyler et al., 2003). Another sample of African American and Hispanic women also cited perceived lack of time as a barrier to physical activity, but actual time commitments were not correlated with perceived lack of time (Heesh and Masse, 2004).

Several environmental factors may promote or discourage activity among ethnic minority women. For instance, knowing people who exercise was positively correlated with physical activity among Native American, Latina, and African American women, and seeing people exercise in the neighborhood was positively correlated with physical activity among Native American and African American women (Eyler et al., 2003). Such findings underscore the importance of social norms in fostering regular activity. Security may also be important; urban African American women who reported greater neighborhood safety were more likely to be physically active than those who perceived less safety (Eyler et al., 2003).

Another explanation for the low activity levels observed in ethnic minority populations may be the displacement of physical activity with sedentary behaviors, such as television viewing. African Americans watch significantly more television than Caucasians (Gorden-Larsen et al., 1999; Sidney et al., 1996). The positive correlation between hours of television viewing and overweight/obesity has been clearly established among Caucasians, African Americans, and Hispanics (e.g., Gortmaker et al., 1996; Hu et al., 2003; Proctor et al., 2003). Television viewing may promote obesity directly by displacing physical activity (e.g., Tucker and Bagwell, 1991), although not all studies have supported this association (e.g., Hu, Li, Colditz, Willet, and Manson 2003). Viewing may also promote weight gain because it burns few calories, is often accompanied by the consumption of high calorie snacks (Van den Bulck, 2000), and promotes food consumption through advertisements.

In sum, available data convincingly demonstrate ethnic disparities in physical activity. However, these data are limited in two important ways. First, most studies rely on self-reported levels of activity, which may not provide valid estimates. Data suggest that self report indices tend to overestimate physical activity levels (Walsh, Hunter, Sirikul, and Gower, 2004), and that the degree of overestimation is larger in individuals with greater percentage body fat (Buchowski, Townsend, Chen, Acra, and Sun, 1999). One study of recently weight-reduced women found that Blacks overestimated their physical activity by a significantly wider margin than Whites (Walsh et al., 2004). Second, many studies limit foci to leisure time physical activity, which excludes physical exertion at work (e.g., manual labor) as well as lifestyle-related activity (e.g., walking rather than driving to work, taking the stairs rather than the elevator). Brownson and colleagues (2001) found that approximately three quarters of lower SES, ethnic minority women reported physical activity sufficient to yield health benefits when *all* types of activity (e.g., occupational duties, household chores, etc.) were taken into account.

Notwithstanding these limitations, the data suggest that interventions designed to promote physical activity in ethnic minority women could potentially prove useful in obesity prevention, especially when informed by the unique barriers to activity within these populations.

ETHNICITY AND DIET

The contribution of cross-ethnic variation in dietary patterns to ethnic disparities in obesity prevalence remains poorly understood (Lin, Bermudez, and Tucker, 2003).

However, numerous studies have described ethnic differences in macronutrient intake which could in part account for ethnic differences in obesity prevalence. For example, several studies have reported differences in dietary fat intake across ethnic groups. Mexican-Americans reported greater consumption of dietary fat compared to non-Hispanic women (Bartholomew, Young, Martin, and Hazuda, 1990). Similarly, Black women consumed significantly more kilocalories, dietary fat grams, and percent calories from fat than Latin-American women (Sanchez-Johnsen et al., 2004). Moreover, sources of dietary fat varied by ethnicity among a sample of Caucasian, Black, and Hispanic women (Kristal, Shattuck, and Patterson, 1999). Black women, for example, consumed less fat from dairy sources than from meats. Amount of fat intake also varied among Hispanics and non-Hispanic Whites, with Hispanics consuming fewer saturated fats and simple sugars and more complex carbohydrates (Bermudez, Falcon, and Tucker, 2000).

There is also data to suggest that Caucasians consume more vegetables than their ethnic minority counterparts. For example, significant differences in the daily frequency of raw vegetable consumption were found between African American and Caucasian women (29% and 51%, respectively) even after adjusting for SES (Madan et al., 2002). Similarly, data from the Third National Health and Nutrition Examination Survey indicated that more than 50% of Mexican-Americans reported adhering to recommended daily guidelines for fruit and vegetable consumption as opposed to fewer than 30% of non-Hispanic Blacks. However, Xie, Gilliland, Yu Fen, and Rockett (2003) reported that the lowest vegetable intake among Hispanic adolescents compared with Whites, Blacks, or Asian adolescents, although those whose parents had higher educational attainment were more likely to meet recommendations for these foods.

In contrast, other studies have reported no association between ethnicity and diet. Patterson, Harlan, Block, and Kahle (1995) found that sources of energy and nutrition among low-income Hispanic-American women were similar to those among African Americans and Caucasians in the 1987 National Health Interview Survey. Nguyen, Larson, Johnson, and Goran (1996) found no association of ethnicity with fat intake and no association of ethnicity with the relationship between fat intake and fat mass among White and Mohawk children, although findings were limited due to a small sample size, consisting of only 56 White and 15 Mohawk children.

To the extent that macronutrient intake differs by ethnicity, marketing and sales practices may represent a contributing factor. Pratt and Pratt (1996), for example, found that magazines geared toward African American readers (e.g., Ebony) were significantly less likely to contain fruit and vegetable advertisements than those directed toward Caucasian readers (e.g., Ladies' Home Journal). Media content analysis on food commercials airing on programs targeted toward African Americans were more likely to contain themes of abundance and flavor than those appearing on shows geared toward a more general audience (Carrillo, Brownell, and Suich, 2004). Finally, two recent studies have demonstrated that, compared to the most popular shows on television, shows specifically targeted toward African American viewers contain significantly more food advertisements per episode (Carrillo et al., 2004; Tirodkar and Jain, 2003).

Cross-ethnic variation in dietary patterns may also relate to the availability and accessibility of healthful food retailers. Block, Scribner, and DeSalvo (2004) found that, on average, predominantly Black neighborhoods contain one more fast food restaurant per square mile than White neighborhoods. Moreover, the proportion of Black residents in a given region was a more powerful predictor of fast food restaurant placement than median household income. Such studies suggest the possibility that variation in the promotion and availability of specific types of food to ethnic minority groups may relate to disparities in dietary patterns related to overweight.

The Prevalence of Dieting across Ethnic Groups

On average, individuals who diet weigh more than those who do not (French and Jeffery, 1994). Although a lack of experimental data prevents insight into the directionality of this relationship, some investigators have suggested that dieting itself may promote weight gain via increased metabolic efficiency (in which dieters require fewer calories to maintain weight over time) as well as an increased risk of developing binge eating disorder (Stice, Cameron, Killen, Hayward, and Barr Taylor, 1999). Indeed, one ongoing prospective cohort study reported that frequent dieters were significantly more likely to become overweight than those who never dieted, independent of height change, maternal BMI, physical activity, and dietary intake (Field and Colditz, 2001).

Available data provide conflicting evidence about potential differences in dieting across ethnic populations. For example, in a large population study (n = 324), Striegel-Moore, Wilfley, Caldwell, Needham, and Brownell (1996) found no differences in either weight loss strategies or reasons for most recent dieting efforts among Black and White dieters. They did find, however, that Black women reported experiencing less social pressure to lose weight, initiated dieting later in life, and were less likely to diet at each developmental milestone, although the causal sequence of these findings was not established.

Many studies investigating cross-ethnic rates of dieting have found that African American and Latina women are less likely to report restrained eating than are Caucasian women (Neff, Sargent, McKeown, Jackson, and Valois, 1997; Neumark-Sztainer, Story, Falkner, Beuhring, and Resnick, 1999; Rucker and Cash 1992; Stevens et al., 1997), although at least one study reported no significant ethnic differences (Schreiber et al., 1996). Moreover, in two studies, Latina and African American adolescents were more likely than Caucasian adolescents to report behaviors associated with weight gain rather than loss, such as snacking, eating when not hungry, and eating in front of the television (McNutt, et al., 1997; Neumark-Sztainer et al., 1999).

Differences in Weight Loss Methods across Ethnic Groups

Ethnic differences in weight loss methods may also contribute to ethnic differences in obesity prevalence. Neff et al. (1997) found that more Caucasian female adolescents reported dieting and exercising to lose weight compared with Hispanic adolescents. Although previous studies showed no differences in BMI among adolescent females in the region where the study was conducted, the investigators did not adjust for BMI, which limits interpretation of these findings. Moreover, Story, French, Resnick, and Blum (1995) found that Black females dieted less frequently than Caucasians, American Indians, and Whites even after adjusting for BMI.

In summary, available data on ethnic differences in dietary intake are conflicting, and mechanisms underlying such differences are not well understood. Importantly, observed differences may partially relate to socio-cultural influences such as differential advertising patterns based on an audiences' ethnicity, as well as the local availability and affordability of healthful foods. These factors may be useful to address within interventions designed to promote healthful dietary behaviors among minority women. Finally, the relation between dieting behaviors and later overweight is incompletely understood across ethnic groups, and additional prospective investigations are recommended to help clarify the causal sequence between the two, should one exist.

Ethnicity and Body Image

Available data suggest important ethnic differences in both perceptual and attitudinal body image. Perceptual body image refers to estimation of current and ideal body sizes; attitudinal body image includes affect, cognitions, and behaviors regarding size or appearance (Rucker and Cash, 1992). Several studies suggest that African American women prefer larger body sizes and are more satisfied with their bodies than are Caucasian women. Comparative data from Hispanic and Native American women are fewer and, as a result, less is known about body size preferences among these populations.

Ideal Size Preferences across Ethnicity

A recent meta-analysis found that non-Whites consistently select a significantly larger ideal body size than Whites (Wildes, Emery and Simmons, 2001). This finding is particularly well-replicated among African American women (e.g., Becker, Yanek, Koffman, and Bronner, 1999; Rucker and Cash, 1992; Thompson, Corwin, and Sargent, 1997), but has also been documented in Hispanic women (Winkleby et al., 1996). In addition, several studies suggest that Caucasian women perceive larger discrepancies between their current and ideal body sizes than do African American women (Altabe, 1998; Gluck and Geliebter, 2002;

Rucker and Cash, 1992). In contrast, another study did not find significant differences between African American, Hispanic, and Caucasian women with respect to perceived discrepancies (Snooks and Hall, 2002). Smaller discrepancies between actual and ideal body size among ethnic minority women may in part reflect ethnic differences in response to perceived social pressures to be thin. For example, in one study, Euro-American males chose thinner ideal female figures than African American males, and a significantly greater proportion of Euro-American males reported wishing their girlfriends would lose weight (Greenberg and LaPorte, 1996). In another investigation, African American women were simply more accurate than Caucasian women in predicting that men preferred average rather than extremely thin figures (Demarest and Allen, 2000).

It is important to note that a preference for larger figures does not represent a categorical idealization of obesity—or even a tolerance for it. In general, results indicate that although African American women, and potentially Hispanic women, prefer body types slightly larger than Caucasians, these ideals still lie in the "normal" size range (Becker et al., 1999; Rucker and Cash, 1992; Winkelby et al., 1996). The finding that young Black men and women rated normal weight figures with low waist-to-hip ratios as more attractive and desirable then either overweight or underweight figures (Singh, 1994) supports this point. For these reasons, more data are required on the impact of the cross-ethnic body size ideals on weight concern and weight control behaviors.

Body Satisfaction across Ethnicity

In addition to differences in perceptional dimensions of body image, ethnic differences have been described in attitudinal body image. Specifically, numerous cross-sectional investigations suggest that Black and African American women exhibit lower levels of body dissatisfaction than their White or Caucasian counterparts (Akan and Grilo, 1995; Altabe, 1998; Barry and Grilo, 2002; Henriques, Calhoun, and Cann, 1996; Felts, Parrillo, Chenier, and Dunn, 1996; Miller et al., 2000; Rucker and Cash, 1992; Schooler, Ward, Merriwether, and Caruthers, 2004; Smith, Thompson, Raczynksi, and Hilner, 1999), although some investigations have found no differences (Demarest and Allen, 2000; Shaw, Ramirez, Trost, Randall, and Stice, 2004).

One potential explanation for greater body satisfaction among African American women is that there may be a culturally based tendency to place greater value on non-weight dimensions of body and self (Bay-Cheng, Zucker, Stewart, and Pomerleau, 2002). Qualitative studies lend support to this hypothesis. Whereas focus groups of White adolescent girls expressed a desire to conform to the predetermined ideal of a tall, thin, and blonde "perfect" girl; African American girls emphasized personal style, grooming, confidence, and "making what you've got work for you" (p. 103) regardless of body shape (Parker et al., 1995; Rubin, Fitts, and Becker, 2003). Likewise, when queried as to what makes a girl "beautiful," African American girls usually cited character, personality, or attitude, while White girls focused almost solely on physical appearance (Duke, 2000). Quantitative studies confirm that appearance may be less strongly associated with overall self-esteem in African American populations. One study found that bogus positive feedback (on social skills, friendliness, and first impression) *increased* and bogus negative feedback *decreased* body satisfaction in Caucasian women but not in African American women (Henriques et al., 1996). Similarly, weekly fluctuations in body esteem were more highly correlated with weekly fluctuations in self-esteem in White women than Black women (Henriques and Calhoun, 1999).

Potential ethnic disparities in body image are less clear cut with respect to Latino women. Although there are some data to suggest that Hispanics are more satisfied than Caucasians with their bodies (Felts et al., 1996), other studies suggest that they do not significantly differ from Caucasians in this respect (Demarest and Allen, 2000; Miller et al., 2000; Shaw et al., 2004), and some investigators have found that Hispanics exhibit even greater body dissatisfaction (Robinson et al., 1996). Most studies are consistent in finding, however, that Hispanics are less satisfied with their bodies than are African Americans (Altabe, 1998; Vander Wal and Thomas, 2004).

Of particular relevance to the present review, investigations controlling for BMI have generally still found that Caucasian women exhibit greater body dissatisfaction than African American women (Gluck and Geliebter, 2002; Wilfley et al., 1996). One potential explanation for the robustness of this finding is that BMI accounts for significantly less of the variance in body satisfaction among African Americans than Caucasians, a theory for which Caldwell, Wilfley, and Brownell (1997) found empirical support in their sample of African American and Caucasian female dieters. These findings have important implications for the continued use of body image measures in which shape and weight comprise the core dimension. Currently, alternate facets of self-presentation, such as skin color, are conspicuously absent from these assessments. There is already data to suggest that such alternative dimensions may be of especial importance in African American populations. For example, African American women were more likely then either Caucasians or Hispanics to identify light skin tone as a salient attribute of their perceived cultural ideal (Altabe, 1998). Indeed, skin tone dissatisfaction was positively correlated with global body dissatisfaction among African American female undergraduates (Bond and Cash, 1992) and global self-worth among overweight African American children (Young-Hyman, Schlundt, Herman-Wenderoth, and Bozylinksi, 2003).

Limitations aside, data suggest that African American women prefer larger ideal body sizes and endorse greater body satisfaction than Caucasian women, both with and without adjusting for BMI. Findings for Latinas are less clear; although several studies indicate that they experience less satisfaction than African Americans, perhaps at levels on par with Caucasians. More data on body image among Native American women are necessary to characterize body satisfaction in that population.

Ethnicity and Disordered Eating

Individuals who endorse disordered eating attitudes and behaviors are significantly more likely to be overweight than those who do not (Fitzgibbon et al., 1998; Striegel-Moore et al., 2000). In particular, women diagnosed with binge eating disorder (BED) demonstrate elevated rates of obesity compared to controls (Bruce and Agras, 1992; Pike, Dohm, Striegel-Moore, Wilfley, and Fairburn, 2001). However, the causal relation between disordered eating and overweight is not established.

The role of ethnicity in the etiology of eating disturbances has attracted significant research attention but yielded markedly discrepant findings. However there are data that suggest eating disorder risk factors that may be unique to ethnic minority women. For example, racial teasing was correlated with disturbed eating behavior and body dissatisfaction among American college students of South Asian descent, even after controlling for self-esteem, BMI, and depressive symptoms (Iyer and Haslam, 2003). Moreover, based on interviews with African American, Latina, and White women, Thompson (1992) proposed that eating problems may initially develop as "survival strategies" (p. 546) for coping with stressful social experiences, several of which disproportionately confront ethnic minority populations, including racism, classism, and poverty.

With respect to the comparative prevalence of disordered eating, a recent meta-analysis reported that Whites experience significantly greater eating disturbance than non-Whites (Wildes, Emery, and Simons, 2001). However, Shaw and colleagues (2004) have observed that early studies (e.g., Abrams, Allen, and Gray, 1993; Gray, Ford, and Kelly, 1987) report greater discrepancies than later studies (e.g., Shaw et al., 2004; Smith, Marcus, Lewis, Fitzgibbon, and Schreiner, 1998). Therefore, it is unclear whether increasingly smaller differences imply an actual increase in eating disorder symptomatology among ethnic minority women or whether they merely reflect recent improvements in study methodology.

In addition, ethnic differences are smallest when clinically significant syndromes are studied in community samples (Wildes, Emery, and Simons, 2001). Given the low base rates of anorexia and bulimia nervosa, extremely large sample sizes would be necessary to reliably detect between-group differences in prevalence. Crago and Shisslak (2003) argue from the extant literature that restricting-type symptoms may be more common among White women, whereas binge eating and purging are equally common across ethnic groups. Indeed, in a recent meta-analysis, O'Neill (2003) determined that although Whites demonstrated significantly more eating pathology (as measured by standard continuous scales) than Blacks overall, bulimic symptoms and rates of binge eating disorder did not differ significantly between the groups. Similarly, whereas several studies have found that Caucasian women score higher than African American women on the Eating Attitudes Test-26 (Akan and Grilo, 1995) and Eating Disorder Inventory (Rhea, 1999)[1], other investigations have reported comparable or even higher rates of binge eating and purging among African American women (Field, Colditz, and Peterson 1997; Le Grange et al., 1998; Story et al., 1995).

In light of investigations citing high levels of disordered eating in Hispanic women, Bay-Cheng and colleagues (2002) warn against the potential peril of a White-not White dichotomy, in which diverse non-White ethnic groups are presumed to comprise a unitary, homogeneous comparison category in relation to Whites. Consistent with findings that Hispanic women may experience greater body dissatisfaction relative to African American women, some investigations have found that Hispanic females' scores on the Eating Disorder Inventory are higher than those of their African American counterparts (Perez, Voelz, Pettit, and Joiner, 2002; Rhea, 1999), and that Hispanics endorse greater binge severity than Whites (Fitzgibbon et al., 1998). Finally, although very little is known about disordered eating among Native American women, there is some evidence that disordered eating is equally (Smith and

[1] The 26 item Eating Attitudes Test (EAT-26) and Eating Disorders Inventory (EDI) are continuous measures that assess disordered eating attitudes and behaviors. Whereas they are not intended to classify subjects into diagnostic categories, both the EAT-26 has and the EDI have been used as the first stage of a two-stage screening process with respondents scoring above a cut score recommended for a screening interview. For both measures, higher scores presumably reflect a greater risk for clinically significant disordered eating or body image disturbance.

Krejci, 1991) or even more prevalent (Croll, Neumark-Sztainer, Story, and Ireland, 2002) among Native American than Caucasian women. Among participants in the 1996 National Eating Disorders Screening Program (a nation-wide educational and screening program held on college campuses), Native Americans did not differ from Caucasians, African Americans, Latinos, and Asians with regard to frequency of food restriction, self-induced vomiting, excessive exercise, or amenorrhea. However, Native Americans were significantly more likely to endorse laxative use and multiple methods of purging than other ethnic groups (Franko, Becker, Thomas, and Herzog, in press).

Binge Eating Disorder

Whereas some data suggest that there are significant differences in BED prevalence across ethnic groups, the nature of these differences is unclear. In a large population-based cohort of young adult women (n = 2,046), Whites were twice as likely as Blacks to meet lifetime criteria for BED (2.7% v. 1.4% respectively); this difference, however, was only marginally statistically significant (OR=2.0, 95% CI [1.0, 3.8]; Striegel-Moore et al., 2003). In contrast, a smaller community sample (n=351) found that significantly more Hispanic women (9.6%) than either Black (3.9%) or White (1.8%) women met criteria for BED (Fitzgibbon et al., 1998). On the other hand, among obese bariatric surgery candidates (n=210) there were no significant differences in the prevalence of either BED or of binge eating between the White group and the ethnic minority group, which was comprised of African Americans and Latinas (Sanchez-Johnson, Dymek, Alverdy, and Le Grange, 2003).

Other investigators have reported important behavioral similarities and attitudinal differences in clinical presentation across ethnic groups. For instance, macronutrient composition of binges did not differ between Black and White women seeking treatment for BED and bulimia nervosa (Gayle, Fitzgibbon, and Martinovich, 2004). However, findings on binge frequency and severity suggest some between group differences. For example, one investigation found that Black women diagnosed with BED binged significantly more frequently than White women (Pike et al., 2001). Another study found that Hispanic women with BED exhibited greater binge severity (e.g., greater frequency, loss of control, and negative affect) than either White or Black women (Fitzgibbon et al., 1998). Moreover, compared to White women diagnosed with BED, Black women with BED reported significantly lower dietary restraint, eating concern, shape concern, and weight concern and were significantly less likely to have a history of bulimia nervosa (Pike et al., 2001). Similarly, Latina and African American gastric bypass surgery candidates, many of whom had BED, endorsed significantly lower levels of disinhibition (vulnerability to disruptions of eating control) than Whites, despite comparable binge eating frequency (Sanchez-Johnsen et al., 2003).

It also appears that factors associated with BED may differ across ethnic groups. For example, whereas BMI and depressive symptoms correlated positively with binge eating severity among White and Hispanic women, neither correlated with binge eating severity among Blacks (Fitzgibbon et al., 1998). Similarly, in a large community sample, discrimination based on physical features was a significant retrospective correlate of BED among White women but not Black women (Striegel-Moore, Dohm, Pike, Wilfley, and Fairburn, 2002).

In summary, available data on ethnic differences in eating pathology raise more questions than answers. While some studies demonstrate greater pathology among Caucasians and Hispanics relative to African Americans, others show no differences, and data on Native Americans are limited. Further research investigating the reliability, etiology, and mediators of ethnic differences in eating disorder symptoms may contribute to understanding of differential risk for obesity.

ETHNICITY AND EXPLANATORY MODELS OF HEALTH

Ethnic differences in explanatory models of health may also impact obesity prevalence by creating group differences in levels of obesity-related distress and motivation for help-seeking.

For example, the value placed on personal responsibility for weight control varies cross-culturally (Becker, 1995). "Anti-fat" attitudes in the U.S. are arguably part of a social ideology that holds individuals accountable for life outcomes. Indeed, a study of Caucasian overweight and non-overweight women demonstrated that endorsement of the Protestant ethic, an ideology characterized by individualism and the notion that failure is caused by moral weaknesses such as self-indulgence and a lack of self-discipline, was associated with lower psychological well-being among very overweight women (Quinn and Crocker, 1999).

Cross-ethnic differences in weight concern have also been reported. For example, Stern, Pugh, Gaskill, and Hazuda (1982) found that Mexican-Americans were more likely than Anglo-Americans to express that Americans were too concerned with losing weight. Moreover, Crandall and Martinez (1996) found that Mexican-American students were significantly less concerned about weight and more accepting of overweight people than American-born students. Similarly, comparisons of weight concern across ethnic groups suggest that Caucasian adolescents and women worry more about their weight than do African Americans (Thompson, Rafiroiu, and Sargent, 2003; Wardle and Marsland, 1990). Additional studies document high rates of weight-loss efforts among Navajo Indians (White et al., 1997) and urban Native Americans (Sherwood, Harnack, and Story, 2000). Finally, one study reported ethnic differences in how women evaluate the effects of obesity on personal health. A comparison of perceived health among Black and White women found that Black women reported similar self-perceptions of their weight, physical fitness level, and eating behaviors as compared with White women despite being more obese, less fit, and consuming higher percentages of fat (Duncan, Anton, Newton, and Perri, 2003).

Several studies also suggest differences in whether obesity is understood as a threat to personal health. Another qualitative study of African American women found that the frame of reference for normal body weight was much wider than the standard classification of weight categories (Gore, 1999). In a qualitative focus group of Hispanic women, few associated obesity with heart disease (Mein and Winkelby, 1998). Massara (1989) described more acceptance and positive social connotations associated with weight gain among Puerto Rican women living in the U.S. than White women, and also more positive views toward overweight and even obesity among Puerto Rican women. Specifically, Puerto Rican women described weight gain and overweight as reflecting positively on a family's economic livelihood and, conversely, weight loss as indicative of illness and economic strain. Finally,

Baughcum et al. (2001) reported that obese, low SES mothers were more concerned about their children being underweight rather than overweight, regardless of the child's actual weight.

Weight related health concerns may also be secondary to body shape ideal concerns. For example, data from a qualitative study of young African American women in which most subjects were obese suggested that being heavy was considered acceptable, and occasionally desirable, to the opposite sex rather than a health concern (Keller and Hargrove, 1992).

Locus of control with respect to weight has predicted confidence in weight loss behavior and response to health education materials (Holt, Clark, and Kreuter, 2001), thus ethnic variations in this construct may be an important consideration for the development of obesity interventions. A large survey of attitudes about weight and appearance in an ethnically diverse sample found significant associations between ethnicity and explanations of weight gain, with no differences in frequency of weight loss efforts. Specifically, Black and Hispanic women were significantly more likely to attribute weight gain to a slow metabolism, carbohydrate cravings, depression, and low self-esteem than Caucasian dieters (Le Grange, Stone, and Brownell, 1998).

Examined together, constructs that have been empirically associated with explanatory models of overweight appear to vary by ethnicity. These constructs include locus of control, personal responsibility for weight status, and health outcomes perceived to be associated with obesity. One main limitation of the research on explanatory models of health across ethnic groups is that the majority of studies have utilized qualitative methods. While these data represent a critical starting point, the use of quantitative methods are highly recommended for the evaluation of these descriptively rich findings in order to inform much needed intervention development.

A second limitation of research on explanatory models of health is the current lack of consensus as to what constitutes a healthful level of weight concern. For example, Heinberg, Thompson, and Matzon (2001) speculate that, in a society in which the majority of individuals are either overweight or obese, a modicum of body dissatisfaction and distress about overweight may actually be beneficial. Specifically, they argue that the relationship between body image distress and healthy weight-loss behaviors can be illustrated with a U-shaped curve such that individuals who are completely satisfied with their bodies may not be motivated to attempt weight loss. Thus alternative strategies, such as emphasis on the health benefits of weight loss, may be needed in order to motivate relatively non-distressed individuals (Heinberg et al., 2001) to achieve and maintain a healthy body weight.

Family Influence

Family influence is one mechanism by which cultural norms transmit and represents a critical area for future investigations of ethnicity and overweight. Family influence may include the effects of parental modeling of eating behavior, the relationship between parental feeding practices and later overweight, and relatedly, the effects of maternal perceptions of child overweight on feeding practices. Although there are few studies that establish these relationships empirically, the effects of family influence represent a critical area of emerging research, particularly given its promise for the development of family-focused interventions.

Whereas the genetics of obesity is one mechanism through which obesity may intergenerationally transmit, not all children born to obese parents become obese (Whitaker, Wright, Pepe, Seidel, and Dietz, 1997). Indeed, there appear to be environmental components to the association between maternal and child BMI (Baughcum et al., 2001) which remain incompletely understood across ethnic groups. In one study, the availability of healthful foods and consistent availability of food in the home was the strongest correlate of dietary intake (Neumark-Sztainer, Wall, Perry, and Story, 2003).

Parental feeding practices may also affect rates of overweight across ethnic groups. An obesity-proneness model developed by Costanzo and Woody (1985) theorizes ways that parental feeding style may render a child more likely to become obese, and has informed much of the research on parental feeding practices and child's weight. This model postulates that parents will be more likely to restrict their daughter's intake when parents especially value or are troubled by eating and appearance, or if the parent perceives the child to be at risk of overweight. Indeed, control over the parent-child feeding interaction has been hypothesized as a likely contributor to later childhood obesity. Increased control over the feeding interaction has been associated with children's lesser ability to self-regulate intake among non-Hispanic White populations (Johnson and Birch, 1994; Satter, 1996), but there is little comparative work among ethnic minority populations.

In addition to excessive parental control, lack of maternal control over feeding has been observed among obese mothers as compared to a matched sample of normal weight mothers in a naturalistic investigation (Wardle, Sanderson, Guthrie, Rapoport, and Plomin, 2002). Whereas obese mothers were no more likely to use food as a reward or comfort or encourage the child to eat more than he or she wanted, obese mothers reported significantly less control over the feeding interaction as compared with non-obese mothers. In contrast, another study did not find a relationship between maternal control over feeding and obesity among their offspring (Saelens, Ernst, and Epstein, 2000), and both studies utilized primarily Caucasian samples. Lack of control over feeding may be associated with a child's ability to self-regulate energy intake resulting in overweight or obesity, although longitudinal data are required to determine the associations between parental feeding style and subsequent weight trajectories.

The design of interventions for the prevention and treatment of overweight in children requires that parents accurately perceive child overweight. A cross-sectional survey of ethnically and socio-economically diverse women (28% were non-white, of whom 88% were non-Hispanic Blacks) found that only 1 in 5 obese mothers correctly identified their overweight children as overweight (Baughcum, Chamberlin, Deeks, Powers and Whitaker, 2000). Moreover, women with less education were even less likely recognize when their children were overweight. A qualitative analysis may provide a more refined understanding of which social norms and cues parents draw from when evaluating the appropriateness of their children's weight for height. Whereas such data are limited, they suggest important differences in how parents might evaluate their child's weight. For example, Jain et al. (2001) found that low income Black and White women did not define overweight or obesity in their children according to standard growth charts used by professionals, but through social indicators such as if the child was teased because of their size or developed limitations in physical activity as a result of excess weight. Similarly, Hispanic adolescents perceived their overweight status relative to their peers' standards rather than health-based standards (Hall, Cousins, and Power, 1991), suggesting a model by which social comparison primarily informs ideas of overweight.

ACCULTURATION

Acculturation is glossed as the bidirectional process of adaptation to a dominant culture (Berry, 2003; Redfield, Linton, and Herskovitz, 1936). Multiple dimensions of this process are likely to impact dietary patterns, physical activity, help-seeking behavior, and body image in ways that have relevance to risk for overweight and obesity. A phenomenon known as the "epidemiologic paradox" describes a decline in health status among immigrant populations as they spend more time in the U.S. (Hernandez and Charney, 1998). This phenomenon has been well-described in Latino groups (Kaplan, Huguet, Newsom, McFarland, 2004) and appears to have application to risk for overweight and obesity as well as to other health behaviors and indices.

Acculturation and Weight Status

A now seminal study on the Pima Indians by Ravussin, Valencia, Esparza, Bennett, and Schulz (1994) illustrates the complicated interplay between genes and environment, and provides an instructive conceptual model for understanding the potential effects of acculturation on weight. In their native Mexico, the Pimas enjoy a traditional diet — including little animal fat and few complex carbohydrates — as well as significant energy expenditure though physical labor. However, upon immigrating to Arizona and adopting a more "Americanized" lifestyle of greater food consumption and less physical activity, the U.S. Pimas now have one of the highest rates of obesity in the world.

Higher levels of acculturation to mainstream U.S. culture have consistently been associated with increased unhealthy body weight among certain ethnic groups. Indeed, length of time in the U.S. and other proxy measures of acculturation (e.g., language use and birth place) were associated with high BMI and obesity among low income, Puerto Rican women (Himmelgreen et al., 2004).

Many studies report on the association between length of time in the U.S. and greater BMI. Greater acculturation, measured by country of birth and primary language spoken, was significantly associated with larger waist circumference among U.S.-born, Spanish speaking women after adjusing for age, education, percent of energy from dietary fat, and physical activity (Sundquist and Winkleby, 2000). Similarly, Kaplan et al. (2004) found a four-fold increase in risk of obesity in longer-term Hispanic immigrants (i.e., 15 years or more) compared to short-term ones (i.e., 5 years or fewer) among, even after adjusting for factors such as health status and differential access to health care. In the same study, obesity rates were highest among women residing in the U.S. for ten years or more (Himmelgreen et al., 2004).

However, not all investigators have replicated these findings, and instead have found assimilation into the majority culture to be associated with lower overall obesity and more favorable body fat distribution among Hispanic women (Hazuda, Mitchell, Haffner, and Stern, 1991). Such findings suggest either that slimmer woman may assimilate more "successfully" to majority culture, and/or those more interested in assimilation become thinner more quickly as a means of adopting dominant social mores. Moreover, a longitudinal study designed to relate child overweight to family factors did not find the expected association between indicators of greater acculturation and overweight in Hispanic-American children (Ariza, Chen, Binns, and Christofell, 2004). Finally, lack of integration into Anglo culture was associated with greater risk for obesity among Mexican women residing in the U.S. (Ayala et al., 2004).

Indeed, some data suggest that dietary changes may help explain Hernandez and Charney's (1998) epidemiologic paradox. For example, in one study, Mexican-Americans born in Mexico consumed significantly less fat and more fiber than those born in the U.S., and more men and women born in Mexico met the dietary guidelines or recommended nutrient intakes than those born in the U.S. (Dixon, Sundquist, and Winkleby, 2000). Likewise, several studies have found that less acculturated Latinas and Hispanics are more likely to consume healthful foods than their more acculturated peers (Bermudez et al., 2000; Garcia-Maas, 1999; Neuhouser, Thompson, Coronado, and Solomon, 2004). Similarly, acculturation to the U.S. among Hispanic adolescents was significantly associated with a higher frequency of fast-food consumption in a longitudinal study (Unger et al., 2004). Finally, a study of primarily low income Latina women found that those born in the U.S. were more likely to consume 2.5 fewer servings of fruits and vegetables per day than foreign-born Latina women who had been in the U.S. for less than four years (Dubowitz, Smith-Warner, Acevedo-Garcia, Subramanian, and Peterson, 2005). Differences within this cohort of women decreased as duration of time in the U.S. increased supporting the relation between longer duration of residence in the U.S. and less healthful dietary patterns.

There has been comparatively less research on African-Americans and acculturation, but one study found an elevated risk of obesity among foreign-born Black and Hispanic immigrants to the U.S. as their length of U.S. residence increased. Although immigrants initially had a significantly lower risk of obesity than their U.S.-born counterparts of equivalent socioeconomic and demographic background, increased length of residence increased obesity risk (Singh and Siahpush, 2002).

Identification with Minority Culture and Obesity Risk

Ethnic minority women must also negotiate body image ideals of the majority culture which may differ substantially from those of their culture of origin. Rejection of such majority views — particularly those which normalize an unattainable, unhealthfully slender ideal — may provide resilience against eating disturbances. For example, Abrams and Stormer (2002) found that African American girls with ethnically heterogeneous friend groups endorsed significantly greater awareness and internalization of the thin ideal than those without mixed friends. Another investigation found that viewing television programs with predominantly Black casts predicted more positive body image among Black women, whereas, in contrast, watching mainstream television (in which casts are predominantly White) predicted poorer body image in White women (Schooler et al., 2004). These findings suggest that greater participation in minority sub-culture may protect ethnic minority women from body image concerns, although not all studies have supported this hypothesis. For example, Akan and Grilo (1995) found that level of assimilation was associated with neither dieting nor body dissatisfaction among African American women.

Conversely, ethnic minority identification could also work to reduce obesity rather than promote it. For example, Root (1990) has argued that negative U.S. media stereotypes of

Hispanic women as being "fat," "powerless," "obedient," and "quiet," may motivate Hispanic women to strive for thinness in order to dissociate themselves from this denigrating characterization.

Acculturation and Eating Disorders

Numerous cross-cultural studies suggest that transnational migration, assimilation, and acculturation may impact risk for disordered eating (Becker, 2003). Likewise, studies have demonstrated that more acculturated Hispanic women exhibit significantly greater eating pathology than their less acculturated peers (Cachelin, Veisel, Striegel-Moore, and Barzegarnazari, 2000; Chamorro and Flores-Ortiz, 2000; Gowen et al., 1999). Similarly, Black women who were more closely identified with "White" cultural values (as measured by the Racial Identity Attitude Scale for Blacks) also had relatively greater eating pathology than Blacks who were less closely identified with White cultural values (Abrams et al., 1993). In one study, body dissatisfaction correlated with bulimic symptoms only among Black and Hispanic females who reported high acculturative stress, suggesting that relatively lower levels of acculturative stress are associated with lower risk for eating disorder symptoms among ethnic minority women (Perez et al., 2002). In contrast, other investigations have found no relationship between acculturation and eating disturbance among African American (Akan and Grilo, 1995) and Mexican American (Joiner and Kashubeck, 1996; Kuba and Harris, 2001) women. And still others find an association between greater traditionality and disordered eating among non-U.S. samples (Becker, Fay, Gilman, and Striegel-Moore, in press; Lake, Staiger, and Glowinski, 2000; Mumford, Whitehouse, and Platts, 1991).

In summary, data suggest that many of the dietary and weight changes observed among ethnic minority women emerge within a process of acculturation. It is particularly difficult to draw comparisons across studies within the literature on acculturation given the multiple proxies for this construct (e.g., language spoken, years of U.S. residence), as well as the extent to which these variables ecologically capture the process. Although the findings on disordered eating and acculturation remain equivocal, from a clinical perspective these data suggest the importance of maintaining a vigilance for these behaviors among women assimilating to a new culture. Finally, the possibility that adherence to certain traditions in some contexts may provide a protective function against eating disturbances and obesity is intriguing, and represents a priority focus for additional research given its potential application to intervention design.

Conclusions

A theoretical framework including socio-culturally specific attitudes and behaviors that contribute to elevated rates of obesity among African-American, Latina, and Native American women in the U.S. is well supported by available data. Overall, data indicate that these ethnic minority women consume more fat, experience poorer access to nutritious foods, and are less physically active than non-Hispanic White women. Moreover, data also suggest that ethnic minority women are more likely to endorse a larger ideal body size, and diet less than

Caucasian women, and that they endorse binge eating more frequently than they endorse restricting behaviors. These ethnic differences are all likely to increase risk for overweight in ethnic minority populations. Finally, culture-specific perceptions of healthy body size, aesthetic ideals, and valuation of non-weight bodily attributes have been documented that likely impact both acceptance of as well as help-seeking behavior for overweight. In addition, many of the attitudes and behaviors that influence dietary patterns evolve in a context of acculturation, a process associated with poor health outcomes, including overweight. These dimensions of socio-cultural contributions to risk suggest some potentially modifiable risk factors that may be specifically incorporated into interventions.

A critical review of the literature suggests several major methodological limitations in characterizing and evaluating socio-cultural contributions to obesity risk. First, most studies that we have identified and reviewed here are cross-sectional, a design which precludes causal inference. Areas where this is particularly true include the still equivocal prospective relationship between dieting and overweight, as well as the impact of acculturation over time on dietary patterns, weight status, and body image. Longitudinal and experimental studies will be essential to establishing causal mechanisms as well as suggesting effective interventions.

Second, assessment of key socio-cultural covariates is problematic for several reasons, primarily because many assessments have not been validated in the study populations of interest. As a result, there may be substantial misclassification of attributes, and interpretations must be made with caution. Moreover, cross-cultural research on eating disorders has strongly suggested that behaviors and symptoms may be driven by context-specific meanings and motivation despite phenomenologic similarities (see also Becker, 2004). Indeed, very few studies have assessed the psychometric properties of commonly used eating disorders assessments among non-White samples (Franko et al., 2004). Franko and colleagues' (2004) recent study of the internal consistency and factor structure of a preliminary children's version of the EDI (EDI-C) on a large community cohort of Black and White adolescent females is a noteworthy exception. They reported significant overlap between Black and White adolescent females on five of the eight EDI-C subscales, but also identified perfectionism as a factor unique to White girls and body satisfaction as a factor unique to Black girls. We support the authors' recommendation for additional work of this nature, particularly among less frequently studied ethnic groups such as Hispanic and American Indian women, in order to assess the validity of measures of eating pathology across ethnicity.

Similar methodological problems exist within the literature on body image. In his recent review of the common pitfalls of body image assessment, Thompson (2004) emphasizes the importance of utilizing measures with sound psychometric properties on the sample of interest. Thus, the fact that many body image assessments were originally validated on majority Caucasian samples calls into question the inferences researchers can make about body image and satisfaction in ethnic minority populations. In addition to norming new measures on diverse reference groups, creative new approaches to body image assessment should be pursued. Specifically, statistical techniques for the conceptualization of latent variables, such as multidimensional scaling, may prove especially useful for uncovering ethnic differences in the salience and relative importance of non-weight-related domains of body image, such as skin tone and hair texture. In addition, qualitative work may lead to a richer understanding of individuals' awareness of and motivation for focusing on alternative dimensions of body image. Finally, with over 50 measures of body image assessment available, each of which may evaluate a potentially distinct and orthogonal aspect of the construct (e.g., satisfaction with specific body part(s), overall satisfaction, or appearance orientation), it is often difficult to compare results across studies (Thompson, 2004). Indeed, in one population-based investigation of the Multidimensional Body Self-Relations Questionnaire, Blacks placed more importance on physical appearance than Whites, yet Blacks were more satisfied with their appearance than Whites (Smith et al., 1999). These findings highlight the importance of assessing multiple dimensions of body image as a means of understanding motivation to achieve or maintain a healthy body weight.

A third methodological limitation is that categorization of ethnicity remains problematic. Although a necessary starting point, the relatively broad, nonspecific terminology referenced in this chapter (i.e., African-American, Latina) is not merely a semantic limitation, but a critical question of validity. Our current nomenclature for describing ethnic differences likely subsumes a variety of subcultures within each group, and therefore remains inadequate to capture their diversity. For example, Latina females may actually identify with any one of many Latin or Central American cultures (e.g., Cuban, Mexican, South American), which renders the generic overarching Latina category comparatively less meaningful. Furthermore, categories such as "Black" or "Latino" imply a cultural homogeneity that almost certainly does not exist with respect to the research questions of interest. Heterogeneity within populations that are often lumped together for research purposes has been thoughtfully discussed elsewhere in the literature (e.g., Story et al., 1999, Wildes, Emery, and Simons, 2001), and is an essential caveat for the current discussion.

A fourth limitation pertains to construct validity, as some of the key constructs of interest discussed here (e.g., SES, body image, and acculturation) are difficult to define and may be context-specific. Indeed, Alvidrez, Azocar, and Miranda (1996) recommend that acculturation be assessed not merely on a single continuum but as independent, orthogonal dimensions, noting that the component processes of acculturation may not occur at the same rate (e.g., assumption of gender roles versus occupational roles).

Fifthly, it is possible that ethnicity acts as a proxy variable for other social categories that account for a greater proportion of the variance in obesity prevalence. As discussed previously, one such social category is SES. For example, in one study, differences in body dissatisfaction between African American and Caucasian female dieters dropped out after controlling for SES (Caldwell et al., 1997). Similarly, Snooks and Hall (2002) found no differences in real-ideal figure discrepancies in a sample of Caucasians, Hispanics, and African Americans, all of whom were middle class.

A final methodological limitation relates to both sampling and use of comparison groups. It is recommended that attempts to over-sample ethnic minority women be made given that the unequal cell sizes (e.g., 80 White women compared to 20 Black women) employed by most studies to date confer decreased statistical power to detect group differences by artificially increasing the variance of the over-represented group. Additionally, while a relatively robust literature has developed to explore several of the issues just reviewed among African-American women, there has been a comparative lack of research on Latina and Native American women. As the 2004 Census data report that Latinos represent the fastest growing minority population in the country (United States Census Bureau Data, 2004), additional work in this area is especially critical.

Although there is a lack of clinical trial data that suggests effective and feasible interventions for overweight and obesity, there exist several promising, culturally informed interventions that suggest workable models for future development. These include the *Pathways* project and "Sister Talk"; commonalities between these programs warrant consideration here. A recent report described the particular lessons learned and unique hardships overcome by one such program, *Pathways*, a multi-center study evaluating the efficacy of a school-based prevention program for obesity in American Indian children (Gittelsohn et al., 2003). The authors underscore the utility of engaging tribal communities and involving families in this initiative, and the concomitant challenges of conducting standardized research in culturally diverse settings.

Another exemplar program is "Sister Talk," a home-based weight control program for Black women available via cable television (Gans et al., 2003). Program developers included community networking with Black women in its design and implementation, and similar to Gittelsohn and colleagues (2003), emphasized the value of assembling an ethnically and socioeconomically diverse research team. "Sister Talk" developers emphasized both the value but also tremendous amount of time that the "formative research" component required, a phase which relied heavily on methods drawn from both cultural anthropology and marketing. The development of this program also enlisted the creative energies and input of Black women in the community. Successful and culturally attuned interventions will likely require thoughtful efforts to mobilize, engage, and collaborate with the target community.

The public health consequences of the present epidemic of obesity and overweight among minority women are serious and unequivocal. Attention to socio-cultural contributions to obesity risk offers promising directions for both research and clinical initiatives. We support the integration of both quantitative and qualitative methods to best characterize the diversity, richness, and reliability of these phenomena. It also appears that community participatory research will be another especially productive strategy for identifying workable, engaging, and culturally salient foci for interventions.

References

Abrams K. K., Allen, L. R., and Gray, J. J. (1993). Disordered eating attitudes and behaviors, psychological adjustment, and ethnic identity: A comparison of black and white female college students. *International Journal of Eating Disorders*, 14, 49-57.

Abrams, L.S. and Stormer, C.C. (2002). Sociocultural variations in the body image perceptions of urban adolescent females. *Journal of Youth and Adolescence*, 31, 443-450.

Ainsworth B.E., Berry, C.B., Schnyder, V.N., Vickers SR. (1992). Leisure-time physical activity and aerobic fitness in African-American young adults. *Journal of Adolescent Health*, 13, 606-611.

Akan, G.E. and Grilo, C.M. (1995). Sociocultural influences on eating attitudes and behaviors, body image, and psychological functioning: A comparison of African American, Asian American, and Caucasian college women. *International Journal of Eating Disorders*, 18, 181-187.

Altabe, M. (1998). Ethnicity and body image: Quantitative and qualitative analysis. *International Journal of Eating Disorders*, 23, 153-159.

Ariza, A.J., Chen, E.H., Binns, H.J., and Kaufer Christoffel, K. (2004). Risk factors for overweight in five-to-six-year-old Hispanic-American children: a pilot study. *Journal of Urban Health*, 81, 150-161.

Atkinson, R.L. and Nitzke, S., W. S. A. (2001). School based programmes on obesity: increase knowledge about nutrition but do not change eating habits by much. *British Medical Journal*, 323, 1018.

Averett S. and Korenman, S. (1999). Black-white differences in social and economic consequences of obesity. *International Journal of Obesity and Related Metabolic Disorders*, 23, 166-173.

Ayala, G.X., Elder, J.P., Campbell, N.R., Slymen, D.J., Roy, N., Engelberg, M., Ganiats, T. (2004). Correlates of body mass index and waist-to-hip ratio among Mexican women in the United States: Implications for intervention development. *Women's Health Issues*, 14, 155-164.

Ballew C. and Sugerman, S.B. (1995). High-risk nutrient intakes among low-income Mexican women in Chicago, Illinois. *Journal of the American Dietetic Association*, 95, 1409-1413.

Ballew, C. and Sugerman, S.B. (1995). High-risk nutrient intakes among low-income Mexican women in Chicago, Illinois. *Journal of the American Dietetic Association*, 95, 1409-1413.

Barry, D. T., and Grilo, C. M. (2002). Eating and body image disturbances in adolescent psychiatric inpatients: Gender and ethnicity patterns. *International Journal of Eating Disorders*, 32, 335-343.

Bartholomew, A.M., Young, E.A., Martin, H.W., Hazuda, H.P. (1990). Food frequency intakes and sociodemographic factors of elderly Mexican Americans and non-Hispanic whites. *Journal of the American Dietetic Association*, 90, 1693-1696.

Baughcum, A., Chamberlin, L.A., Deeks, C.M., Powers, S.W., and Whitaker, R. (2000). Maternal perceptions of overweight preschool children. *Pediatrics*, 106, 1380-1386.

Baughcum, A.E., Powers, S.W., Johnson, S.B., Chamberlin, L.A., Deeks, C.M., Jain, A., and Whitaker, R. (2001). Maternal feeding practices and their relationship to overweight in early childhood. *Journal of Developmental and Behavioral Pediatrics*, 22, 391-408.

Bay-Cheng, L. Y., Zucker, A. N., Stewart, A. J., and Pomerleau, C. S. (2002). Linking femininity, weight concern, and mental health among Latina, black, and white women. *Psychology of Women Quarterly*, 26, 36-45.

Beaumont, J., Lang, T., Leather, S., and Mucklow, C. (1995). Report from the policy sub-group to the Nutrition Task Force Low Income Project Team of the Department of Health. Radlett, Hertfordshire: Institute of Grocery Distribution.

Becker, A.E. (1995). *Body, Self, and Society: The View from Fiji*. Philadephia: University of Pennsylvania Press.

Becker, A.E. (2003). Eating disorders and social transition. *Primary Psychiatry*, 10, 75-79.

Becker, A.E. (2004). New global perspectives on eating disorders. *Culture, Medicine, and Psychiatry*, 28, 433-437.

Becker, A.E., Fay, K., Gilman, S.E., Striegel-Moore, R. (in press). Facets of acculturation and their diverse relations to body shape concern in Fiji. *International Journal of Eating Disorders*.

Becker, A.E., Keel, P., Anderson-Fye, E., Thomas, J.J. (2004). Genes and/or jeans?: genetic and socio-cultural contributions to risk for eating disorders. *Journal of Addictive Disorders*, 23, 81-103.

Becker, D. M., Yanek, L. R., Koffman, D. M., and Bronner, Y. C. (1999). Body image preferences among urban African Americans and whites from low income communities. *Ethnicity and Disease,* 9, 377-386.

Beech, B.M., Kumanyika, S.K., Baranowski, T., Davis, M., Robinson, T.R., Sherwood, N.W., Taylor, W.C., Relyea, G., Zhou, A., Pratt, C., Owens, A., and Thompson, N.S. (2004). Parental cultural perspectives in relation to weight-related behaviors and concerns of African-American girls. *Obesity Research*, 12 (Suppl 1), 7S-19S.

Bermudez, O.I., Falcon, L.M., and Tucker, K.L. (2000). Intake and food sources of macronutrients among older Hispanic adults: association with ethnicity, acculturation, and length of residence in the United States. *Journal of the American Dietetic Association*, 100, 665-673.

Berrigan, D., Dodd, K., Troiano, R.P., Krebs Smith, S.M., and Ballard Barbash, R. (2003). Patterns of health behavior in U.S. adults. *Preventive Medicine*, 36, 615-623.

Berry, J.W. (2003). Conceptual approaches to acculturation. In K.M. Chun, P.B. Organista, and G. Marin (Eds.) *Acculturation: advances in theory, measurement, and applied research*. Washington, D.C.: American Psychological Association.

Beutler, L.E., Brown, M.T., Crother, L., Booker, K., and Seabrook, M.K. (1996). The dilemma of factitious demographic distinctions in psychological research. *Journal of Consulting and Clinical Psychology*, 64, 892-902.

Block, G., Rosenberger, W.F., and Patterson, B.H. (1988). Calories, fat, and cholesterol: intake patterns in the U.S. population by race, sex, and age. *American Journal of Public Health*, 78, 1150-1155.

Block, J.P., Scribner, R.A., and DeSalvo, K.B. (2004). Fast food, race/ethnicity, and income: a geographic analysis. *American Journal of Preventive Medicine*, 27, 211-217, L.

Bond, S., and Cash, T. F. (1992). Black beauty: Skin color and body images among African American college women. *Journal of Applied Social Psychology*, 22, 874-888.

Bouchard (Ed.), *The genetics of obesity* (pp. 223-233). Boca Raton, FL: CRC Press.

Bouchard, C. B. (1994). Genetics of obesity: Overview and research direction. In C.B.

Bradley, C. B., McMurray, R. G., Harrell, J. S., and Deng, S. (2000). Changes in common activities of 3rd through 10th graders: the CHIC Study. *Medicine and Science in Sports and Exercise*, 32, 2071-2078.

Bradley, C. B., McMurray, R. G., Harrell, J. S., and Deng, S. (2000). Changes in common activities of 3rd through 10th graders: the CHIC Study. *Medicine and Science in Sports and Exercise*, 32, 2071-2078.

Bray, G.A. and Popkin, B.M. (1998). Dietary fat does affect obesity! *American Journal of Clinical Nutrition*, 68, 1157-1173.

Bray, G.A. (2004). Medical consequences of obesity. *Journal of Clinical Endocrinology and Metabolism*, 89, 2583-2589.

Bray, G.A. (2004). Obesity is a chronic, relapsing neurochemical disease. *International Journal of Obesity*, 28(1): 34-38.

Brooks, G.A., Butte, N.F., Rand, W.M., Flatt, J.P., Caballero, B. (2004). Chronicle of the Institute of Medicine physical activity recommendation: how a physical activity recommendation came to be among dietary recommendations. *American Journal of Clinical Nutrition*, 79, 921S-930S.

Brownell, K. D. (1994). Get slim with higher taxes (OpEd). *New York Times*, December 15, A-29.

Brownson, R. C., Eyler, A. A., King, A. C., Brown, D. R., Shyu, Y. L., and Sallis, J. F. (2001). Patterns and correlates of physical activity among US women 40 years and older. *American Journal of Public Health*, 90, 264-270.

Bruce, B., and Agras, W. S. (1992). Binge eating in females: A population-based investigation. *International Journal of Eating Disorders,* 7, 115-119.

Buchowski, M. S., Townsend, K. M., Chen, K. Y., Acra, S. A., and Sun, M. (1999). Energy expenditure determined by self-reported physical activity is related to body fatness. *Obesity Research*, 7, 23-33.

Cachelin, F. M., Veisel, C., Striegel-Moore, R. H., and Barzegarnazari, E. (2000). Disordered eating, acculturation, and treatment-seeking in a community sample of Hispanic, Asian, black and white women. *Psychology of Women Quarterly*, 24, 244-253.

Caldwell, M. B., Brownell, K. D., and Wilfley, D. E. (1997). Relationship of weight, body dissatisfaction, and self-esteem in African American and white female dieters. *International Journal of Eating Disorders*, 22, 127-130.

Carnell, S., Edwards, C., Croker, H., Boniface, D., and Wardle, J. (2005). Parental perceptions of overweight in 3-5 year olds. *International Journal of Obesity*, 29, 353-355.

Carrillo, J. L., Brownell, K. D., and Suich, A. E. (2004, November). *Food marketing to ethnic populations: Environmental contributions to the obesity epidemic.* Poster session presented at the annual convention of the Association for the Advancement of Behavior Therapy, New Orleans, LA.

Chamorro, R. and Flores-Ortiz, Y. (2000). Acculturation and disordered eating patterns among Mexican American women. *International Journal of Eating Disorders*, 28, 125-129.

Colditz, G.A. (1999). Economic costs of obesity and inactivity. *Medicine Science Sports and Exercise*, 31 (11 Suppl), S663-667.

Costanzo, P.R. and Woody, E.Z. (1985). Domain-specific parenting styles and their impact on the child's development of particular deviance: The example of obesity proneness. *Journal of Social and Clinical Psychology*, 3, 425-445.

Crago, M., and Shisslak, C. M. (2003). Ethnic differences in dieting, binge eating, and purging behaviors among American females: A review. *Eating Disorders*, 11, 289-304.

Crandall, C.S. and Martinez, R. (1996). Culture, ideology, and anti-fat attitudes. *Personality and Social Psychology Bulletin*, 22, 1165-1176.

Crespo, C. J., Smit, E., Andersen, R. E., Carter-Pokras, O., and Ainsworth, B. E. (2000). Race/ethnicity, social class and their relation to physical activity during leisure time: Results from the third National Health and Nutrition Examination Survey, 1988-1994. *American Journal of Preventive Medicine*, 18, 46-53.

Croll, J., Neumark-Sztainer, D., Story, M., and Ireland, M. (2002). Prevalence and risk and protective factors related to disordered eating behaviors among adolescents: Relationship to gender and ethnicity. *Journal of Adolescent Health*, 31, 166-175.

Cummins, S., and Macintyre, S. (2002). "Food deserts" — evidence and assumption in health policy making. *British Medical Journal*, 325, 436-438.

Demarest, J., and Allen, R. (2000). Body image: Gender, ethnic, and age differences. *Journal of Social Psychology*, 140, 465-472.

DeNavas-Walt, C., Proctor, B. D., and Mills, R. J. (2004). Income, Poverty, and Health Insurance Coverage in the United States: 2003. In *Current Population Reports*, U.S. Census Bureau, p. 60-226. Washington, D.C.: U.S. Government Printing Office.

Devine, C.M., Wolfe, W.S., Frongillo, E.A., and Bisogni, C.A. (1999). Life-course events and experiences: association with fruit and vegetable consumption in three ethnic groups. *Journal of the American Dietetic Association*, 99, 309-314.

Dixon, L.B., Sundquist, J., and Winkleby, M. (2000). Differences in energy, nutrient, and food intakes in a U.S. sample of Mexican-American women and men: findings from the Third National Health and Nutrition Examination Survey, 1988-1994. *American Journal of Epidemiology*, 152, 548-557.

Dowda, M., Ainsworth, B. E., Addy, C. L., Saunders, R., and Riner, W. (2003). Correlates of physical activity among U.S. young adults, 18-30 years of age, from NHANES III. *Annals of Behavioral Medicine*, 26, 15-23.

Dubowitz, T., Smith-Warner, S.A., Acevedo-Garcia, D., Subramanian, S.V., and Peterson, K.E. (2005, November). *Nativity and duration of time in the United States: Differences in fruit and vegetable intake among low-income postpartum women.* Paper presented at the Annual Meeting of the American Public Health Association, Philadelphia, PA.

Duke, L. (2000). Black in a blonde world: Race and girls' interpretations of the feminine ideal in teen magazines. *Journalism and Mass Communication Quarterly*, 77, 367-392.

Duncan, G.E., Anton, S.D., Newton, R.L., and Perri, M.G. (2003). Comparison of perceived health to physiological measures of health in Black and White women. *Preventive Medicine*, 36, 624-628.

Dykes, J., Brunner, E. J., Martikainen, P. T., and Wardle, J. (2004). Socioeconomic gradient in body size among women: The role of dietary restraint, disinhibition and hunger in the Whitehall II study. *International Journal of Obesity*, 28, 262-268.

Ettner, S. L., and Grzywacz, J. G. (2003). Socioeconomic status and health among Californians: An examination of multiple pathways. *American Journal of Public Health*, 93, 44-444.

Eyler, A. A., Baker, E., Cromer, L., King, A. C., Brownson, R. C., and Donatelle, R. J. (1998). Physical activity and minority women: A qualitative study. *Health Education and Behavior*, 25, 640-652.

Eyler, A. A., Matson-Koffman, D., Young, D. R., Wilcox, S., Wilbur, J., Thompson, J. L., Sanderson, B., and Evenson, K. R. (2003). Quantitative study of correlates of physical activity in women from diverse racial/ethnic groups: The women's cardiovascular health network project summary and conclusions. *American Journal of Preventive Medicine*, 25, 93-103.

Felts, W. M., Parrillo, A. V., Chenier, T., and Dunn, P. (1996). Adolescents' perceptions of relative weight and self-reported weight loss activities: Analysis of 1990 YRBS national data. *Journal of Adolescent Health*, 18, 20-26.

Field, A. E., Colditz, G. A., and Peterson, K. E. (1997). Racial/ethnic and gender differences in concern with weight and in bulimic behaviors among adolescents. *Obesity Research*, 5, 447-454.

Field, A.E. and Colditz, G.A. (2001). Frequent dieting and the development of obesity among children and adolescents. *Nutrition*, 17, 355-356.

Field, A.E., Coakley, E.H., Must, A., Spadano, J.L., Laird, N., Dietz, W.H., et al. (2001). Impact of overweight on risk of developing common chronic diseases during a 10-year period. *Archives of Internal Medicine*, 161, 1581-1586.

Finkelstein, E.A., Fiebelkorn, I.C., Wang, G. (2003). National medical spending attributable to overweight and obesity: how much, and who's paying? *Health Affairs*, Jan-Jun (Suppl) Web Exclusives, W3-219-26.

Fitzgibbon, M. L., Spring, B., Avellone, M. E., Blackman, L. R., Pingitore, R., and Stolley, M. R. (1998). Correlates of binge eating in Hispanic, black, and white women. *International Journal of Eating Disorders*, 24, 43-52.

Flegal, K.M., Carroll, M.D., Kuczmarski, R.J., and Johnson, C.L. (1998). Overweight and obesity in the United States: Prevalence and trends, 1960-1994. *International Journal of Obesity and Related Metabolic Disorders*, 2, 39-47.

Flegal, K.M., Carroll, M.D., Ogden, C.L., and Johnson, C.L. (2002). Prevalence and trends in obesity among US adults, 1999-2000. *Journal of the American Medical Association*, 288, 1723-1727.

Franko, D. L., Becker, A. E., Thomas, J. J., and Herzog, D. B. (in press). Cross-ethnic differences in eating disorder symptoms and related distress. *International Journal of Eating Disorders.*

Franko, D.L., Striegel-Moore, R.H., Barton, B.A., Schumann, B.C., Garner, D.M., *et al.* (2004). Measuring eating concerns in Black and White adolescent girls. *International Journal of Eating Disorders*, 35, 179-189

French, S.A. and Jeffery, R.W. (1994). Consequences of dieting to lose weight: Effects on physical and mental health. *Health Psychology*, 13, 195-212.

Friedman, M.A., and Brownell, K.D. (1995). Psychological correlates of obesity: Moving to the next research generation. *Psychological Bulletin*, 117, 3-20.

Gable, S., and Lutz, S. (2001). Nutrition socialization experiences of children in the Head Start Program. *Journal of the American Dietetic Association*, 101, 572-7.

Gans, K.M., Kumanyika, S.K., Lovell, H.J., Risica, P.M., Goldman, R., and Odoms-Young, A., *et al.* (2003). The development of SisterTalk: a cable TV-delivered weight control program for black women. *Preventive Medicine*, 37, 654-667.

Garcia-Maas, L.D. (1999). Intergenerational analysis of dietary practices and health perceptions of Hispanic women and their adult daughters. *Journal of Transcultural Nursing*, 10, 213-219.

Gayle, J. L., Fitzgibbon, M. L., and Martinovich, Z. (2004). A preliminary analysis of binge episodes: Comparison of a treatment-seeking sample of black and white women. *Eating Behaviors*, 5, 303-313.

Gittelsohn, J., Davis, S.M., Steckler, A., Ethelbah, B., Clay, T., and Metcalfe, L. *et al.* (2003). Pathways: lessons learned and future directions for school-based interventions among American Indians. *Preventive Medicine*, 37, S107-112.

Gluck, M. E., and Geliebter, A. (2002). Racial/ethnic differences in body image and eating behaviors. *Eating Behaviors*, 3, 143-151.

Gordon-Larsen, P., McMurray, R. G., Popkin, B. M. (1999). Adolescent physical activity and inactivity vary by ethnicity: The National Longitudinal Study of Adolescent Health. *Journal of Pediatrics*, 135, 301-306.

Gore, S.V. (1999). African-America women's perceptions of weight: paradigm shift for advanced practice. *Holistic Nursing Practice*, 13, 71-79.

Gortmaker, S.L., Must, A., Perrin, J.M. (1993). Social and economic consequences of overweight in adolescence and young adulthood. *New England Journal of Medicine*, 329, 1008-1012.

Gortmaker, S. L., Must, A., Sobol, A. M., Perterson, K., Colditz, G. A., and Dietz, W. H. (1996). Television viewing as a cause of increasing obesity among children in the United States, 1986-1990. *Archives of Pediatrics and Adolescent Medicine*, 150, 356-362.

Gray, J. J., Ford, K., and Kelly, L. M. (1987). The prevalence of bulimia in a black college population. *International Journal of Eating Disorders*, 6, 733-740.

Greenberg, D. R., and LaPorte, D. J. (1996). Racial differences in body type preferences of men for women. *International Journal of Eating Disorders*, 19, 275-278.

Guinn, B., Semper, T., Jorgensen, L. (1997). Mexican-American female adolescent self-esteem: The effect of body image, exercise behavior, and body fatness. *Hispanic Journal of Behavioral Sciences*, 19, 517-526.

Haiman, C. and Devlin, M.J. (1999). Binge eating before the onset of dieting: a distinct subgroup of bulimia nervosa? *International Journal of Eating Disorders*, 25, 151-157.

Hall, S.K., Cousins, J.H., and Power, T.G. (1991). Self-concept and perceptions of attractiveness and body size among Mexican-America mothers and daughters. *International Journal of Obesity*, 15, 567-575.

Harnack, L., Sherwood, N., and Story, M. (1999). Diet and physical activity patterns of urban American Indian women. *American Journal of Health Promotion*, 13, 233-236.

Hazuda, H.P., Mitchell, B.D., Haffner, S.M., and Stern, M.P. (1991). Obesity in Mexican-American subgroups: Findings from the San Antonio Heart Study. *American Journal of Clinical Nutrition*, 53, 1529S-1534S.

Hedley, A.A., Ogden, C.L., Johnson, C.L., Carroll, M.D., Curtin, L.R., and Flegal, K.M. (2004). Prevalence of overweight and obesity among US children, adolescents, and adults, 1999-2002. *Journal of the American Medical Association*, 292, 1304.

Heesh, K. C., and Masse, L. C. (2004). Lack of time for physical activity: Perception or reality for African American and Hispanic women? *Women and Health*, 39, 45-62.

Heinberg, L.J., Thompson, J., Matzon, J.L. (2001). Body dissatisfaction as a motivator for health lifestyle change: Is some distress beneficial? In R. H. Striegel-Moore and L. Smolak, (Eds), *Eating disorders: Innovative directions in research and practice.* (pp. 215-232). Washington, D. C.: American Psychological Association.

Henriques, G. R., Calhoun, L. G., and Cann, A. (1996). Ethnic differences in women's body satisfaction: An experimental investigation. *Journal of Social Psychology*, 136, 689-697.

Henriques, G. R., and Calhoun, L. G. (1999). Gender and ethnic differences in the relationship between body esteem and self-esteem. *Journal of Psychology*, 133, 357-368.

Hernandez, D.J. and Charney, E. (Eds). (1998). *From Generation to Generation: The Health and Well-Being of Children in Immigrant Families.* National Research Council and Institute of Medicine.

Hewett, J.K. (1997). The genetics of obesity: what the genetic studies told us about environment. *Behavior Genetics*, 27, 353-357.

Himmelgreen, D.A., Perez-Escamilla, R., Martinez, D., Bretnall, A., Eells, B., Peng, Y., and Bermudez, A. (2004). The longer you stay, the bigger you get: Length of time and language use in the U.S. are associated with obesity in Puerto Rican women. *American Journal of Physical Anthropology*, 125, 90-96.

Holt, C.L., Clark, E.M., and Kreuter, M.W. (2001). Weight locus of control and weight-related attitudes and behaviors in an overweight population. *Addictive Behavior*, 26, 329-340.

Horodynski, M.A., Hoerr, A., and Coleman, G. (2004). Nutrition education aimed at toddlers: A pilot program for rural, low-income families. *Family and Community Health*, 27, 103-113.

Hovey, J.D. (1998). Acculturative stress, depression, and suicidal ideation among Mexican-American adolescents: Implications for the development of suicide prevention programs in schools. *Psychological Reports*, 83, 249-250.

Hu, F. B., Li, T. Y., Colditz, G. A., Willet, W. C., and Manson, J. E. (2003). Television watching and other sedentary behaviors in relation to risk of obesity and type 2 diabetes mellitus in women. *Journal of the American Medical Association*, 289, 1785-1791.

Hunter, G. R., Kekes-Szabo, T., Snyder, S., *et al.* (1997). Fat distribution, physical activity, and cardiovascular risk factors. *Medicine and Science in Sports and Exercise*, 29, 362-369.

Iyer, D.S., and Haskam, N. (2003). Body image and eating disturbance among South Asian American women: The role of racial teasing. *International Journal of Eating Disorders*, 34, 142-147.

Jain, A., Sherman, S.N., Chamberlin, D.L., Carter, Y., Powers, S.W., Whitaker. R.C. (2001). Why don't low-income mothers worry about their preschoolers being overweight? *Pediatrics*, 107, 1138-1146.

Jain, A., Sherman, S.S., Chamberlin, L.A., and Whitaker, R.C. (2004). Mothers misunderstand questions on a feeding questionnaire. *Appetite*, 42, 249-254.

Johnson, S.L. and Birch, L.L. (1994). Parents' and children's adiposity and eating style. *Pediatrics*, 94, 653-661.

Joiner, G.W., and Kashubeck, S. (1996). Acculturation, body image, self-esteem, and eating disorder symptomatology in adolescent Mexican American women. *Psychology of Women Quarterly*, 20, 419-435.

Kaplan, M.S., Huguet, N., Newsom, J.T., Bentson, H., and McFarland, B.H. (2004). *American Journal of Preventive Medicine*, 27, 323-326.

Keller, C.S. and Hargrove, H.J. (1992). Factors contributing to perception of health in young Black women. *Perceptual and Motor Skills*, 75, 706.

Keller, K.L., Pietrobelli, A.S., Must, S.S., and Faith, M.S. (2002). Genetics of eating and its relation to obesity. *Current Atherosclerosis Reports*, 4, 176-182.

Kimm, S.S., Obarzanek, E., Barton, B.A., Aston, C.E., Similo, S.L., Morrison, J.A. *et al.* (1996). Race, SES, and obesity in 9-10-year-old girls: the NHLBI Growth and Health Study. *Annals of Epidemiology*, 6, 266-275.

King, A. C., Castro, C., Wilcox, S., Eyler, A. A., Sallis, J. F., and Brownson, R. C. (2000). Personal and environmental factors associated with physical inactivity among different racial-ethnic groups of U.S. middle-aged and older-aged women. *Health Psychology*, 19, 354-364.

Kristal, A.R., Shattuck, A.L., and Patterson, R.E. (1999). Differences in fat-related dietary patterns between black, Hispanic and White women: results from the Women's Health Trial Feasibility Study in Minority Populations. *Public Health Nutrition*, 2, 253-262.

Kuba, S.A., and Harris, D.J. (2001). Eating disturbances in women of color: An exploratory study of contextual factors in the development of disordered eating in Mexican-American women. *Health Care for Women International*, 22, 281-298.

Lake, A.J., Staiger, P.K., and Glowinski, H. (2000). Effect of Western culture on women's attitudes to eating and perceptions of body shape. *International Journal of Eating Disorders*, 27, 83-89.

Le Grange, D., Louw, J., Breen, A., and Katzman, M. A. (2004). The meaning of 'self-starvation' in impoverished black adolescents in South Africa. *Culture, Medicine, and Psychiatry*, 28, 439-61.

Le Grange, D., Stone, A.A., and Brownell, K.D. (1998). Eating disturbances in white and minority female dieters. *International Journal of Eating Disorders*, 24, 395-403.

Lee, R. E., and Cubbin, C. (2002). Neighborhood context and youth cardiovascular health behaviors. *American Journal of Public Health*, 92, 428-436.

Lee, S. (1995). Self-starvation in context: Towards a culturally sensitive understanding of anorexia nervosa. *Social Science and Medicine*, 41, 25-36.

Lee, S., Lee, A. M., Ngai, E., Lee, D. T. S., and Wing, Y. K. (2001). Rationales for food refusal in Chinese patients with anorexia nervosa. *International Journal of Eating Disorders*, 29, 224-229.

Lin, H., Bermudez, O.I., and Tucker, K.L. (2003). Dietary patterns of Hispanic elders are associated with acculturation and obesity. *American Society for Nutritional Sciences*, 133, 3651-3657.

Lindquist, C. H., Reynolds, K. D., and Goran, M. I. (1999). Sociocultural determinants of physical activity among children. *Preventive Medicine*, 29, 305-312.

Madan, A.K., Barden, C.B., Beech, B., Fay, K., Sintich, M., and Beech, D.J. (2002). Self-reported differences in daily raw vegetable intake by ethnicity in a breast cancer screening program. *Journal of the National Medical Association*, 94, 894-900.

Maes, H.H.M., Neale, M.C., and Eaves, L.J. (1997). Genetic and environmental factors in relative body weight and adiposity. *Behavior Genetics*, 27, 325-351.

Massara, E.B. (1989). *Que gordita! A study of weight among women in a Puerto Rican community*. New York: AMS Press.

McNutt, S.W., Hu, Y., Schreiber, G.B., Crawford, P.B., Obarzanek, E., Mellin, L. (1997). A longitudinal study of the dietary practices of black and white girls 9 and 10 years old at enrollment: the NHLBI Growth and Health Study. *Journal of Adolescent Health,* 20, 27–37.

Mein, S. and Winkleby, M.A. (1998). Concerns and misconceptions about cardiovascular disease risk factors: A focus group evaluation with low-income Hispanic women. *Hispanic Journal of Behavioral Sciences*, 20, 192-211.

Miller, K. J., Gleaves, D. H., Hirsh, T. G., Green, B. A., Snow, A. C., and Corbett, C. C. (2000). Comparisons of body image dimensions by race/ethnicity and gender in a university population. *International Journal of Eating Disorders*, 27, 310-316.

Molnar, B. E., Gortmaker, S. L., Bull, F. C., and Buka, S. L. (2004). Unsafe to play? Neighborhood disorder and lack of safety predict reduced physical activity among urban children and adolescents. *American Journal of Health Promotion*, 18, 378-386.

Morland, K., Wing, S., and Roux, A. D. (2002). The contextual effect of the local food environment on residents' diets: The atherosclerosis risk in communities study. *American Journal of Public Health*, 92, 1761-1767.

Morland, K., Wing, S., Diez Roux, A., and Poole, C. (2002). Neighborhood characteristics associated with location of food store and food service places. *American Journal of Preventive Medicine*, 22, 23-29.

Mumford, D.B., Whitehouse, A.M., Platts, M. (1991). Sociocultural correlates of eating disorders among Asian schoolgirls in Bradford. *British Journal of Psychiatry*, 158, 222-228.

Nasser, M., Katzman, M.A., and Gordon, R.A. (Eds) (2001). *Eating disorders and cultures in transition*. New York, NY, US: Brunner-Routledge.

Neff, L.J., Sargent, R.G., McKeown, R.E., Jackson, K.L., and Valois, R.F. (1997). Black-White differences in body size perceptions and weight management practices among adolescent females. *Journal of Adolescent Health*, 20, 459-465.

Neuhouser, M.L., Thompson, B., Coronado, G.D., and Solomon, C.C. (2004). Higher fat intake and lower fruit and vegetable intake are associated with greater acculturation among Mexicans living in Washington State. *Journal of the American Dietetic Association*, 104, 51-57.

Neumark-Sztainer D., Story, M., Resnick, M.D., Blum, R.W. (1997). Psychosocial concerns and weight control behaviors among overweight and nonoverweight Native American adolescents. *Journal of the American Dietetic Association*, 97, 598-604.

Neumark-Sztainer, D., Story, M., Falkner, N.H., Beuhring, T., Resnick, M.D. (1999). Sociodemographic and personal characteristics of adolescents engaged in weight loss and weight/muscle gain behaviors: Who is doing what? *Preventive Medicine*, 28, 40-50.

Neumark-Sztainer, D., Wall, M., Perry, C., and Story, M. (2003). Correlates of fruit and vegetable intake among adolescents: Findings from Project EAT. *Preventive Medicine*, 37, 198-208.

Newton, R.L., and Perri, M.G. (1997). Health beliefs and obesity: effects of race and obesity status. *Annals of Behavioral Medicine*, 19, S105.

Nguyen, V.T., Larson, D.E., Johnson, R.K., Goran, M.I. (1996). Fat intake and adiposity in children of lean and obese parents. *American Journal of Clinical Nutrition*, 63, 507-513.

Nicholson, B.L. (1997). The influence of pre-emigration and post-emigration stressors on mental health: A study of southeast Asian refugees. *Social Work Research*, 21, 19-31.

O'Neill, S. K. (2003). African American women and eating disturbances: A meta-analysis. *The Journal of Black Psychology*, 29, 3-16.

Parker, S., Nichter, M., Nichter, M., Vuckovic, N., Sims, C., and Ritenbaugh, C. (1995). Body image and weight concerns among African American and White adolescent females: Differences that make a difference. *Human Organization*, 54, 103-114.

Parks, S. E., Housemann, R. A., Brownson, R. C. (2003). Differential correlates of physical activity in urban and rural adults of various socioeconomic backgrounds in the United States. *Journal of Epidemiology and Community Health*, 57, 29-35.

Patterson, B.H., Harlan, L.C., Block, G., and Kahle, L. (1995). Food choices of whites, blacks, and Hispanics: data from the 1987 National Health Interview Survey. *Nutrition and Cancer*, 23, 105-119.

Patterson, T.L., Rupp, J.W., Sallis, J.F., Atikins, C.J., and Nader, P.R. (1988). Aggregation of dietary calories, fats, and sodium in Mexican-American and Anglo families. *American Journal of Preventive Medicine*, 4, 75-82.

Pawson, I.G., Martorell, R., and Mendoza, F.E. (1991). Prevalence of overweight and obesity in U.S. Hispanic populations. *American Journal of Clinical Nutrition*, 53, 1522S-1528S.

Perez, M., Voelz, Z.R., Pettit, J.W., and Joiner, T.E. (2002). The role of acculturative stress and body dissatisfaction in predicting bulimic symptomatology across ethnic groups. *International Journal of Eating Disorders*, 31, 442-454.

Perry, A. C., Rosenblatt, E. B., and Wang, X. (2004). Physical, behavioral, and body image characteristics in a tri-racial group of adolescent girls. *Obesity Research*, 12, 1670-1679.

Pike, K. M., Dohm, F-A., Striegel-Moore, R. H., Wilfley, D. E., and Fairburn, C. G. (2001). A comparison of Black and White women with binge eating disorder. *American Journal of Psychiatry*, 158, 1455-1460.

Pratt, C.A., and Pratt, C.B. (1996). Nutritional Advertisements in Consumer Magazines: Health implication for African Americans. *Journal of Black Studies*, 26, 504-523.

Presnell, K. and Stice, E. (2003). An experimental test of the effect of weight-loss dieting on bulimic pathology: Tipping the scales in a different direction. *Journal of Abnormal Psychology*, 112, 166-170.

Price, R. A. (2002). Genetics and common obesities: Background, current status, strategies and future prospects. In T. A. Wadden and A. J. Stunkard (Eds.), *Handbook of obesity treatment* (pp. 73-94). New York: Guilford Press.

Proctor, M. H., Moore, L. L., Gao, D., Cupples, L. A., Bradlee, M. L., Hood, M. Y., and Ellison, R. C. (2003). Television viewing and change in body fat from preschool to early adolescence: The Framingham Children's Study. *International Journal of Obesity*, 27, 827-833.

Puhl, R. and Brownell, K.D. (2001). Bias, discrimination, and obesity. *Obesity Research*, 9, 88-805.

Quinn, D.M. and Crocker, J. (1999). When ideology hurts: Effects of belief in the Protestant ethic and feeling overweight on the psychological well-being of women. *Journal of Personality and Social Psychology*, 77, 402-414.

Ravussin, E., Valencia, M.E., Esparza, J., Bennett, P.H., and Schulz, L.O. (1994). Effects of a traditional lifestyle on obesity in the Pima Indians. *Diabetes Care*, 17, 1067-1074.

Redfield, R., Linton, R., and Herskovitz, M.J. (1936). Memorandum for the study of acculturation. *American Anthropologist*, 38, 149-152.

Rhea, D. J. (1999). Eating disorder behaviors of ethnically diverse urban female adolescent athletes and non-athletes. *Journal of Adolescence*, 22, 379-388.

Robinson, T. N., Killen, J. D., Litt, I. F., Hammer, L. D., Wilson, D. M., Haydel, K. F., Hayward, C., and Barr Taylor, C. (1996). Ethnicity and body dissatisfaction: Are Hispanic and Asian girls at increased risk for eating disorders? *Journal of Adolescent Health*, 19, 384-393.

Root, M.P. (1990). Disordered eating in women of color. *Sex Roles*, 22 (7-8), 525-536.

Rubin, L. R., Fitts, M. L., and Becker, A. E. (2003). "Whatever feels good in my soul": Body ethics and aesthetics among African American and Latina women. *Culture, Medicine and Psychiatry*, 49-75.

Rucker, C.E., and Cash, T.F. (1992). Body images, body-size perceptions, and eating behaviors among African-American and White college women. *International Journal of Eating Disorders*, 12, 291-300.

Saelens, B.E., Ernst, M.M., and Epstein, LH. (2000). Maternal child feeding practices and obesity: A discordant sibling analysis. *International Journal of Eating Disorders*, 27, 459-463.

Sanchez-Johnsen, L. A. P., Dymek, M., Alverdy, J., and le Grange, D. (2003). Binge eating and eating-related cognitions in ethnically diverse obese women. *Obesity Research*, 11, 1002-1009.

Sanchez-Johnsen, L.A., Fitzgibbon, M.L., Martinovich, Z., Stolley, M.R., Dyer, A.R., and Van Horn, L. (2004). Ethnic differences in correlates of obesity between Latin-American and black women. *Obesity Research*, 12, 652-60.

Satter, E.M. (1996). Internal regulation and the evolution of normal growth as the basis for prevention of obesity in children. *Journal of the American Dietetic Association*, 98, 860-864.

Schooler, D., Ward, L. M., Merriwether, A., and Caruthers, A. (2004). Who's that girl: Television's role in the body image development of young white and black women. *Psychology of Women Quarterly*, 28, 38-47.

Schreiber, G.B., Robins, M., Striegel-Moore, R., Obarzanek, E., Morrison, J.A., and Wright, D.J. (1996). Weight modification efforts reported by black and white preadolescent girls: National Heart, Lung, and Blood Institute Growth and Health Survey. *Pediatrics*, 98, 63-70.

Schulsinger, F. (1986). *New England Journal of Medicine*, 314, 193-198.

Seidell, J.C. (1998). Societal and personal costs of obesity. *Experimental and Clinical Endocrinology and Diabetes*, 106 (Suppl 2), 7-9.

Shaw, H., Ramirez, L., Trost, A., Randall, P., and Stice, E. (2004). Body image and eating disturbances across ethnic groups: More similarities than differences. *Psychology of Addictive Behaviors*, 18, 12-18.

Shea, S., Stein, A.D., Basch, C.E., Lantigua, R., Maylahn, C., Strogatz, D.S., *et al.* (1991). Independent associations of educational attainment and ethnicity with behavioral risk factors for cardiovascular disease. *American Journal of Epidemiology*, 134, 567-582.

Sherwood, N.E., Harnack, L., and Story, M. (2000). Weight-loss practices, nutrition beliefs, and weight-loss program preferences of urban American Indian women. *Journal of the American Dietetic Association*, 100, 442-446.

Sidney, S., Sternfeld, B., Haskell, W.L., Jacobs, D.B., Chesney, M.A., and Hulley, S.B. (1996). Television viewing and cardiovascular risk factors in young adults: the CARDIA study. *Annals of Epidemiology*, 6, 154-159.

Singh, D. (1994). Body fat distribution and perception of desirable female body shape by young black men and women. *International Journal of Eating Disorders*, 16, 289-294.

Singh, G.K., and Siahpush, H.J. (2002). Ethnic-immigrant differentials in health behaviors, morbidity, and cause-specific mortality in the United States: an analysis of two national data bases. *Human Biology*, 74 , 83-109.

Smith, D. E., Marcus, M. D., Lewis, C., Fitzgibbon, M., Schreiner, P. (1998). Prevalence of binge eating disorder, obesity, and depression in a biracial cohort of young adults. *Annals of Behavioral Medicine*, 20, 227-232.

Smith, D. E., Thompson, J. K., Raczynski, J. M., and Hilner, J. E. (1999). Body image among men and women in a biracial cohort: The CARDIA study. *International Journal of Eating Disorders*, 25, 71-82.

Smith, J. E., and Krejci, J. (1991). Minorities join the majority: Eating disturbances among Hispanic and Native American youth. *International Journal of Eating Disorders*, 10, 179-186.

Smolak, L., and Striegel-Moore, R. H. (2001). Challenging the myth of the golden girl: Ethnicity and eating disorders. In R. H. Striegel-Moore and L. Smolak, (Eds), *Eating disorders: Innovative directions in research and practice.* (pp. 215-232). Washington, D. C.: American Psychological Association.

Snooks, M. K., and Hall, S. K. (2002). Relationship of body size, body image, and self-esteem in African American, European American, and Mexican American middle-class women. *Health Care for Women International*, 23, 460-466.

Sobal, J., and Stunkard, A. J. (1989). Socioeconomic status and obesity: A review of the literature. *Psychological Bulletin*, 105, 260-275.

South Asian adolescents. *International Journal of Eating Disorders*, 37, 161-165.

Stern, M.P., Pugh, J.A., Gaskill, S.P. and Hazuda, H.P. (1982). Knowledge, attitudes, and behaviors related to obesity and dieting in Mexican American and Anglos: the San Antonio Heart Study. *American Journal of Epidemiology*, 115, 917-928.

Stevens, J., Alexandrov, A.A., Smirnova, S.G., Deev, A.D., Gershunskaya, Y.B., and Davis, C.E., *et al.* (1997). Comparison of attitudes and behaviors related to nutrition, body size, dieting, and hunger in Russian, black-American, and white-American adolescents. *Obesity Research*, 5, 227-236.

Stice, E., Cameron, R.P., Killen, J.D., Hayward, C., and Barr Taylor, C. (1999). Naturalistic weight reduction efforts prospectively predicted growth in relative weight and onset of obesity among female adolescents. *Journal of Consulting and Clinical Psychology*, 67, 967-974.

Story, M., French, S.A., Resnick, M.D., and Blum, R.W. (1995). Ethnic/racial and socioeconomic differences in dieting behaviors and body image perceptions in adolescents. *International Journal of Eating Disorders*, 18, 173-179.

Striegel-Moore, R. H., Dohm, F. A., Kraemer, H. C., Taylor, C. B., Daniels, S., Crawford, P. B., and Schreiber, G. B. (2003). Eating disorders in white and black women. *American Journal of Psychiatry*, 160, 1326-1331.

Striegel-Moore, R. H., Dohm, F. A., Pike, K. M., Wilfley, D. E., and Fairburn, C. G. (2002). Abuse, bullying, and discrimination as risk factors for binge eating disorder. *American Journal of Psychiatry*, 159, 1902-1907.

Striegel-Moore, R., H., Schreiber, G. B., Lo, A., Crawford, P., Obarzanek, E., and Rodin, J. (2000). Eating disorder symptoms in a cohort of 11 to 16-year-old black and white girls: The NHLBI growth and health study. *International Journal of Eating Disorders*, 27, 49-66.

Striegel-Moore, R.H., Wilfley, D.E., Caldwell, M.B., Needham, M.L., and Brownell, K.D. (1996). Weight-related attitudes and behaviors of women who diet to lose weight: a comparison of black dieters and white dieters. *Obesity Research*, 4, 109-116.

Stunkard, A.J., Soresen, T.I., Hanis, C., Teasdale, T.W., Chakraborty, R., Schul, W.J., and

Sundquist, J. and Winkleby, M. (2000). Country of birth, acculturation status and abdominal obesity in a national sample of Mexican-American women and men. *International Journal of Epidemiology*, 29, 470-477.

Tareen, A., Hodes, M., and Rangel, L. (2005). Non-fat-phobic anorexia nervosa in British

Thompson, B. W. (1992). 'A way outa no way': Eating problems among African-American, Latina, and white women. *Gender and Society*, 6, 546-561.

Thompson, J. K. (2004). The (mis)measurement of body image: Ten strategies to improve assessment for applied and research purposes. *Body Image*, 1, 7-14.

Thompson, S. H., Corwin, S. J., Sargent, R. G. (1997). Ideal body size beliefs and weight concerns of fourth-grade children. *International Journal of Eating Disorders*, 21, 279-284.

Thompson, S.H., Rafiroiu, A.C., and Sargent, R.G. (2003). Examining gender, racial, and age differences in weight concern among third, fifth, eight, and eleventh graders. *Eating Behavior*, 3, 307-323.

Tillotson, J.L., Bartsch, G.E., Gorder, D., Grandits, G.A., and Stamler, J. (1997). Group and nutrient intakes at baseline in the Multiple Risk Factor Intervention Trial. *American Journal of Clinical Nutrition*, 65 (Suppl 1), 228S-257S.

Tirodkar, M.A., and Jain, A. (2003). Food messages on African American television shows. *American Journal of Public Health*, 93, 439-441.

Tucker, L. A., and Bagwell, M. (1991). Television viewing and obesity in adult females. *American Journal of Public Health*. 81, 908-911.

Unger, J.B., Reynolds, K., Shakib, S., Spruijt-Metz, Sun, P., Johnson Anderson, C. (2004). Acculturation, physical activity, and fast-food consumption among Asian-American and Hispanic adolescents. *Journal of Community Health*, 29, 467-481.

United States Census Bureau Data (2004, March). United States Interim Projections by Age, Sex, Race, and Hispanic Origin. Retrieved September 3, 2005, from the United States Census Bureau Web site: *http:// www. census.gov /ipc/ www/usinterimproj/*

Valdez R, Gonzalez-Villalpando C, Mitchell BD, Haffner SM, Stern MP. (1995) Differential impact of obesity in related populations. *Obesity Research*, 3 (Suppl 2), 223s-232s.

Van den Bulck, J. (2000). Is television bad for your health? Behavior and body image of the adolescent "couch potato." *Journal of Youth and Adolescence*, 29, 273-288.

Vander Wal, J. S., and Thomas, N. (2004). Predictors of body image dissatisfaction and disturbed eating attitudes and behaviors in African American and Hispanic girls. *Eating Behaviors*, 4, 291-301.

Walsh, M. C., Hunter, G. R., Sirikul, B., and Gower, B. A. (2004). Comparison of self-reported with objectively assessed energy expenditure in black and white women before and after weight loss. *American Journal of Clinical Nutrition*, 79, 1013-1019.

Wardle, J., and Marsland, L. (1990). Adolescent concerns about weight and eating; a social-development perspective. *Journal of Psychosomatic Research*, 34, 377-391.

Wardle, J., Sanderson, S., Guthrie, C.A., Rapoport, L., and Plomin, R. (2002). Parental feeding style and inter-generational transmission of obesity risk. *Obesity Research*, 10, 453-462.

Weber J.L., Lytle L., Gittelsohn, J., Cunningham-Sabo, L., Heller, K., Anliker, J.A., Stevens, J., Hurley, J, Ring, K. (1995). Validity of self-reported dietary intake at school meals by American Indian children: the Pathways study. *Journal of the American Dietetic Association*, 104, 746-752.

Whitaker, R.C., Wright, J.A., Pepe, M.S., Seidel, K.D., and Dietz, W.H. (1997). Predicting obesity in young adulthood from childhood and parental obesity. *New England Journal of Medicine*, 337, 869-873.

White, L.L., Ballew, C., Gilbert, T.J., Mendlein, J.M., Mokdad, A.H., Strauss, K.F. (1997). Weight, body image, and weight control practices of Navajo Indians: Findings from the Navajo Health and Nutrition Survey. *Journal of Nutrition*, 127 (Suppl 10), 2094S-2098S.

Wilcox, S., Castro, C., King, A. C., Housemann, R., and Brownson, R. C. (2000). Determinants of leisure time physical activity in rural compared with urban older and ethnically diverse women in the United States. *Journal of Epidemiology and Community Health*, 54, 667-672.

Wildes, J.E., Emery, R.E., and Simons, A.D. (2001). The role of ethnicity and culture in the development of eating disturbance and body dissatisfaction: A meta-analytic review. *Clinical Psychology Review*, 21, 521-555.

Wilfley, D. E., Schreiber, G. B., Pike, K. M., Striegel-Moore, R. H., Wright, D. J., and Rodin, J. (1996). Eating disturbance and body image: A comparison of a community sample of adult Black and White women. *International Journal of Eating Disorders*, 20, 377-387.

Wilson, D. K., Kirtland, K. A., Ainsworth, B. E., and Addy, C. L. (2004). Socioeconomic status and perceptions of access and safety for physical activity. *Annals of Behavioral Medicine*, 28, 20-28.

Winkleby, M. A., Gardner, C. D., and Barr Taylor, C. (1996). The influence of gender and socioeconomic factors on Hispanic/white differences in body mass index. *Preventive Medicine*, 25, 203-211.

Wolf A.M. and Colditz, G.A. (1998). Current estimates of the economic cost of obesity in the United States. *Obesity Research.* 6, 97–106.

Wolf, A. W., Gortmaker, S. L., Cheung, L., Gray, H. M., Herzog, D. B., and Colditz, G. A. (1993). Activity, inactivity, and obesity: racial, ethnic, and age differences among schoolgirls. *American Journal of Public Health*, 83, 1625-1627.

Wrigley, N., Warm, D., Margetts, B., and Whelan, A. (2002). Assessing the impact of improved retail access on diet in a 'food desert': A preliminary report. *Urban Studies*, 39, 2061-2082.

Xie, B., Gilliland, F.D., Yu Fen, L., and Rockett, H.R.H. (2003). Effects of ethnicity, family income, and education on dietary intake among adolescents. *Preventive Medicine*, 36, 30-40.

Young-Hyman, D., Schlundt, D. G., Herman-Wenderoth, L., and Bozylinksi, K. (2003). Obesity, appearance, and psychosocial adaptation in young African American children. *Journal of Pediatric Psychology*, 28, 463-472.

Zhang, Q., and Wang, Y. (2004). Trends in the association between obesity and socioeconomic status in U.S. adults: 1971-2000. *Obesity Research*, 12, 1622-1632.

In: New Developments in Obesity Research
Editor: Lawrence F. Ditmier, pp. 121-152

ISBN 1-60021-296-4

Chapter 5

Determinants of the Global Obesity Epidemic: Contributions of Early Life, Nutrition, Physical Activity, Television, Smoking, and Economic and Social Change

Christina O. Stubbs
Public Health Nutritionist, Accredited Practicing Dietitian, Brisbane, Australia[1]

Abstract

Since 1980, the prevalence of obesity in many developed countries has doubled or tripled, and many developing countries are following the same trend. The World Health Organisation describes this as a serious global epidemic and a major contributor to the global burden of chronic disease and disability. In the last 25 years there has been a widespread environmental change resulting in a small positive disturbance in energy balance for a large proportion of the global population.

This chapter reviews current evidence regarding the underlying determinants of this epidemic. The mechanisms of energy balance are described, and limitations in interpretation of energy intake data from dietary surveys and energy expenditure data from physical activity surveys are discussed.

There is no evidence of a decrease in population energy expenditure as a factor in the obesity epidemic. On the contrary, both theoretical considerations and objective evidence demonstrate that per capita energy expenditure increases as mean BMI increases.

Data from Europe, the United States and Australia show an increase in self-reported energy intake and an increase in population food supply energy concurrently with the dramatic increase in the prevalence of overweight and obesity in children and adults since 1980. Increased consumption of sweetened drinks, snack foods and fast food with an increase in portion sizes contribute to increased energy intake. Absolute fat intake has changed little, but fat as a percentage of energy has decreased due to an increase in

[1] Address for correspondence: Christina Stubbs. Email: Christina.stubbs@gmail.com

carbohydrate intake and total energy intake. Economic evidence also supports the role of low-cost, energy-dense but nutrient-poor foods as major contributors to this epidemic.

Population physical activity level has also decreased, but over a much longer period than the recent epidemic of obesity and a large proportion of adults and children in developed countries are insufficiently active for good health. Physical inactivity increases the susceptibility of individuals to obesogenic changes in the food supply. Increasing physical activity is associated with many health benefits apart from its role in obesity prevention.

The increased prevalence of ex-smokers in developed countries may have contributed to the increased prevalence of adult obesity, but evidence for this is conflicting, and it is unlikely to have contributed to the epidemic in children or in developing countries.

Comprehensive efforts are required by governments, industry and communities working together to address these determinants and create the economic, physical and social environments that support individuals to choose healthy foods and an active lifestyle.

INTRODUCTION

The global prevalence of overweight and obesity amongst both children and adults has increased to epidemic proportions, according to the World Health Organization (WHO). Across the globe, both high-income and low-income countries are experiencing an obesity epidemic, and the associated increase in the burden of preventable chronic diseases, including cardiovascular disease, diabetes, musculo-skeletal disorders, endocrine disorders and cancer. [1] Each year, at least 2.6 million people globally die prematurely as a result of being overweight or obese. [2]

In 1995, Prentice and Jebb published their paper examining the role of "gluttony or sloth" on the prevalence of obesity in Britain [3], which has been widely cited as evidence that the obesity epidemic has occurred despite minimal or no increase in per capita energy intake and/or energy from the food supply. There are vested interests that support this view. According to the International Obesity Taskforce (IOTF), the food industry seeks to focus on inactivity and promote sports to divert attention from the role of foods and drinks. [4] Prentice and Jebb have subsequently asserted that their paper has been "wilfully misquoted" as implying that diet plays an unimportant role in the obesity epidemic. [5]

The IOTF considers that the causes of the obesity epidemic are twofold: an abundance of energy-dense foods and drinks, leading to a pervasive passive over-consumption of energy; and an environment that limits opportunities for physical activity, leading to an almost universal sedentary state.

Under normal physiological circumstances, energy intake (food and drink consumption) is determined by energy expenditure and food supply, and the body finely regulates this balance to maintain a stable weight. Weight gain occurs when energy intake exceeds energy expenditure on average. Over time, this imbalance leads to overweight and then to obesity.

Evidence suggests that individual susceptibility to obesity is strongly influenced by genetic variations. [6] However, the rapid increase in the global prevalence of obesity in both children and adults since 1980 [7] cannot be explained by genetic susceptibility. In many countries, the majority of the adult population and a quarter of children and adolescents are now overweight. This dramatic increase can only be explained by substantial changes in the environment of the late 20th century affecting energy intake, energy expenditure, or both. This chapter will explore these environmental determinants of obesity.

THE CHANGING GLOBAL PREVALENCE OF OBESITY

Body Mass Index (BMI), defined as weight in kilograms divided by the square of height in metres (kg/m^2), is the widely accepted proxy measure of fatness and associated health risks used to determine the prevalence of overweight and obesity. There is a continuous association between BMI and risk of a range of diseases, with the lowest risk at a BMI of 21. [1] However, for convenience, the risk associated with BMI is frequently treated as a categorical variable.

For adults, the WHO classifies BMI within the range of 18.5 to less than 25 as healthy weight (average risk), 25 to less than 30 as overweight (increased risk), and greater than 30 as obese (substantially increased risk). [8] When describing population prevalence, the term overweight is frequently used to refer to BMI ≥25. Throughout this chapter the following terms will be used except where otherwise specified.

Overweight:	BMI ≥25.
Overweight but not obese:	BMI ≥25 but < 30
Obese:	BMI ≥ 30

For children, a variety of different methods of defining overweight and obesity have been used. The International Obesity Taskforce (IOTF) has defined age-specific BMI cut-offs for overweight and obesity equivalent to the adult cut-offs, based on data from six different reference populations. [9] This method provides a standard benchmark which is most useful for monitoring and evaluating changes in populations. The US Centres for Disease Control (CDC) method defines "at risk of overweight" as BMI > 85th percentile for age, and "overweight" as BMI > 95th percentile, based on US population dataprior to the obesity epidemic. [10] For consistency, these classifications will be referred to in this chapter as overweight and obese respectively.

The WHO has estimated that more than one billion adults worldwide are overweight (BMI ≥ 25) and at least 300 million are obese (BMI ≥ 30). The prevalence of obesity varies from under 5% in China, Japan and some African nations to greater than 25% in urban Samoa and the USA. However, even in countries with low prevalence of obesity, urban populations may have a much higher prevalence. [2]

The "nutrition transition" has been described as the shift in human diet and health patterns from a high prevalence of undernutrition and infectious diseases to a high prevalence of obesity and non-communicable degenerative diseases, with an increased consumption of fats, sugars, animal products, a decrease in cereal intake and fibre, and an increasingly sedentary lifestyle. [11] This transition is associated with economic and demographic shifts from subsistence agriculture to increasing urbanization and globalisation.

In developed western market economies, this transition has occurred over the last few centuries. However in developing countries, it is occurring much more rapidly, and undernutrition and obesity coexist and pose a double health burden to these nations.

Data on the prevalence of overweight is limited in many parts of the world. Reliable, representative, objective data based on measured weight and height are expensive to obtain. Self-reported data are easier to obtain but under-estimate the true prevalence of overweight. Comparison with measured data shows that individuals under-estimate their weight and overestimate their height. [12]. In a community-based study in Sydney in 1997, males under-reported their weight by 1.4 kg and females by 3 kg. The degree of under-reporting was related to weight; for those above the median weight, males under-reported by 1.7kg and females by 4.5kg. [13] However, self-reported BMI is useful to monitor trends in the prevalence of overweight and obesity. [14]

The most comprehensive national prevalence data available are from the USA. Objective measured data representative of the US adult population aged 20-75 are available from the National Health and Nutrition Examination Survey (NHANES) conducted by the National Centre for Health Statistics. This dataset shows that the mean BMI of US adults increased from 25.3 in 1971-74 to 28.0 in 1999-2000. [15] The WHO recommends that for optimum health, the median BMI of the population should be in the range 21 to 23, while the range for individuals should be 18.5-24.9. [16]

In 1960-62, the prevalence of overweight (BMI ≥25) in the USA was 48% of men and 39% of women, while the prevalence of obesity was 10% and 15% respectively. [7] By 1971-74, this had changed little, but by 2000, 65 % of adults were overweight and 30% were obese. [15] Changes in the prevalence of overweight and obesity for US adults from 1971 to 2000 are shown in Figure 1. [15]

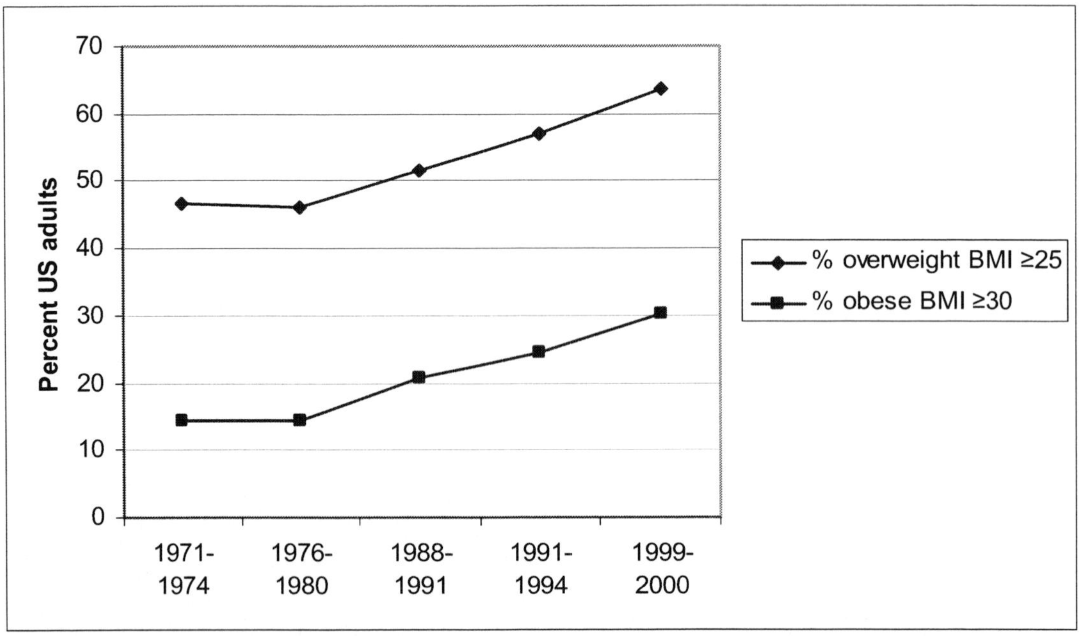

Figure 1: Trends in prevalence of overweight in US adults (Jolliffe 2004)

Similar changes in prevalence of overweight and obesity have occurred amongst US children since 1980, as shown in Figure 2. [17]

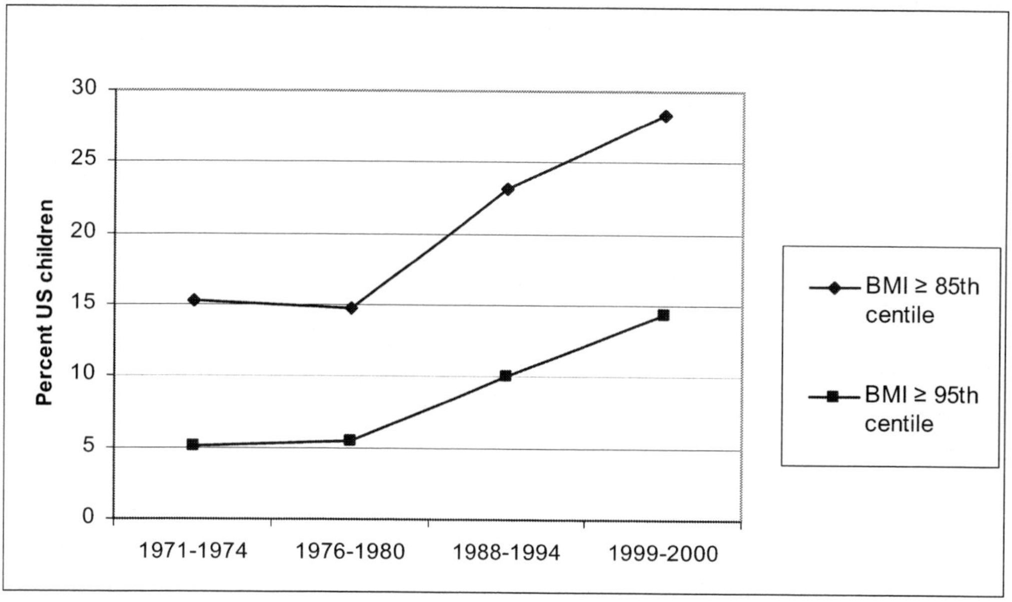

Figure 2: Prevalence of overweight in US children (Jolliffe 2004)

These comparable, objective measures of the obesity epidemic indicate that the prevalence of overweight and obesity in the USA was stable from 1960 to the late 1970s. However, there was a dramatic increase from 1980 to 2000. During these two decades, the prevalence of adult overweight (BMI ≥25) increased by 37%, while the prevalence of obesity increased by 112%.

In addition, not only are there more obese children and adults, there are more severely obese children and adults. That is, the population distribution of BMI has shifted to the right, and has also become more skewed to the right as shown in Figure 3. [15]

In other developed countries there has also been a dramatic increase in prevalence of overweight since 1980. In Australia, based on surveys of measured height and weight in state capital cities, the mean adult BMI increased from 25.5 to 27.2 for men, and from 24.3 to 26.8 for women aged 25–64 between 1983 and 1995. [18] The AusDiab study conducted nationally in 1999-2000 found that 67% of males and 52% of females aged over 25 in Australia were overweight (BMI ≥ 25) and 19% and 22% respectively were obese. When assessed by waist circumference (waist ≥102cm in men or ≥ 88cm in women), the prevalence of obesity was even higher at 26% in men and 34% in women. [19] This is a 250% increase in prevalence of adult obesity compared to a 1980 survey. Furthermore, low response rates in the AusDiab survey may be responsible for a selection bias that has resulted in an under estimate of the prevalence. [20]

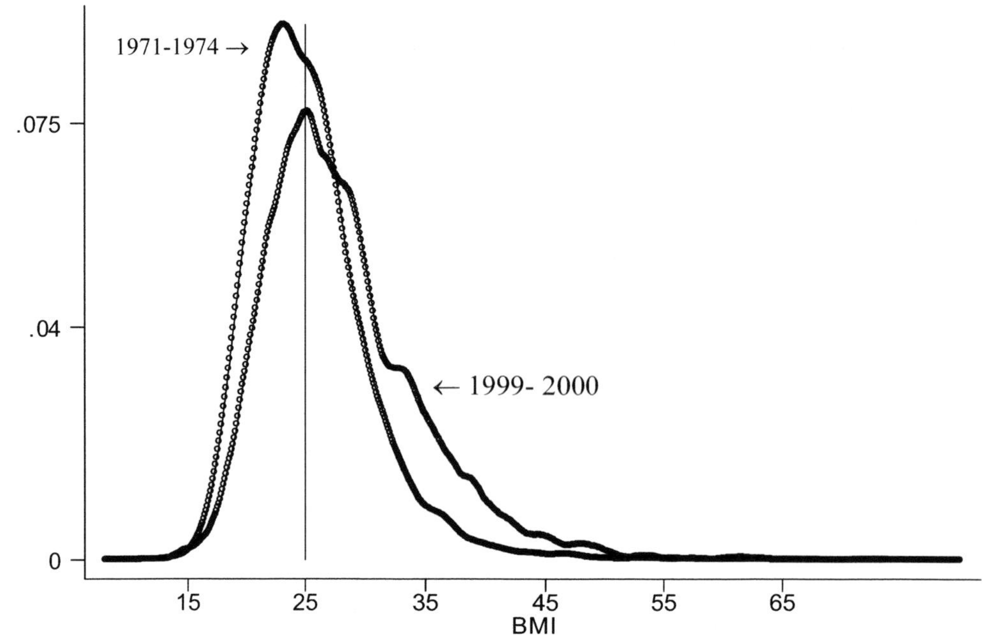

Figure 3: Distribution of BMI in US adults, 1971-74 and 1999-2000 (Jolliffe 2004). Reproduced by kind permission of the Population Association of America.

The prevalence of overweight but not obese, based on the IOTF standards, in Australian children increased from 10% in 1985 to 15.4% in 1995 and the rate of obesity increased from 1.3% to 4.8% [21] but the increase from 1969 to 1985 was much smaller. [22] Although national data are not available for subsequent years, state estimates indicate that the prevalence of overweight continues to increase. In South Australia, rates of overweight in four-year-old children rose from 12.9% to 21.5% (girls) and 10.6% to 18.4% (boys) between 1995 and 2002. [23]

Studies in the United Kingdom (UK), Spain and Australia have found that central adiposity measured by waist circumference have also increased markedly in young people. The Australian study suggests that the prevalence of obesity in children may be higher when based on waist circumference than on BMI. [24] This is of particular concern because waist circumference is strongly associated with cardiovascular risk markers in both adults and children.

Even amongst developed nations, rates of obesity vary considerably. Amongst adults in Europe, measured prevalence of obesity amongst women varies from 10% in Denmark to 38% in Greece. [25] The Health Behaviour in School-aged Children study compared prevalence rates across Europe using the international standard definitions of overweight and obesity. Overweight (pre-obese + obese) rates varied from 5.1% in Lithuania to 25.4% in Malta, which was comparable to the United States. Obesity rates varied from 0.4% in Lithuania to 7.9% in Malta, based on self-reported height and weight. [26] While Eastern European countries make up the five countries with the lowest prevalence of childhood overweight, the next five include Netherlands, Poland, Belgium and Switzerland. Rates are highest in southern Europe, where recent estimates suggest that 36% of 9-year-olds in mainland Italy, and 39% of 12-year-olds in Crete were overweight. [10]

The increased prevalence of childhood obesity is of particular concern because of its implication for the future prevalence of chronic diseases, as this cohort of children grows older. The risk of adult obesity increases with higher childhood and adolescent BMI and with increasing age. [27] The current high prevalence of overweight and chronic diseases in adults occurs in a cohort that had a much lower prevalence of obesity in childhood, which raises grave concerns for the future health of the current generation of children.

The rapid increase in the rates of overweight and obesity across the world since 1980, combined with the great variability in rates between countries, provide convincing evidence that the cause of this epidemic lies in recent changes in the physical, social and economic environment.

SOCIOECONOMIC DETERMINANTS

The social and physical environments in which people live and work have direct effects on their health. A social gradient in health outcomes is well-established, particularly for cardiovascular disease. [28] Although the obesity epidemic now affects all strata of many industrialized societies to the extent that overweight is now becoming the "normal" body type, obesity is more prevalent amongst children and adults from low socioeconomic groups, while in low-income countries, it is more prevalent amongst the wealthier sections of society. [2, 10]

A range of economic, environmental and social determinants, such as income, education, occupation and food supply underlie the behavioral determinants of health, which in turn lead to the biological factors which are associated with disease.

Disadvantage acts through early life experiences, education, employment, income, living conditions, food supply and the physical environment to make healthy lifestyles less attainable for many individuals.

Groups with low income, low levels of education and those in low status occupations are less likely than those with higher socioeconomic status (SES) to have food purchasing, preparation and consumption patterns consistent with dietary guidelines. [29, 30] Compared to that of higher socioeconomic groups, the diet of lower SES groups includes more cheap, energy-dense foods rich in fats and carbohydrates, but fewer foods of low energy density rich in protective nutrients, such as fruit and vegetables. [31]

Food insecurity, or the limited or uncertain access to nutritionally adequate, safe and culturally acceptable foods, is paradoxically associated with obesity. In the 1998-1999 California Women's Health Survey, obesity was twice as prevalent amongst food-insecure women (31%) compared with food secure women (16%) and the relationship was strongest in non-whites. [32]

Hitchman et al in a qualitative study found that food poverty in modern Britain consists of nutritional rather than caloric inadequacy, and is caused by the interaction of factors ranging from income and family structure through to transport availability and the nature of modern food retailing. [33] Access to healthy food was determined not only by income but also by the location of stores, their pricing policies and transport to stores. Small local stores often have higher prices and a smaller range of products than the larger more distant supermarkets, which are less accessible to those dependent on public transport. In Australia in

2000 and 2004, healthy food cost 14% more in remote areas and 30% more in the very remote areas compared with major cities. In addition, availability of healthy food declined with remoteness. [34, 35]

It is not clear how the relative contributions of education, occupation and income contribute to the socioeconomic gradient in obesity. However, these factors are themselves inter-related and are strongly determined by childhood socioeconomic factors. Low SES groups are also differentially disadvantaged by the intergenerational effects of poor maternal diet, and poor infant and childhood growth and development. [10] However, cross-sectional studies of obesity and adult social class may obscure the effects of childhood social and economic conditions.

There is no clear association between SES in early life and childhood BMI, but a consistent strong inverse relationship is reported between early life SES and adult BMI. [36] A large longitudinal study of the 1958 birth cohort in Britain (all persons born in England, Scotland and Wales during a particular week in March 1958) examined the effect of social class at several ages with obesity to the age of 33. Fatness and social class were unrelated in childhood, but a social gradient was evident for both sexes in early adulthood, and for girls from age 16, with a higher risk of adult obesity for those from lower social classes in childhood. [37] The association remained after controlling for parental obesity, which is a major determinant of childhood obesity. Educational attainment was also strongly associated with obesity at age 33.

In contrast to western industrialized nations, obesity in developing countries is more prevalent amongst urban people and higher socioeconomic groups. For example, in China, obesity rates vary from under 5% in rural areas to over 20% in urban areas [2] In India, the odds ratio for obesity was 1.68 for high-status women compared to low-status, and the prevalence of obesity amongst women in Karnataka state was 3% overall compared to 15% in large cities. [38]

Intergenerational and Early Life Factors

Early life factors which are associated with risk of overweight and obesity in later childhood or adult life include parental overweight, birth weight, socioeconomic factors, timing or rate of maturation, physical activity, dietary factors including early feeding, and other behavioural factors. [36]

There is a large body of evidence suggesting that the foetal environment significantly influences health in adulthood. [39] Children whose parents are overweight are at greater risk of obesity, and this relationship is strongest for maternal weight. [10] Since maternal pre-pregnancy weight and weight gain during pregnancy are both associated with birth-weight, it is likely that these factors are in the same causal pathway, mediated through the intrauterine environment.

Many epidemiological studies have consistently demonstrated a direct relationship between birth-weight and adult BMI, with an increment of 0.5 to 0.7 kg/m2 in adult BMI for each 1 kg increment in birth-weight. [40] Some studies show that this is a J or U-shaped relationship, with both high and low birth-weight babies at increased risk of child or adult obesity. [10]

Low birth weight rather than high birth weight is associated with increased risk of cardiovascular disease in adulthood [41], and low birth-weight appears to be more strongly associated with abdominal obesity than with BMI. [40] This has been confirmed using dual-energy X-ray absorptiometry (DEXA) to objectively measure body composition by Kensara *et al*, who compared two groups of 64-72 year old men who had either low or high birth-weight. There was no difference in BMI between the two groups, but the low birth-weight men had higher fat mass, higher trunk-to-limb fat ratio, and lower lean mass, muscle mass and muscle-to-fat ratio. [42]

A similar finding in childhood was reported by Wells *et al* using foot-to-foot impedance to measure body composition in a longitudinal study of 9-year old Brazilian boys. This study found that birth-weight was directly associated with later height and lean mass, but not fatness. [43]

Babies born to mothers with diabetes during pregnancy have a higher birth-weight and a greater risk of obesity in childhood, after 4-5 years. The risk of obesity is associated with diabetes independently of maternal weight and birth-weight, suggesting it is due to the intrauterine environment. [10]

Growth in early childhood is also associated with later risk of obesity, but the effect varies with age. BMI growth charts show that BMI normally increases during the first year of life, then decreases to about the age of 5 and then increases throughout the rest of childhood and adolescence. The nadir of BMI is known as the adiposity rebound.

There is a substantial body of evidence confirming a relationship between early adiposity rebound and later risk of obesity, with early rebounders gaining fat mass at up to three times the rate of late rebounders. [44] The average age of rebound may have decreased. In a study of children born before 1960, less than 30% of children had an age of rebound of 5.1 years or less, compared to 37% of children in a study of children born in 1991. [44]

Studies of the role of infant feeding in subsequent obesity have shown conflicting results, but two systematic reviews have found that breastfeeding shows a small but consistent protective effect against obesity, and that there is a dose relationship with the duration of breastfeeding. [45, 46]

The WHO recommends exclusive breastfeeding to around 6 months of age. [47] However definitions of breastfeeding across studies are inconsistent, with many studies reporting the effect of any breastfeeding rather than the effect of exclusive breastfeeding. This is likely to underestimate the effect of breastfeeding. The power of studies to detect the effect of breastfeeding is also limited by the small proportion of mothers in developed countries who practice exclusive breastfeeding. In Australia, up to 90% of mothers initiate breastfeeding, but by three months a third of these have stopped breastfeeding or introduced other foods [48] and breastfeeding rates in the US are lower.

Results may also be confounded by maternal BMI, maternal smoking and socioeconomic class. In studies which adjusted for these factors the protective effect of breastfeeding remained but the effect was considerably smaller. [46]

Women who are overweight or obese before pregnancy are less likely to initiate breastfeeding and discontinue breastfeeding earlier than women in the healthy weight range. Obese women have been shown to have heavier infants at one year but method of feeding may confound this relationship. A large, prospective study from the Danish National Birth Cohort studied these variables simultaneously and confirmed that maternal pre-pregnant BMI was positively associated with infant weight gain at one year. Shorter durations of breastfeeding and earlier introduction of complementary foods were also associated with infant weight gain, and there was an interaction between these factors. [49].

Persistence of overweight or obesity from childhood to adulthood has been demonstrated in a number of studies, with the strongest correlation for obesity in adolescence. [10] Although many overweight or obese adults were not overweight as children, the increased prevalence of obesity in children suggests that the prevalence of adult obesity and chronic disease will be much higher when this generation of children become adults, with dire consequences for already struggling health systems.

There may be an intergenerational cohort effect occurring, with increased maternal BMI and prevalence of maternal diabetes since 1980 contributing to the rapid increase in early childhood obesity observed in recent years.

ENERGY BALANCE

Since energy cannot be created or destroyed, energy intake is equal to energy expenditure plus any change in body energy stores. Weight gain or loss is associated with changes in lean mass, consisting of muscle and organs, as well as fat mass.

Stable bodyweight in adults indicates that energy expenditure is equal to energy intake (in balance). Only a small, persistent positive energy imbalance is required to cause weight gain in adults, or excessive weight gain in children, which over time progressively increases BMI. Energy expenditure is closely correlated with body mass, and will gradually increase due to increased body mass, until equilibrium is once again reached.

Experimental studies of overfeeding with objective measures of energy expenditure and body composition have shown that the energy cost of weight gain associated with overfeeding is approximately 45-55 MJ/kg weight gain (12,000kcal). [50, 51] About two-thirds of excess energy intake is stored as body energy (lean mass and fat mass) and there is a substantial increase in energy expenditure. [50]

In addition to the increase in resting energy expenditure (REE) observed in controlled experimental environments, there is evidence from both animal and human studies that levels of spontaneous physical activity, including fidgeting and activities of daily living, increase in response to overfeeding in free-living individuals, and that this is one of the mechanisms which maintain our finely-tuned energy balance. [52, 53]

A large longitudinal study of mid-life Australian women found that average weight gain over five years was 0.5kg per year. [54] This is equivalent to an energy gap of about 70kJ/day (16 Calories/d), based on 50MJ/kg weight gain. This seems a very small imbalance to account for the obesity epidemic. However, this small continuous per capita gap is in addition to the increased energy expenditure due to increased per capita body mass. Thus the overall increase in per capita daily energy expenditure between surveys would be larger than this.

A 5kg difference in body weight is associated with a difference in estimated energy expenditure of approximately 340kJ (80 kcal) (see table 2). Comparison of the Australian National Nutrition Surveys in 1983 and 1995 found that mean body mass of adults increased by around 6kg, and self-reported energy intake increased by 3-4% or around 350kJ. [18]

ENERGY EXPENDITURE

Total daily energy expenditure is made up of resting energy expenditure (basal metabolic rate plus the thermic effect of food), and activity energy expenditure. Resting energy expenditure is the largest component and is mainly related to lean body mass. Activity energy expenditure depends on both the amount of physical activity and body mass to be moved. Therefore both total mass and lean mass contribute to total energy expenditure.

Terms and Definitions

- *Total Daily Energy Expenditure* (TEE) is the amount of energy expended by an individual in one day.
- *Basal Metabolic Rate* (BMR) is energy expended in the rested and fasted state.
- *The Thermic Effect of Food* (TEF) is the energy used in digestion of food (estimated at ~10% TEE).
- *Resting Energy Expenditure* (REE) or *Resting Metabolic Rate* (RMR) is the energy expended by the body in the resting state to maintain itself and digest food i.e., REE = BMR + TEF (or REE = BMR x 1.1)
- *Activity Energy Expenditure* (AEE) is the energy expended by the body in movement, including both the energy expended in sedentary or obligatory activities and that expended in discretionary activity. AEE is the only aspect of energy expenditure which is under conscious control. AEE makes up 20-40% of TEE, depending on physical activity level.
- TEE = REE + AEE
- *Physical Activity Level* (PAL) is expressed as the ratio of TEE divided by REE.
- PAL = TEE/REE
- TEE = REE x PAL

Energy expenditure can be calculated for individuals by measuring changes in respiration of oxygen and carbon dioxide. Traditionally this was done by indirect calorimetry, which restricts the activity of participants. However, energy expenditure of free-living individuals can now be measured using the doubly-labelled water (DLW) technique. This method involves the administration of water labelled with the stable isotopes deuterium and oxygen-18. Carbon dioxide production can be measured by the difference in elimination of these two isotopes. This technique has been validated against indirect calorimetry and is accurate to within 1%. [55]

A large review of objective studies of energy expenditure using DLW found that resting energy expenditure, activity-related energy expenditure and total energy expenditure are all substantially increased in obesity. For BMI categories ranging from <25 to >35, total energy expenditure ranged from 9.5 to 13.5 MJ/d (2,260 to 3,200 kcal/d) for women and 12.9 to 17.5 MJ/d (3,070 to 4,160 kcal/d) for men. [56] This contradicts the widely-held view that obesity is due to 'low metabolism' and is maintained despite a low level of food intake.

Several equations have been developed to predict REE (also referred to as Resting Metabolic Rate, RMR) based on weight, gender, height and age, or more simply on weight and gender. The factors of gender and age account for the major variations in lean body mass. In general, such equations tend to underestimate energy expenditure in the obese.

A systematic review of studies which validated predictive resting energy equations by indirect calorimetry found that the Mifflin-St Jeor equation was the most reliable for healthy obese and non-obese adults. [57]

Table 1: Mifflin-St Jeor equations for predicting REE [58]

	REE (kcal/d)
Men	(9.99 x weight [kg]) + (6.25 x height [cm])– (4.92 x age) +5
Women	(10 x weight [kg]) + (6.25 x height [cm])– (5 x age) - 161.

AEE includes energy used in sedentary activities, as well as energy used in voluntary activities. Both of these are affected by body mass, since it takes more energy to move a larger body than a smaller body.

However, there is no reliable method of measuring energy expenditure at a population level. Surveys of self-reported leisure-time physical activity are sometimes reported as proxies for energy expenditure, but energy expenditure depends on several factors in addition to physical activity level, and leisure-time physical activity is in turn, a proxy for actual activity level.

Over the last century, it is thought that physical activity levels have declined due to increased mechanization and ownership of automobiles, television and labour-saving devices. However, population mean height and weight have increased and the consequent increases in per capita REE and sedentary AEE could be expected to counteract any decline in AEE due to decreased obligatory and voluntary physical activity. This possibility is supported by a study using objective measures to compare energy expenditure between African-American and Nigerian adults from rural and urban sites. The Nigerians had significantly lower height, weight and BMI than the American subjects, but the Americans had significantly higher total energy expenditure, resting energy expenditure and activity energy expenditure [59].

Table 2: Energy expenditure for a hypothetical women at varying BMI and physical activity levels, compared with observed TEE and PAL

For a hypothetical woman, age 50, height 164cm			**Estimated TEE (kJ)**			**Observed (Prentice 1996)**		
Weight kg	BMI	REE kJ	@ PAL 1.5	@ PAL 1.6	@ PAL 1.7	BMI	TEE	PAL
57	21.2	4973	7459	7956	8454	21.3	9510	1.67
70	26.0	5519	8278	8830	9382			
75	27.9	5729	8593	9166	9739	27.6	10210	1.73
85	31.6	6149	9223	9838	10453	32.1	11360	1.73
120	44.6	7619	9853	10510	12952	43.6	13510	1.65

Table 2 illustrates the theoretical relationship between BMI and TEE at varying levels of physical activity (PAL) using the Mifflin-St Jeor equation to calculate REE. TEE increases with both BMI and PAL. TEE is underestimated compared with the measured TEE reported by Prentice. [56]

A decrease in PAL from 1.7 to 1.6 for a 70kg woman would lead to a reduction in energy expenditure of around 550kJ (130kcal). If energy intake remained the same, weight would increase to around 76kg, when energy expenditure would again equal energy intake.

Analysis of a large combined dataset of objectively measured energy expenditure used to determine the US Dietary Reference Intakes (DRI) for macronutrients shows that the observed mean physical activity level (TEE/BMR) was 1.7 for adults in both the healthy weight range (BMI <25) and for overweight or obese adults (mean BMI 30). [60, 61] Prentice reported that PAL values for the moderately obese form a similar distribution to those of the non-obese, but those for the massively obese are skewed to the left. [56]

As shown in table 2, estimated TEE at a PAL of 1.7 would vary from 8454 kJ at a BMI of 22 to 10450 kJ at a BMI of 32. Even if physical activity fell from very active (PAL 1.8) at a BMI of 22 to inactive (PAL 1.5) at a BMI of 32, estimated TEE would still increase from 8950 to 9853.

This exercise demonstrates that a decline in per capita energy expenditure concurrently with increasing mean population body mass is theoretically unlikely and is not supported by objective evidence. However, energy expenditure per kg body mass does decline with increasing BMI, even at the same PAL.

Energy Intake: The Contribution of Nutrition

All energy intake is derived from food and drinks. Carbohydrate provides 16 kilojoules (kJ) per gram, protein 17kJ, alcohol 29kJ and fat 37kJ per gram. (one Calorie or kcalorie = 4.2kJ). Water, fibre and micronutrients do not yield any significant energy.

There has been much debate about the role of energy and fat intake in the obesity epidemic. [62, 63] In the 1990s there was a prevailing view that fat intake was a major determinant of weight gain. However, failure to distinguish between fat as a percentage of total energy intake and absolute fat intake has caused some confusion in this debate.

In 1996, Heini and Weinsier noted that the contribution of fat to total energy intake of US adults had decreased by 11% between 1976-1980 and 1988-1991, based on the NHANES survey. This was described as "the American paradox" and cited as evidence that the obesity epidemic may be due to decreased physical activity related energy expenditure and not to changes in eating patterns. [64]

While reducing fat intake is recognised as an important strategy to reduce energy intake in successful weight loss interventions [63], there is little evidence that population fat intake is associated with the obesity epidemic independently of energy intake. [62] A decrease in fat intake as a proportion of energy has been demonstrated in the United States and Australia, but the proportional decline is due more to an increase in total energy intake than to any decrease in absolute fat intake. [18, 65]

Interpretation of dietary intake studies is complex as there is evidence that a systematic bias occurs in dietary self-report. Energy intake in population studies is based on either interviewer or self-completed diet records or recalls or food-frequency questionnaires. [66] When self-reported intake using a variety of dietary assessment instruments has been compared against energy expenditure based on DLW measurements, energy intake is typically under-reported by up to 26%. [55]

In a weight-stable individual, energy expenditure is equal to energy intake. When energy expenditures are measured objectively, TEE/RMR gives an estimate of physical activity level. However, when reported energy intake (REI) is divided by estimated RMR (calculated from weight, gender, height and age), the result provides an estimate of the degree of accuracy of food intake reporting.

Under-reporting is not random. Overweight individuals under-report their dietary intake to a greater extent than normal-weight individuals. [67, 68] Socially undesirable foods (such as those high in fat or sugar) and foods eaten between meals are under-reported to a greater extent than other foods. This introduces a systematic bias resulting in the paradoxical observation that obese individuals appear to eat less energy, fat and sugar than lean individuals, although objective measures of energy expenditure demonstrate that the opposite is true. [56] Techniques have been developed to reduce this bias by excluding the most blatant under-reporters, based on REI/BMR. Inclusion of only those respondents within 1.4 standard deviations of mean REI/BMR has been found to yield the most reliable results. [69] Failure to exclude under-reporters results in spurious associations which have often been reported uncritically.

Despite its shortcomings, self-reported energy intake is useful to monitor trends over time when used carefully. The NHANES data from the US provides comparable estimates of self-reported energy intake over time. Between 1971 and 2000 energy intake significantly increased by 7% in men and 22% in women ($p>0.01$). [70] However, all of this increase occurred between 1980 and 2000, concurrently with the increase in prevalence of overweight (See Figure 4). Since the extent of under-reporting is likely to have increased in parallel with the increasing BMI of the population, these figures are likely to under-estimate the increase in energy intake.

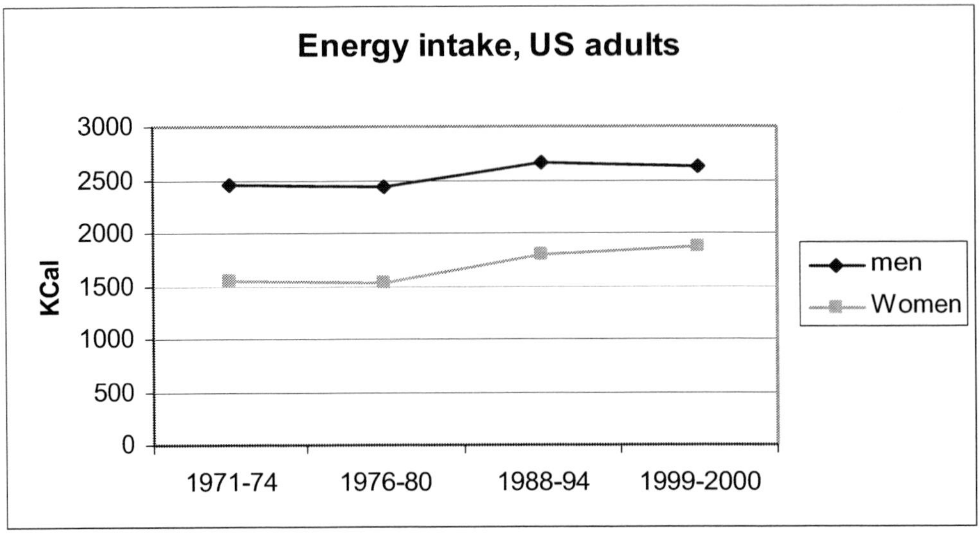

Figure 4: Energy intake for US adults (Wright 2004)

Analyses of other US food consumption surveys also indicate that energy intake has increased, and describe the changes in eating patterns of Americans. Energy intake from food consumed at home decreased by 11% to 21% between 1977 and 1996, and energy intake from restaurant and fast foods increased by between 91% and 208%, depending on age group. [71]

In Australia, where the national prevalence of obesity tripled amongst children and doubled amongst adults between the early 1980s and 1995, the mean energy consumption of Australian adults living in capital cities increased significantly by around 3–4% (about 350kJ) between National Nutrition Surveys in 1983 and 1995. Between 1985 and 1995, mean energy intake increased significantly by 11% for girls and 15% for boys aged 10–15 years. [18]

For adults, increased energy intake was derived from significantly increased total carbohydrate and there was a small but significant reduction in fat intake. For children, total carbohydrate intake (including sugars) and protein intake increased significantly, while fat intake did not change significantly. [18]

When examined by food groups, the increased energy consumption in both adults and children was mainly due to increased consumption of sugar-sweetened drinks, cereal-based foods (including cakes, biscuits, pies, pizza and some desserts), and confectionary. [18] For both children and adults there was a significant decrease in intake of fats and oils used as spreads or added to food after cooking, but this was offset by an increase in intake of fat from other sources, such as cereal-based foods and confectionary. [18]

The level of under-reporting, as measured by the ratio of reported energy intake to estimated REE was consistent across these surveys and the ability of this study to detect changes in food group consumption that are consistent with changes in food supply and promotion suggests that the results are reliable. The consistency between the changes in food group and nutrient consumption between Australia and the United States also point to common factors underlying the change in food consumption.

Prentice and Jebb reported that total average energy intake in Britain had decreased between 1970 and 1990, [3] based on the UK National Food Survey, which measures household expenditure on food rather than food intake per se. Prior to 1992, this survey did not include expenditure on food consumed away from home, confectionary, soft drinks or alcohol. Consumption of these foods has been increasing in recent decades while consumption of food prepared at home has been falling. Subsequent to 1994 when all of these were included there is still a small downward trend reported in UK energy intake. [72] This survey is not comparable to surveys in other countries which use self-reported intake, and since it does not adjust for under-reporting, as discussed above, the apparent decline in energy intake is likely to be artefactual.

An alternative method of assessing population energy intake is by estimates of food disappearance, or the amount of energy available in a national food supply, divided by the population to give the available energy per capita. This method takes into account food production, imports and exports, animal feed and losses due to processing and preparation. It tends to overestimate energy intake since it does not account for wastage, but is useful to establish time trends or to compare populations.

Data on food supply in the US indicates that per capita energy supply increased by 15% between 1970 and 1994. [73] In Europe, data from the WHO MONICA study indicate that per capita energy supply increased from the early 1980s to the mid 1990s in most high-income countries and China, in parallel with increased population BMI. In contrast, both energy supply and population BMI decreased in Eastern European countries. Trends in total energy supply per capita explained 41% of the between-population differences in trends in BMI. [74] Trends in supply of total fat and total sweeteners were also correlated with trends in BMI.

Energy density is the energy content per unit weight of foods, drinks or meals. Fat and sugar are positively related to energy density, while water and fibre are negatively related. Solid foods are considered separately from beverages. Substantial evidence from experimental feeding studies demonstrates that higher energy density of foods promotes over-consumption. [75] Food intake in humans appears to be regulated by weight and volume rather than by energy. A small increase in weight of an energy-dense food results in a larger increase in energy intake. Energy-dense diets are associated with higher consumption of grains, fats and sweets and lower consumption of vegetables and fruit, and with higher energy intakes. [76]

US data regarding trends in food purchasing and preparation suggest that Americans are eating more meals outside the home, relying more heavily on convenience foods, and consuming larger food portions. [77] Between the late 1970s and mid 1990s, the number of fast food restaurants in the US more than doubled and consumption of food prepared away from home increased from 18% to 32% of total energy intake. [78] In a cross-sectional analysis of state-level prevalence of obesity in the US and the number of fast-food outlets, the number of residents per fast-food restaurant was strongly correlated with state prevalence of obesity. [79]

Fast foods have a very high energy density. [75] Fast-food consumption is associated with higher energy intake and poorer dietary quality amongst children [80] and adults. [81] In a 15-year prospective analysis, baseline fast-food consumption and change in fast-food consumption were both directly correlated with weight gain and risk of insulin resistance. Over 15 years, frequent fast food consumers gained 4.5kg more than infrequent fast food consumers after adjustment for other lifestyle and dietary factors. [81]

Portion sizes in family-type restaurants, takeaway and fast-food outlets in the US began to increase in the 1970s, rose sharply in the 1980s, and have continued to increase in parallel with increasing population BMI. [82] A graph published by Young and Nestle [82] of the number of new, larger portions introduced over time from 1970 to 1999 looks remarkably similar to the graphs of prevalence of overweight in the US shown in Figure 1 and Figure 2. The difference between actual portion sizes in 2000 compared to standard portion sizes defined by the US Department of Agriculture and the Food and Drug Administration varies from zero for white bread to 700% for cookies. [82]

The marketing strategy of providing larger portions or “meal deals” to appeal to consumers’ desire for value for money has been described as “upsizing”. Meal deals in four major fast food chains in Australia provided an average 23% increase in energy for a 12% increase in cost. [83] The single meal deals examined provided 52% and 77% respectively of male and female daily energy intakes reported in the 1995 Australian National Nutrition Survey.

The association between portion size and energy intake amongst adults and older children has been demonstrated in laboratory settings. [84-86] In restaurants, customers who purchase larger portions also consume more energy, and rate the larger meal as better value for money. [86] In both laboratory and real life situations, consumers do not compensate for larger main courses by reducing side dishes, or by reducing energy intake at a subsequent meal.

Individuals under-report energy intake to a greater extent when consuming large portions. [77]

Portion size and energy density have independent effects, and combine to affect energy intake. Consuming large portions of low-energy-dense foods such as salads and most vegetables does not increase energy intake and may reduce it. [84] Dietary patterns based on high intakes of low-energy dense foods such as fruit, vegetables and reduced-fat dairy foods, and low intakes of energy dense foods such as fast food and sweetened drinks, are associated with smaller weight gains in longitudinal studies. [87]

In the 1995 Australian National Nutrition Survey, "non-core"[2] foods and beverages, which are recommended for occasional consumption only, made up 41% of total energy intake for children and adolescents. [88] However, the expected association between consumption of non-core foods and overweight was not observed except in young children, probably due to greater under-reporting in older children and failure to exclude under-reporters. [69]

Consumption of sweetened drinks (soda, soft-drinks, juices, sports drinks) has increased dramatically over the past few decades, particularly amongst children and young people. [89, 90] Consumption of sweetened drinks is associated with an increased risk of obesity in at least one longitudinal study. [91] In a randomised controlled trial, a targeted school based education programme produced a modest reduction in the number of carbonated drinks consumed, which was associated with a reduction in the number of overweight and obese children. [92]

In summary, evidence from ecological data in several countries suggests that per capita energy intake has increased in parallel with population BMI. Evidence from self-reported intake studies is conflicting, due to a systematic bias in energy intake reporting. Studies which have attempted to account for this bias have demonstrated an increase in energy intake. The widespread belief in the 1990s that fat intake was the major cause of obesity has also caused confusion, particularly when fat intake was considered independently of energy intake. There is good evidence that increased consumption of carbohydrate and energy-dense foods particularly sugars, sweetened drinks, snacks and fast foods, increased consumption of food prepared away from home and larger portion sizes have all contributed to increased energy intake. The pattern of these dietary changes is consistent with geographic variations in prevalence of obesity and the timeline of the obesity epidemic.

The Contribution of Physical Activity

Physical activity includes all bodily movement from fidgeting to running. It is a complex behaviour and difficult to measure. Activity energy expenditure (AEE) can be assessed at an individual level by measuring total energy expenditure (TEE) (eg, using doubly labelled water) and deducting resting energy expenditure (REE). AEE includes sedentary activities as well as obligatory and voluntary physical activity. Estimates of these activities can be made using accelerometers (movement sensors) or pedometers (step counters). At a population

[2] Core foods in the Australian Guide to Healthy Eating (NHMRC 2003) include water and the nutrient-dense recommended five food groups- fruit, vegetables, grains, meat and alternatives, milk and alternatives. Non-core foods are energy-dense, high in fat, sugar or salt and are recommended for occasional use only

level, physical activity measurement usually relies on self-report of voluntary or leisure-time activity and suffers from methodological difficulties similar to nutrition. [93]

A strong relationship has been demonstrated cross-sectionally and longitudinally between physical activity measured in a variety of different ways and a range of health outcomes, including fitness, cardiovascular disease, diabetes, bone health, mental health, musculoskeletal disorders, and some cancers. However, the relationship with obesity is not so clear. [94, 95]

In weight control studies, physical activity is an effective adjunct to dietary management for weight loss and is particularly important for maintenance of weight loss. [94]

It appears anecdotally that the level of obligatory physical activity required for survival activities, transport and activities of daily living has declined markedly over the last century in industrialised countries, and more recently with increasing prosperity and urbanisation in developing countries. It has been estimated that the difference in physical activity between a pre-modern lifestyle (based on 19th century Sydney) and a modern sedentary office worker is equivalent to walking 8-16km per day. [96] However, there are no long-term comparable measures of physical activity to confirm this, and most of these activity changes occurred in industrialised nations before 1980.

Population-level studies generally rely on self-reported leisure-time physical activity, including responses to questions about time spent walking and in other moderate and vigorous activity. 150 minutes of moderate physical activity per week is considered "sufficient activity" for health for adults. [94, 95]

In the US, participation in leisure-time physical activity measured in the Behavioural Risk Factor Surveillance System (BRFSS) remained unchanged from 1990-1998. [97] Prevalence of leisure-time inactivity declined significantly from 30% in 1994 to 24% in 2004. [98]. However, these reports do not include non leisure-time activity such as transport, occupational or house or yard work activity. Of employed adults aged 18 or over in the 1990 BRFSS survey, half of those who reported doing no leisure-time activity also reported doing at least one hour of hard occupational activity per day. [99]

The difficulties of estimating physical activity from self-reported data are highlighted by a report from the European Physical Activity Surveillance System. The first stage has indicated that there is no correlation between physical activity measures obtained using the International Physical Activity Questionnaire (IPAQ) and previous national estimates of physical activity using a variety of different instruments. [100]

Australian national surveys in 1997, 1999 and 2000, using an instrument similar to IPAQ, indicate that the proportion of the population that met the criterion for sufficient physical activity for health (150 minutes per week of moderate physical activity) declined from 62% in 1997 to 57% in 2000. [101] The questions in these surveys were not confined to leisure-time activity.

It has been suggested 45-60 minutes of moderate activity daily, or a PAL of 1.7, is required to prevent the transition to overweight. [102] Estimates of the amount of physical activity required to prevent overweight must also take into consideration energy intake, and this estimate was made in the context of an abundant energy-dense food supply.

Many, but not all cross sectional data based on self-report indicate that obese individuals tend to have lower levels of physical activity than normal-weight individuals. [95] The relationship may depend on how physical activity is measured.

The energy cost of performing the same amount of physical activity is related to the body mass of the individual. A study using accelerometry to measure physical activity and DLW to measure energy expenditure in 18 adolescents found that obese adolescents were less physically active than their normal-weight peers were, and physical activity was inversely related to % body fat. However there was no difference between the two groups in total energy expenditure or physical activity-related energy expenditure adjusted for fat-free mass. [103]

Studies which are based on objective measures of energy expenditure have not consistently shown that overweight is associated with lower levels of physical activity. The National Academies of Sciences has published a large combined dataset of objectively measured energy expenditure used to determine the US Dietary Reference Intakes (DRI) for macronutrients. [104] Analysis of this dataset reveals that the mean PAL (measured TEE/ estimated BMR) was 1.73 for 538 adults in the healthy weight range (BMI <25) and 1.73 for 443 overweight or obese adults (mean BMI 30). [60]

Prentice also reported a PAL of around 1.7 for obese and non-obese individuals, with somewhat lower levels amongst the severely obese. [56]

Cross-sectional studies showing an association between physical activity levels and obesity cannot determine whether inactivity causes obesity or obesity causes inactivity. Prospective observational population studies of the effect on body weight of physical activity measured at baseline are few and the results are inconsistent. [102, 105, 106]

The Copenhagen City Heart Study examined over 5000 adults at three time points over 15 years, using measured BMI and self-reported physical activity. The design accounted for the potential confounding factors of initial BMI and activity level and previous weight gain. [106] This large and elegant study found that initial BMI predicted later activity level, but no evidence that initial inactivity predicts later obesity.

A study of 138 5-year old Pima Indian children found that at 5 years, obesity was associated with reduced participation in sport and increased television viewing, but not with PAL calculated from energy expenditure measured by DLW. After follow-up at age 10, obesity was associated with reduced participation in sport and increased television viewing and also with a reduced PAL. None of the baseline variables measured, including AEE and TEE predicted body fat at 10 years after adjusting for body fat at 5 years. However, obesity predicted later inactivity. [107]

Few longitudinal studies have attempted to determine the relative contribution of energy intake and expenditure to subsequent risk of obesity using objective measures. Stunkard et al found that in infants, measured energy intake at 3 months but not measured energy expenditure determined weight gain between 3 months and 12 months. [108] Subsequent follow-up of these infants confirmed that this relationship continued at two years. [109] Body weight at two years was also predicted positively by sucking behaviour and total energy expenditure, and negatively by family income.

Tataranni et al found that amongst Pima Indian adults, baseline total energy intake (calculated from energy expenditure measured by DLW), was a predictor of weight gain over four months or more. [110] However, baseline AEE and PAL were not related to weight gain. PAL was lower in obese subjects and the authors concluded that this was secondary to obesity.

It is clear that a sedentary lifestyle is widespread in developed societies, and that this is related to physical, technological and economic environmental conditions. [111] However, the evidence also suggests that physical inactivity may be a consequence of obesity. Physical activity has a profound impact on health, [95] independently of its role in obesity, and community-wide interventions to promote more active lifestyles could have a major impact on health outcomes.

THE CONTRIBUTION OF TELEVISION VIEWING

Evidence from epidemiological, cross-sectional and experimental studies suggests that television viewing contributes to obesity in children. Many cross-sectional studies have reported an association between television viewing and childhood obesity. [112] Similar associations have been reported in adults. [113] Longitudinal studies in pre-school children [114, 115] , adolescents [116, 117] and adults [118] have reported that television viewing predicts later obesity. Two longitudinal studies have reported the effect of childhood television viewing on adult risk of obesity. [119, 120]

A follow-up of the British 1970 birth cohort found that weekend television viewing at 5 years but not weekday hours of television predicted higher BMI at 30 years when adjusted for TV viewing and activity level at 10 years, sex, socioeconomic status, parental BMI, and birth weight. Each additional hour of TV watched on weekends at 5 years increased risk of adult obesity by 7%. [120]

A New Zealand study which followed 1000 individuals from birth to 26 years reported that childhood and adolescent television viewing predicted adult BMI, cardiorespiratory fitness, cigarette smoking and raised serum cholesterol. This relationship was independent of television viewing at age 26, childhood socioeconomic status, BMI at age 5 years, parental BMI, parental smoking, and physical activity at age 15. 17% of overweight at 26 years could be attributed to more than 2 hours of television viewing daily between the ages of 5 and 15. [119]

School interventions which reduced television-viewing time, have been shown to be effective in reducing the prevalence of obesity, [121, 122] whereas school interventions which have not targeted television viewing have generally failed to show an effect on BMI.

These studies suggest that the relationship between television viewing and obesity is causal. Potential mechanisms include: [112]

- Reduced energy expenditure from the displacement of physical activity.
- Decreased metabolic rate during viewing.
- Increased dietary intake from eating during viewing or from the effects of food advertising.

Television viewing is associated with inactivity and reduced fitness, but several studies have reported that the correlation between television viewing time and overweight is independent of the level of physical activity. [113, 123] The longitudinal studies described above adjusted for adolescent physical activity. Television viewing may act partially through displacement of physical activity, but it is not the only mechanism.

A study using indirect calorimetry to objectively measure energy expenditure and physical activity level found that while television viewing was related to body composition, there was no relationship with total energy expenditure, activity energy expenditure or activity level. [124] Reduced metabolic rate is an unlikely mechanism for the effect of television viewing.

The strongest evidence relates to the effect of television viewing on eating behaviour. [125] A systematic review commissioned by the UK Food Standards Agency of the effects of food promotion on children found that the food industry spends many millions of dollars annually advertising to children, and most of these advertisements are for pre-sweetened breakfast cereals, sweetened drinks, confectionary, savoury snacks, or fast food outlets. [126]

Ecological evidence from the USA, Australia and Europe shows a significant correlation between national levels of advertising on children's television in 10 countries, and the prevalence of childhood overweight. For advertising of energy-dense foods, the correlation was 0.86 ($p<.005$), while there was a weaker negative correlation with advertising of healthier foods (-0.56, $p<0.10$). The highest prevalence of obesogenic advertising was in the USA and Australia, with the lowest in Sweden and the Netherlands. [127]

The UK review found good evidence that food promotion affects children's preferences, purchase behaviour and consumption. This effect is independent of other factors and operates at both a brand and category level. These findings are likely to understate the effect that food promotion has on children. The literature focuses principally on television advertising but the cumulative effect of this combined with other forms of promotion and marketing is likely to be significantly greater. In addition, the studies have only looked at *direct* effects on individual children, and understate *indirect* influences. [126]

The Contribution of Smoking Cessation

Smoking prevalence has declined markedly in some industrialised countries in the past 25 years. Quitting smoking is associated with an increased risk of weight gain, [128, 129] and it has been hypothesised that the increased number of ex-smokers may contribute to the increased prevalence of obesity. [129] There is evidence from animal and human studies that nicotine increases energy expenditure and reduces appetite. [130, 131] Several studies have shown an association between cigarettes smoked per day and overweight and obesity amongst former smokers. [132]

Epidemiological analysis of trends in smoking and obesity in New Zealand and Australia in the 1980s to mid 1990s concluded that changes in the prevalence of smoking explained little, if any of the change in prevalence of obesity. [133, 134] In the USA, Flegal et al estimated that reduction in prevalence of smoking contributed between a quarter (men) and a sixth (women) of the increase in prevalence of overweight over 10 years in the 1980s. [129]

In the WHO MONICA project, trends in prevalence of ex-smokers explained around 10% of the population trends in mean BMI in Europe, the USA, Australia and China from the early 1980s to the mid 1990s. Energy supply combined with prevalence of ex-smokers explained 69% of the between-country differences in change of prevalence of overweight. [74]

In an economic model of the US obesity epidemic, Chou estimated that the increase in the real price of cigarettes has a significant effect on the probability of obesity. [135] However, the decline in smoking prevalence in the US between 1960 and 1978, when the prevalence of obesity was stable, was similar to the decline in 1978-2000, which casts some doubt on the role of this factor.

In some developing countries, increases in the prevalence of obesity may have occurred at the same time as increases in smoking prevalence. There is also no evidence to suggest that increasing prevalence of ex-smokers contributes to the obesity epidemic amongst children.

THE CONTRIBUTION OF ECONOMIC AND SOCIAL CHANGE

Technological changes including industrialization, ownership of televisions, computers, private vehicles and labour-saving devices have undoubtedly contributed to a decrease in obligatory physical activity. However, with the exception of the widespread adoption of computers in the home and workplace, these changes predate the obesity epidemic which began in developed nations around 1980. There is evidence that physical activity levels in the US based on time use diaries fell between 1965 and 1975 but have remained fairly constant since. [136]

Advances in food technology have reduced the relative cost of producing and distributing processed foods high in sugar and fat, while the cost of primary production and distribution of basic fresh foods such as fruit, vegetables, meat and milk has not decreased to the same extent. According to economic theory, a decrease in the price of food is likely to lead to an increase in consumption. [136]

From 1960 to 1980 food prices in the US rose slightly faster than the rate of inflation whereas from 1980 to 2000, food prices rose less than the Consumer Price Index (CPI). The relative price of energy-dense foods and drinks decreased from the 1980s compared to less energy-dense foods such as fruit and vegetables. [136] In Australia, the CPI for food in Queensland rose by 24% between 1998 and 2004, while the cost of basic fruit and vegetables rose by 33% and the cost of soft-drinks (soda), water and juices fell by 3%. [35]

An economic analysis of factors underlying the trends in obesity in the US found that the per capita increase in the numbers of restaurants accounted for 61% of the increase in BMI and 65% of the increase in the percentage of the population who are obese. [135]

The relationship between socioeconomic status and obesity in developed nations suggests that economic factors influence food selection. Computer modelling of the French national diet examined the relationship between the cost and energy density of modelled diets. The results suggest that consumers on a low food budget are more likely to select an energy-dense diet. However, if cost is not an issue, selecting an energy-dense diet induced only a moderate reduction in diet cost. [137]

The increased participation of women in the labour force has increased family incomes but rendered many families “time-poor”. This may partly explain the increased demand for and consumption of fast food and convenience foods. Although women’s participation in the workforce began to rise in the late 1960s, it increased more rapidly in the late 1970s. In the mid 1970s, some employers still expected women to resign when they married. Female participation in the Australian labour force rose from 43.6% in 1979 to 55.9% in 2003. [138] This does not capture the increase in hours worked. Formal long-day child-care to enable mothers to work has become more widespread and accessible during the 1980s and early 1990s in Australia.

In summary, economic evidence suggests that lower relative food costs, particularly for processed energy-dense foods and sweetened drinks, and the proliferation of fast food restaurants offering large portions of low-cost energy-dense foods have made a significant contribution to the obesity epidemic. Social, workforce, technological and broader economic developments have contributed to these changes in the food economy.

CONCLUSION

The global prevalence of obesity was relatively stable during the 1960s and 1970s, but has dramatically increased since 1980. The explanation for this increase lies in changes in the physical, economic and social environment which correspond with this time-frame.

There is no evidence of a decrease in population energy expenditure as a factor in the obesity epidemic. On the contrary, both theoretical considerations and objective evidence demonstrate that per capita energy expenditure increases as mean BMI increases. An increase in energy supply and consumption associated with decreased real cost and increased availability and convenience of energy-dense foods and drinks appears to be the primary driver of the obesity epidemic.

Population physical activity level has also decreased, but over a much longer period than the recent epidemic of obesity. Activity energy expenditure is directly associated with body mass but physical activity levels in the very obese tend to be lower than the non-obese. Widespread low levels of activity may have contributed by increasing susceptibility of the population to a cheap, energy dense food supply. There are many other benefits of physical activity and healthy eating besides weight management, and both physical activity and nutrition must be addressed together to improve the health of populations.

The increased prevalence of ex-smokers in developed countries may have contributed a small proportion of the increased prevalence of adult obesity, but evidence for this is conflicting, and it is unlikely to have contributed to the epidemic in children or in developing countries.

These results have clear implications that policies and interventions to prevent obesity must focus on the food supply and social and economic influences on eating behaviour, in addition to influences on physical activity. Information alone will not make a difference, except to the most advantaged sections of society. The obesity epidemic can only be turned around when comprehensive efforts are made by governments, industry and communities working together to address these determinants and create the economic, physical and social environments that support individuals to choose healthy foods and an active lifestyle.

REFERENCES

[1] WHO, *The World Health Report 2002: Reducing risks, promoting healthy life.* 2003, World Health Organization: Geneva.

[2] WHO, *Preventing chronic diseases: a vital investment: WHO global report.* 2005, World Health Organization: Geneva.

[3] Prentice, AM and Jebb, SA, Obesity in Britain: gluttony or sloth? *BMJ*, 1995. **311**(7002): 437-439.

[4] IOTF and EASO, *Obesity in Europe*. 2002, International Obesity Task Force and European Association for the Study of Obesity: London.

[5] Prentice, AM and Jebb, S, Energy intake/physical activity interactions in the homeostasis of body weight regulation. *Nutr Rev*, 2004. **62**(7 Pt 2): S98-104.

[6] Maes, HH, Neale, MC, and Eaves, LJ, Genetic and environmental factors in relative body weight and human adiposity. *Behav Genet*, 1997. **27**(4): 325-351.

[7] Flegal, KM, The obesity epidemic in children and adults: current evidence and research issues. *Med. Sci. Sports Exerc.*, 1999. **31**(11 Suppl): S509-S514.

[8] WHO, *Obesity: Preventing and Managing the Global Epidemic, Report of a WHO Consultation on Obesity*. 1998, World Health Organisation: Geneva.

[9] Cole, TJ, Bellizzi, MC, Flegal, KM, and Dietz, WH, Establishing a standard definition for child overweight and obesity worldwide: international survey. *BMJ*, 2000. **320**(7244): 1240-.

[10] Lobstein, T, Baur, L, Uauy, R, and IASO International Obesity Taskforce, Obesity in children and young people: a crisis in public health. *Obesity Reviews*, 2004. **5**(Suppl. 1): 4-85.

[11] Popkin, BM, An overview on the nutrition transition and its health implications: The Bellagio meeting. *Public Health Nutrition*, 2002. **5**(1A): 93-103.

[12] DelPrete, LR, Caldwell, M, English, C, Banspach, SW, and Lefebvre, C, Self-reported and measured weights and heights of participants in community-based weight loss programs. *J Am Diet Assoc*, 1992. **92**(12): 1483-1486.

[13] Flood, V, Webb, K, Lazarus, W, and Pang, G, Use of self-report to monitor overweight and obesity in populations: some issues for consideration. *Aust N Z J Public Health*, 2000. **24**(1): 96-99.

[14] Australian Institute of Health and Welfare (AIHW), *A Growing Problem: Trends and Patterns in Overweight and Obesity among Adults in Australia, 1980 to 2001*. 2003, AIHW: Canberra.

[15] Jolliffe, D, Continuous and robust measures of the overweight epidemic: 1971-2000. *Demography*, 2004. **41**(2): 303-314.

[16] WHO, *Diet, nutrition and the prevention of chronic diseases*. 2003, World Health Organization: Geneva.

[17] Jolliffe, D, Extent of overweight among US children and adolescents from 1971 to 2000. *Int J Obes*, 2004. **28**(1): 4-9.

[18] Cook, T, Rutishauser, IHE, and Seelig, M, *Comparable data on food and nutrient intake and physical measurements from the 1983, 1985 and 1995 national nutrition surveys*. 2001, Commonwealth Department of Health and Aged Care: Canberra.

[19] Cameron, A, Welborn, T, Zimmet, P, Dunstan, D, Owen, N, et al, Overweight and obesity in Australia: the 1999-2000 Australian Diabetes, Obesity and Lifestyle Study (AusDiab). *MJA*, 2003. **178**: 427-432.

[20] Coyne, TJ, Findlay, MG, Ibiebele, TI, and Firman, DW, Overweight and obesity in Australia: an underestimate of the true prevalence? *MJA*, 2004. **180**(2): 93-94.

[21] Magarey, AM, Daniels, LA, and Boulton, TJ, Prevalence of overweight and obesity in Australian children and adolescents: reassessment of 1985 and 1995 data against new standard international definitions. *MJA*, 2001. **174**(11): 561-4.

[22] Booth, ML, Chey, T, Wake, M, Norton, K, Hesketh, K, et al, Change in the prevalence of overweight and obesity among young Australians, 1969-1997. *Am J Clin Nutr*, 2003. **77**(1): 29-36.

[23] Vaska, VL and Volkmer, R, Increasing prevalence of obesity in South Australian 4-year-olds: 1995–2002. *J Paediatr Child Health*, 2004. **40**: 353-355.

[24] Garnett, SP, Cowell, CT, Baur, LA, Shrewsbury, VA, Chan, A, et al, Increasing central adiposity: the Nepean longitudinal study of young people aged 7-8 to 12-13. *Int J Obesity*, 2005. **29**(11): 1353-1360.

[25] IOTF, *EU Platform on Diet, Physical Activity and Health: International Obesity Taskforce EU Platform Briefing Paper*. 2005, International Obesity Task Force and European Association for the Study of Obesity: Brussels.

[26] Janssen, I, Katzmarzyk, PT, Boyce, WF, Vereecken, C, Mulvihill, C, et al, Comparison of overweight and obesity prevalence in school-aged youth from 34 countries and their relationships with physical activity and dietary patterns. *Obesity Reviews*, 2005. **6**(2): 123-132.

[27] Bray, G, Predicting obesity in adults from childhood and adolescent weight. *Am J Clin Nutr*, 2002. **76**: 497-8.

[28] Wilkinson, R and Marmot, M, eds. *The Social Determinants of Health: The Solid Facts*. 1998, WHO: Europe.

[29] Turrell, G, Educational differences in dietary guideline food practices: are they associated with educational differences in food and nutrition knowledge? *Aust J Nutr Diet*, 1997. **54**: 25-33.

[30] Mishra, G, Ball, K, Patterson, A, Brown, W, Hodge, A, et al, Socio-demographic inequalities in the diets of mid-aged Australian women. *Eur J Clin Nutr*, 2005. **59**: 185-195.

[31] James, WPT, Nelson, M, Ralph, A, and Leather, S, Socioeconomic determinants of health: The contribution of nutrition to inequalities in health. *BMJ*, 1997. **314**(7093): 1545-.

[32] Adams, AK, Harvey, HE, and Prince, RJ, Association of maternal smoking with overweight at age 3 y in American Indian children. *Am J Clin Nutr*, 2005. **82**(2): 393-398.

[33] Hitchman, C, Christie, I, Harrison, M, and Lang, T, *Inconvenience food: the struggle to eat well on a low income*. 2002 http://www.demos.co.uk/catalogue/inconveniencefood/ (accessed November 2005), Demos: London.

[34] Lee, AJ, Darcy, AM, Leonard, D, Groos, AD, Stubbs, CO, et al, Food availability, cost disparity and improvement in relation to accessibility and remoteness in Queensland. *Aust N Z J Public Health*, 2002. **26**(3): 266-72.

[35] Queensland Health, *The 2004 Healthy Food Access Basket (HFAB) Survey*. 2005, Queensland Health: Brisbane.

[36] Parsons, TJ, Power, C, Logan, S, and Summerbell, CD, Childhood predictors of adult obesity: a systematic review. *Int J Obes Relat Metab Disord.*, 1999. **23**(Suppl 8): S1-S107.

[37] Power, C, Manor, O, and Matthews, S, Child to adult socioeconomic conditions and obesity in a national cohort. *Int J Obes*, 2003. **27**: 1081-1086.

[38] Griffiths, P and Bentley, M, Women of higher socio-economic status are more likely to be overweight in Karnataka, India. *Eur J Clin Nutr*, 2005. **59**(10): 1217-1220.

[39] Barker, D, *Mothers and Babies and Health in Later Life*. 2nd ed. 1998, Edinburgh: Churchill Livingstone.
[40] Oken, E and Gillman, MW, Fetal Origins of Obesity. *Obes Res*, 2003. **11**(4): 496-506.
[41] Barker, DJ, Fetal origins of cardiovascular disease. *Annals of Medicine*, 1999. **31**(Suppl. 1): 3-6.
[42] Kensara, OA, Wootton, SA, Phillips, DI, Patel, M, Jackson, AA, et al, Fetal programming of body composition: relation between birth weight and body composition measured with dual-energy X-ray absorptiometry and anthropometric methods in older Englishmen. *Am J Clin Nutr*, 2005. **82**(5): 980-987.
[43] Wells, JCK, Hallal, PC, Wright, A, Singhal, A, and Victora, CG, Fetal, infant and childhood growth: relationships with body composition in Brazilian boys aged 9 years. *Int J Obes*, 2005. **29**(10): 1192-1198.
[44] Taylor, RW, Grant, AM, Goulding, A, and Williams, SM, Early adiposity rebound: review of papers linking this to subsequent obesity in children and adults. *Current Opinion in Clinical Nutrition & Metabolic Care*, 2005. **8**(6): 607-612.
[45] Arenz, S, Ruckerl, R, Koletzko, B, and von Kries, R, Breast-feeding and childhood obesity--a systematic review. *Int J Obes Relat Metab Disord*, 2004. **28**(10): 1247-1256.
[46] Owen, CG, Martin, RM, Whincup, PH, Smith, GD, and Cook, DG, Effect of Infant Feeding on the Risk of Obesity Across the Life Course: A Quantitative Review of Published Evidence. *Pediatrics*, 2005. **115**(5): 1367-1377.
[47] WHO, *The optimal duration of exclusive breastfeeding*. 2001, World Health Organization: Geneva.
[48] NHMRC, *Dietary Guidelines for Children and Adolescents in Australia*. 2003, National Health & Medical Research Council: Canberra.
[49] Baker, JL, Michaelsen, KF, Rasmussen, KM, and Sorensen, TI, Maternal prepregnant body mass index, duration of breastfeeding, and timing of complementary food introduction are associated with infant weight gain. *Am J Clin Nutr*, 2004. **80**(6): 1579-1588.
[50] Tremblay, A, Despres, J, Theriault, G, Fournier, G, and Bouchard, C, Overfeeding and energy expenditure in humans. *Am J Clin Nutr*, 1992. **56**(5): 857-862.
[51] Votruba, SB, Blanc, S, and Schoeller, DA, Pattern and cost of weight gain in previously obese women. *Am J Physiol Endocrinol Metab*, 2002. **282**(4): E923-930.
[52] Thorburn, AW and Proietto, J, Biological determinants of spontaneous physical activity. *Obesity Reviews*, 2000. **1**(2): 87-94.
[53] Levine, JA and Kotz, CM, NEAT - non-exercise activity thermogenesis - egocentric & geocentric environmental factors vs. biological regulation. *Acta Physiologica Scandinavica*, 2005. **184**(4): 309-318.
[54] Brown, WJ, Williams, L, Ford, JH, Ball, K, and Dobson, AJ, Identifying the Energy Gap: Magnitude and Determinants of 5-Year Weight Gain in Midage Women. *Obes Res*, 2005. **13**(8): 1431-1441.
[55] Trabulsi, J and Schoeller, DA, Evaluation of dietary assessment instruments against doubly labeled water, a biomarker of habitual energy intake. *Am J Physiol Endocrinol Metab. 2001 Nov;281(5):*, 2001. **281**(5): 891-899.
[56] Prentice, AM, Black, AE, Coward, WA, and Cole, TJ, Energy expenditure in overweight and obese adults in affluent societies: an analysis of 319 doubly-labelled water measurements. *Eur J Clin Nutr*, 1996. **50**(2): 93-7.

[57] Frankenfield, D, Roth-Yousey, L, Compher, C, and Evidence Analysis Working Group, Comparison of predictive equations for resting metabolic rate in healthy non-obese and obese adults: a systematic review. *J Am Diet Assoc*, 2005. **105**: 775-789.

[58] Mifflin, MD, St Jeor, ST, Hill, LA, Scott, BJ, Daugherty, SA, et al, A new predictive equation for resting energy expenditure in healthy individuals. *Am J Clin Nutr*, 1990. **51**(2): 241-247.

[59] Luke, A, Durazo-Arvizu, RA, Rotimi, CN, Iams, H, Schoeller, DA, et al, Activity energy expenditure and adiposity among black adults in Nigeria and the United States. *Am J Clin Nutr*, 2002. **75**(6): 1045-1050.

[60] Institute of Medicine of the National Academies, *Doubly labelled water dataset*. 2002 http://www.iom.edu/subpage.asp?id=7292 (accessed November 2005), National Academies of Sciences.

[61] Food and Nutrition Board, *Dietary Reference Intakes for Energy, Carbohydrate, Fiber, Fat, Fatty Acids, Cholesterol, Protein, and Amino Acids (Macronutrients) (2005)*. 2002, Institute of Medicine of the National Academies: Washington D.C.

[62] Willett, WC, Dietary fat plays a major role in obesity: no. *Obes Rev*, 2002. **3**(2): 59-68.

[63] Astrup, A, Dietary fat is a major player in obesity- but not the only one. *Obes Rev*, 2002. **3**: 57-58.

[64] Heini, A and Weinsier, R, Divergent trends in obesity and fat intake patterns: the American paradox. *Am J Med*, 1997. **102**(3): 259-64.

[65] Wright, JD, Kennedy-Stephenson, J, Wang, CY, McDonnell, MA, and National Center for Health Statistics CDC, Trends in intake of energy and macronutrients- United States, 1971-2000. *MMWR Weekly*, 2004. **53**(04): 80-82.

[66] Rutishauser, IHE, Dietary intake measurements. *Public Health Nutr*, 2005. **8**(Special Issue 1): 1100-7.

[67] Heitmann, BL and Lissner, L, Dietary underreporting by obese individuals- is it specific or non-specific? *BMJ*, 1995. **311**: 986-89.

[68] Zhang, J, Temme, EHM, Sasaki, S, and Kesteloot, H, Under- and Overreporting of Energy Intake using Urinary Cations as Biomarkers: Relation to Body Mass Index. *Am. J. Epidemiol.*, 2000. **152**(5): 453-462.

[69] Huang, TT-K, Roberts, SB, Howarth, NC, and McCrory, MA, Effect of Screening Out Implausible Energy Intake Reports on Relationships between Diet and BMI. *Obes Res*, 2005. **13**(7): 1205-1217.

[70] Wright, JD, Kennedy-Stephenson, J, Wang, CY, McDowell, MA, Johnson, CL, et al, Trends in intake of energy and macronutrients- United States, 1971-2000. *MMWR weekly*, 2004. **53**(04): 80-82.

[71] Nielsen, SJ, Siega-Riz, AM, and Popkin, BM, Trends in Energy Intake in U.S. between 1977 and 1996: Similar Shifts Seen across Age Groups. *Obes Res*, 2002. **10**(5): 370-378.

[72] DEFRA, *Family Food 2003-04- Report on the Expenditure & Food Survey*. 2005 http://statistics.defra.gov.uk/esg/publications/efs/default.asp (accessed November 2005), National Statistics: London.

[73] Harnack, LJ, Jeffery, RW, and Boutelle, KN, Temporal trends in energy intake in the United States: an ecologic perspective. *Am J Clin Nutr*, 2000. **71**(6): 1478-1484.

[74] Silventoinen, K, Sans, S, Tolonen, H, Monterde, D, Kuulasmaa, K, et al, Trends in obesity and energy supply in the WHO MONICA Project. *Int J Obes*, 2004. **28**(5): 710-718.

[75] Prentice, AM and Jebb, SA, Fast foods, energy density and obesity: a possible mechanistic link. *Obesity Reviews*, 2003. **4**(4): 187-194.

[76] Drewnowski, A and Darmon, N, The economics of obesity: dietary energy density and energy cost. *Am J Clin Nutr*, 2005. **82**(1): 265S-273.

[77] Harnack, L, Steffen, L, Arnett, DK, Gao, S, and Luepker, RV, Accuracy of estimation of large food portions. *J Am Diet Assoc*, 2004. **104**(5): 804-806.

[78] Guthrie, JF, Lin, BH, and Frazao, E, Role of food prepared away from home in the American diet, 1977-78 versus 1994-96: changes and consequences. *J Nutr Educ Behav*, 2002. **34**(3): 140-150.

[79] Maddock, J, The relationship between obesity and the prevalence of fast food restaurants: state-level analysis. *Am J Health Promot.*, 2004. **19**(2): 137-143.

[80] Bowman, SA, Gortmaker, SL, Ebbeling, CB, Pereira, MA, and Ludwig, DS, Effects of Fast-Food Consumption on Energy Intake and Diet Quality Among Children in a National Household Survey. *Pediatrics*, 2004. **113**(1): 112-118.

[81] Pereira, MA, Kartashov, AI, Ebbeling, CB, Van Horn, L, Slattery, M, et al, Fast-food habits, weight gain, and insulin resistance (the CARDIA study): 15-year prospective analysis. *Lancet*, 2005. **365**(9453): 36-42.

[82] Young, LR and Nestle, M, The contribution of expanding portion sizes to the US obesity epidemic. *Am J Public Health*, 2002. **92**(2): 246-249.

[83] Cameron-Smith, D, Bilsborough, SA, and Crowe, TC, Upsizing Australia's waistline: the dangers of "meal deals". *MJA*, 2002. **177**(11/12): 686-689.

[84] Ello-Martin, JA, Ledikwe, JH, and Rolls, BJ, The influence of food portion size and energy density on energy intake: implications for weight management *Am J Clin Nutr*, 2005. **82**(1): 236S-241.

[85] Rolls, BJ, Morris, EL, and Roe, LS, Portion size of food affects energy intake in normal-weight and overweight men and women. *Am J Clin Nutr*, 2002. **76**(6): 1207-1213.

[86] Diliberti, N, Bordi, PL, Conklin, MT, Roe, LS, and Rolls, BJ, Increased Portion Size Leads to Increased Energy Intake in a Restaurant Meal. *Obes Res*, 2004. **12**(3): 562-568.

[87] Newby, P, Muller, D, Hallfrisch, J, Qiao, N, Andres, R, et al, Dietary patterns and changes in body mass index and waist circumference in adults. *Am J Clin Nutr*, 2003. **77**(6): 1417-1425.

[88] Bell, AC, Kremer, PJ, Magarey, AM, and Swinburn, BA, Contribution of 'noncore' foods and beverages to the energy intake and weight status of Australian children. *European Journal of Clinical Nutrition*, 2005. **59**(5): 639-645.

[89] St-Onge, M-P, Keller, KL, and Heymsfield, SB, Changes in childhood food consumption patterns: a cause for concern in light of increasing body weights. *Am J Clin Nutr*, 2003. **78**(6): 1068-1073.

[90] Nielsen, SJ and Popkin, BM, Patterns and trends in food portion sizes, 1977-1998. *JAMA*, 2003. **289**(4): 450-453.

[91] Ludwig, DS, Peterson, KE, and Gortmaker, SL, Relation between consumption of sugar-sweetened drinks and childhood obesity: a prospective, observational analysis. *Lancet*, 2001. **357**(9255): 505-8.

[92] James, J, Thomas, P, Cavan, D, and Kerr, D, Preventing childhood obesity by reducing consumption of carbonated drinks: cluster randomised controlled trial. *BMJ*, 2004: bmj.38077.458438.EE.

[93] Sallis, JF and Owen, N, *Physical Activity and Behavioural Medicine*. Behavioural Medicine and Health Psychology, ed. JR Turner. 1999, Thousand Oaks: Sage Publications.

[94] Centres for Disease Control and Prevention, *Physical Activity and Health: a report of the Surgeon General*. 1996, US Dept of Health and Human Services.

[95] UK Department of Health, *At least five a week: Evidence on the impact of physical activity and its relationship to health. A report from the Chief Medical Officer*. 2004, UK Department of Health: London.

[96] Eggar, GJ, Vogels, N, and Westertep, KR, Estimating historical changes in physical activity levels. *Med J Aust*, 2001. **175**: 635-6.

[97] Centers for Disease Control and Prevention (CDC), Physical Activity Trends --- United States, 1990--1998. *MMWR Morb Mortal Wkly Rep*, 2001. **50**(9): 166-169.

[98] Centers for Disease Control and Prevention (CDC), Trends in Leisure-Time Physical Inactivity by Age, Sex, and Race/Ethnicity --- United States, 1994--2004. *MMWR Morb Mortal Wkly Rep*, 2005. **54**(39): 991-994.

[99] Centers for Disease Control and Prevention (CDC), Prevalence of leisure-time and occupational physical activity among employed adults--United States, 1990. *MMWR Morb Mortal Wkly Rep*, 2000. **49**(19): 420-424.

[100] Rutten, A, Ziemainz, H, Schena, F, Stahl, T, Stiggelbout, M, et al, Using different physical activity measurements in eight European countries. Results of the European Physical Activity Surveillance System (EUPASS) time series survey. *Public Health Nutr*, 2003. **6**(4): 371-376.

[101] Bauman, A, Bellew, B, Vita, P, Brown, W, and Owen, N, *Getting Australia active: towards better practice for the promotion of physical activity*. 2002, National Public Health Partnership: Canberra.

[102] Saris, WHM, Blair, SN, van Baak, MA, Eaton, SB, Davies, PSW, et al, How much physical activity is enough to prevent unhealthy weight gain? Outcome of the IASO 1st Stock Conference and consensus statement. *Obesity Reviews*, 2003. **4**(2): 101-114.

[103] Ekelund, U, Aman, J, Yngve, A, Renman, C, Westerterp, K, et al, Physical activity but not energy expenditure is reduced in obese adolescents: a case-control study. *Am J Clin Nutr*, 2002. **76**(5): 935-941.

[104] Food and Nutrition Board, *Dietary Reference Intakes for Energy, Carbohydrate, Fiber, Fat, Fatty Acids, Cholesterol, Protein, and Amino Acids (Macronutrients)*. 2005, Institute of Medicine of the National Academies: Washington D.C.

[105] Blair, SN, LaMonte, MJ, and Nichaman, MZ, The evolution of physical activity recommendations: how much is enough? *Am J Clin Nutr*, 2004. **79**(5): 913S-920.

[106] Petersen, L, Schnohr, P, and Sørensen, TIA, Longitudinal study of the long-term relation between physical activity and obesity in adults. *Int J Obes*, 2004. **28**(1): 105-112.

[107] Salbe, AD, Weyer, C, Harper, I, Lindsay, RS, Ravussin, E, et al, Assessing Risk Factors for Obesity Between Childhood and Adolescence: II. Energy Metabolism and Physical Activity. *Pediatrics*, 2002. **110**(2): 307-314.

[108] Stunkard, AJ, Berkowitz, RI, Stallings, VA, and Schoeller, DA, Energy intake, not energy output, is a determinant of body size in infants. *Am J Clin Nutr*, 1999. **69**(3): 524-530.

[109] Stunkard, AJ, Berkowitz, RI, Schoeller, D, Maislin, G, and Stallinlgs, VA, Predictors of body size in the first 2 y of life: a high-risk study of human obesity. *Int J Obes Relat Metab Disord*, 2004. **28**(4): 503-513.

[110] Tataranni, PA, Harper, IT, Snitker, S, Parigi, AD, Vozarova, B, et al, Body weight gain in free-living Pima Indians: effect of energy intake vs expenditure. *Int J Obes Relat Metab Disord*, 2003. **16**: 16.

[111] James, W, A public health approach to the problem of obesity. *Int J Obesity*, 1995. **19**(Suppl. 3): S37-S45.

[112] Robinson, TN, Television viewing and childhood obesity. *Pediatr Clin North Am*, 2001. **48**(4): 1017-25.

[113] Salmon, J, Bauman, A, Crawford, D, Timperio, A, and Owen, N, The association between television viewing and overweight among Australian adults participating in varying levels of leisure-time physical activity. *Int J Obes Relat Metab Disord*, 2000. **24**(5): 600-6.

[114] Proctor, MH, Moore, LL, Gao, D, Cupples, LA, Bradlee, ML, et al, Television viewing and change in body fat from preschool to early adolescence: The Framingham Children's Study. *Int J Obes Relat Metab Disord*, 2003. **27**(7): 827-33.

[115] Jago, R, Baranowski, T, Baranowski, JC, Thompson, D, and Greaves, KA, BMI from 3-6 years is predicted by TV viewing and physical activity, not diet. *Int J Obes*, 2005. **29**: 557-564.

[116] Berkey, CS, Rockett, HRH, Gillman, MW, and Colditz, GA, One-Year Changes in Activity and in Inactivity Among 10- to 15-Year-Old Boys and Girls: Relationship to Change in Body Mass Index. *Pediatrics*, 2003. **111**(4): 836-843.

[117] Kaur, H, Choi, WS, Mayo, MS, and Jo Harris, K, Duration of television watching is associated with increased body mass index. *The Journal of Pediatrics*, 2003. **143**(4): 506-511.

[118] Hu, FB, Li, TY, Colditz, G, Willett, WC, and Manson, JE, Television Watching and Other Sedentary Behaviors in Relation to Risk of Obesity and Type 2 Diabetes Mellitus in Women. *JAMA*, 2003. **289**(14): 1785-1791.

[119] Hancox, RJ, Milne, BJ, and Poulton, R, Association between child and adolescent television viewing and adult health: a longitudinal birth cohort study. *The Lancet*, 2004. **364**(9430): 257-262.

[120] Viner, RM and Cole, TJ, Television viewing in early childhood predicts adult body mass index. *J Pediatr*, 2005. **147**(4): 429-435.

[121] Robinson, TN, Reducing children's television viewing to prevent obesity: a randomized controlled trial. *JAMA*, 1999. **282**(16): 1561-7.

[122] Gortmaker, S, Peterson, K, Wiecha, J, Sobol, AM, Dixit, S, et al, Reducing obesity via a school-based interdisciplinary intervention among youth: Planet Health. *Arch Pediatr Adolesc Med*, 1999. **153**(4): 409-418.

[123] Trost, SG, Sirard, JR, Dowda, M, Pfeiffer, KA, and Pate, RR, Physical activity in overweight and nonoverweight preschool children. *Int J Obes Relat Metab Disord*, 2003. **27**(7): 834-9.

[124] Grund, A, Krause, H, Siewers, M, Rieckert, H, and Muller, MJ, Is TV viewing an index of physical activity and fitness in overweight and normal weight children? *Public Health Nutr*, 2001. **4**(6): 1245-51.

[125] Ludwig, DS and Gortmaker, SL, Programming obesity in childhood. *The Lancet*, 2004. **364**(9430): 226-227.

[126] Hastings, GB, Stead, M, McDermott, L, Forsyth, A, MacKintosh, AM, et al, *Review of Research on the Effects of Food Promotion to Children - Final Report and Appendices. Prepared for the Food Standards Agency*. 2003, Institute for Social Marketing: Stirling.

[127] Lobstein, T and Dibb, S, Evidence of a possible link between obesogenic food advertising and child overweight. *Obesity Reviews*, 2005. **6**(3): 203-208.

[128] Koh-Banerjee, P, Chu, N-F, Spiegelman, D, Rosner, B, Colditz, GA, et al, Prospective study of the effect of changes in dietary intake, physical activity, alcohol consumption, and smoking with a 9-y gain in waist circumference among 16 587 US men. *Am J Clin Nutr*, 2003. **78**: 719-727.

[129] Flegal, KM, Troiano, RP, Pamuk, ER, Kuczmarski, RJ, and Campbell, SM, The Influence of Smoking Cessation on the Prevalence of Overweight in the United States. *New England Journal of Medicine*, 1995. **333**(18): 1165-1170.

[130] Hofstetter, A, Schutz, Y, Jequier, E, and Wahren, J, Increased 24-hour energy expenditure in cigarette smokers. *N Engl J Med*, 1986. **314**: 79-82.

[131] Hughes, JR and Hatsukami, DK, Effects of three doses of transdermal nicotine on post-cessation eating, hunger and weight. *J Subst Abuse*, 1997. **9**: 151-159.

[132] John, U, Hanke, M, Rumpf, H-J, and Thyrian, JR, Smoking status, cigarettes per day, and their relationship to overweight and obesity among former and current smokers in a national adult general population sample. *Int J Obes*, 2005. **29**(10): 1289-1294.

[133] Boyle, CA, Dobson, A, Egger, G, and Magnus, P, Can the increasing weight of Australians be explained by the decreasing prevalence of cigarette smoking? *Int J Obes Relat Metab Disord.*, 1994. **18**(1): 55-60.

[134] Simmons, G, Jackson, R, Swinburn, B, and Yee, RL, The increasing prevalence of obesity in New Zealand: is it related to recent trends in smoking and physical activity? *N Z Med J*, 1996. **109**(1018): 90-92.

[135] Chou, S-Y, Grossman, M, and Saffer, H, An economic analysis of adult obesity: results from the Behavioral Risk Factor Surveillance System. *J Health Economics*, 2004. **23**: 565-87.

[136] Finkelstein, EA, Ruhm, CJ, and Kosa, KM, Economic causes and consequences of obesity. *Annu Rev Public Health*, 2005. **26**: 239-257.

[137] Darmon, N, Ferguson, E, and Briend, A, Do economic constraints encourage the selection of energy-dense diets? *Appetite*, 2003. **41**: 315-322.

[138] Australian Bureau of Statistics, *Australian Labour Market Statistics (cat. no. 6105.0)*. 2005, Australian Bureau of Statistics: Canberra.

In: New Developments in Obesity Research
Editor: Lawrence F. Ditmier, pp. 153-174

ISBN 1-60021-296-4

Chapter 6

ADVANCES IN THE IMMUNOLOGY OF OBESITY

Mary Madeline Rogge*

Indiana University School of Nursing, Indianapolis, IN

ABSTRACT

At the same time the incidence of obesity has been increasing dramatically world-wide, new evidence has begun to shed light on obesity as an immunologic disorder. Like macrophages, adipocytes secrete a wide array of cytokines that mediate the inflammatory response, and the increased risk obesity poses for developing heart disease, stroke, and diabetes are closely linked to elevated levels of adipocyte-secreted immune modulators. New scientific developments have also revealed that high levels of several immunologic factors are precursors to weight gain, and point to chronic-low grade inflammation as a contributing factor in the onset and progression of obesity. Both infectious organisms and environmental chemical agents are capable of inducing lipogenesis and adipogenesis by stimulating pro-inflammatory cytokine signaling, and theoretical models of how obesity and inflammation intersect are being developed. These models offer promising approaches for research to extend our understanding of the current obesity epidemic.

INTRODUCTION

In 1982, a research team reported for the first time, that canine distemper virus caused obesity in normally lean mice infected with the virus. [1] Mice inoculated with the virus became nearly twice as fat as the control mice, and demonstrated a higher number of adipocytes (hyperplasia) and larger adipocyte cell size (hypertrophy) than the controls. The pancreases of the infected mice also showed islet cell hyperplasia and hypertrophy. No viral antigens were detected in the hypothalamus or other regions of the brain, pointing to a peripheral, rather than central, defect in energy regulation as the cause of the obesity. The researchers concluded, "The present findings raise the possibility that virus infection also plays a role in some cases of obesity in children and in some cases of severe spontaneous obesity in adults."[1] The researchers' serendipitous findings seem to have attracted little attention at the time. However, these findings take on added significance when viewed within the context of other research over the past two decades, which has revealed preadipocytes and adipocytes are highly metabolically active cells and are closely affiliated with the immune system. Adipocytes secrete an array of chemical products that exert autocrine, paracrine, and systemic effects, and one of the most intriguing areas of obesity research to emerge has been the role of adipocytes in innate immunity and inflammation. Accumulating evidence points to obesity as a disease of chronic inflammation.[2-5] In both genetic and dietary-induced obesity, particularly central obesity, inflammatory markers are elevated and tend to be positively correlated with measures of adiposity. Furthermore, the elaboration of signaling molecules, now called adipocytokines or adipokines, from adipose tissue may provide the nexus between obesity and its co-morbidities.

What is often overlooked in the research examining the relationships between adipokines and diabetes or cardiovascular disease is the evidence suggesting obesity itself may be the consequence of immunologic processes. In the following chapter, general features of innate immunity and inflammation will be reviewed, including the metabolic effects of inflammation, and the parallel secretory functions of adipocytes will be described. The linkages between adipose immune mediators and common complications of obesity will be explored, as well as the modulating effects of weight loss or exercise. Finally, non-dietary environmental factors that may contribute to obesity as an immunologic disorder will be discussed, along with the pathways that may be involved, and a model of obesity as an immunogenic disorder will be presented, with recommendations for future research.

INNATE IMMUNITY

Innate immunity is the inborn, antigen-nonspecific defense a host is able to mount against injury and infection. In order to provide protection against microbes to which the body has not previously been exposed, innate immunity depends on recognizing pathogen-associated structures common to many different types of microorganisms, rather than responding to specific antigens. This first line of defense is composed of phagocytic cells, including neutrophils, monocytes, and macrophages; T lymphocytes, especially natural killer cells; proteins of the complement cascade; and chemical mediators, including pro-inflammatory cytokines and acute phase proteins.

Central to innate immunity is the inflammatory process, which is initiated when toll-like receptors on host cells bind with characteristic molecules from pathogens, called pathogen-associated molecular patterns (PAMPs), that distinguish pathogens from host cells. Among the more common pathogen-associated molecular patterns are lipopolysaccharide (LPS) from the cell walls of gram-negative bacteria, peptidoglycan and lipoteichoic acids from the cell walls of gram-positive bacteria, bacterial DNA, glucans from fungi cell walls, and double-stranded RNA from many viruses. [6-7] Once toll-like receptors bind with PAMPs, inactive nuclear factor kappa-B (NF-KB) in the cytoplasm of mononuclear cells is activated and translocates to the nucleus, where it is able to regulate the transcription of more than 150

* (317) 274-7665; mmrogge@iupui.edu

genes. Activation of NF-KB in mononuclear cells, particularly macrophages, increases the secretion of the pro-inflammatory cytokines tumor necrosis factor-alpha (TNF-α), interleukin-1 (IL-1), interleukin-6 (IL-6), and inducible nitric oxide synthase (iNOS). (8-9) These molecules initiate and coordinate immune processes (Figure 1) to recognize and isolate foreign substances, to target them for phagocytosis and destruction by heat (fever) or chemical means, and to promote healing of injured tissues. To date, ten different types of toll-like receptors have been identified, capable of initiating responses to PAMPs, and different types of receptors may need to be activated in specific combinations to adapt immune responses to particular types of environmental challenges through the cytokines they evoke. [10-11]

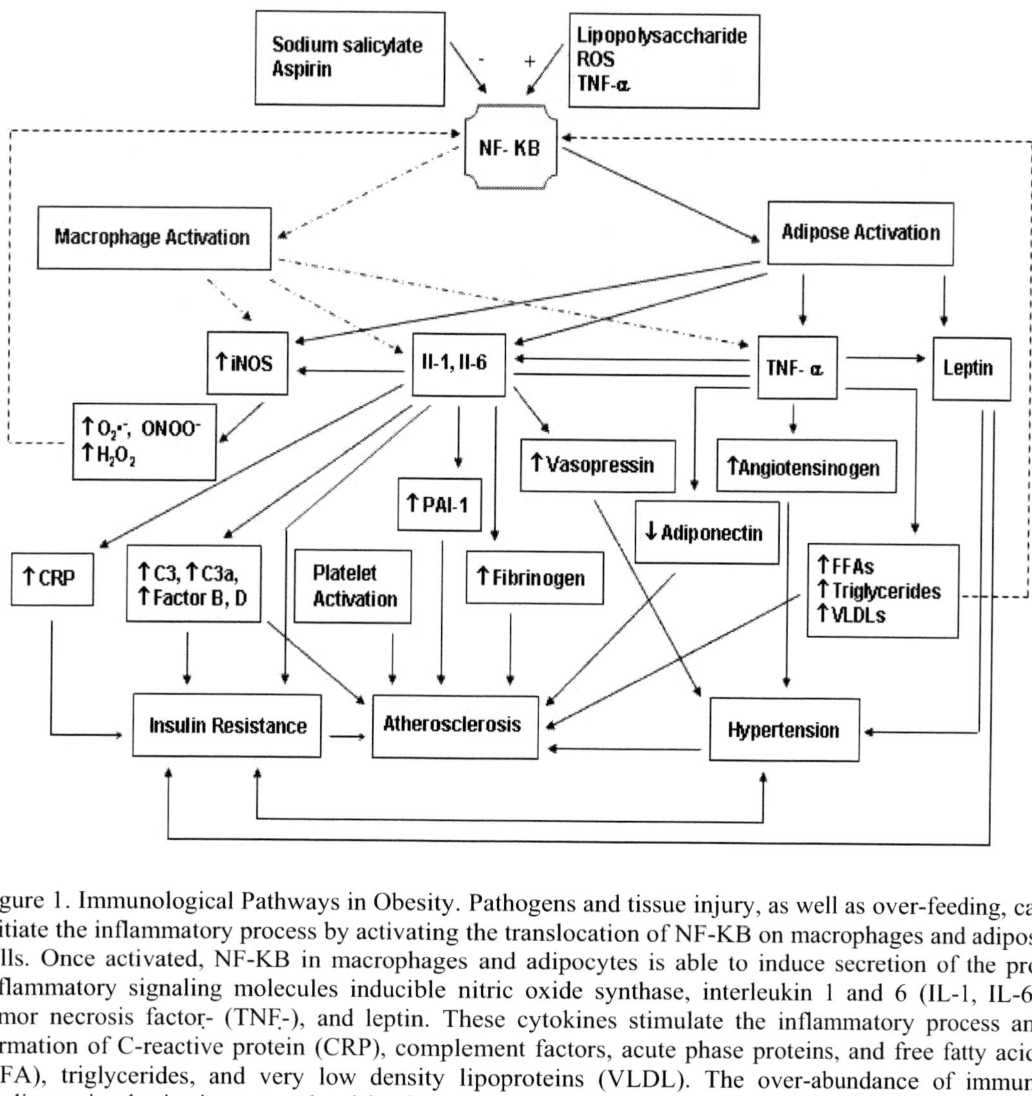

Figure 1. Immunological Pathways in Obesity. Pathogens and tissue injury, as well as over-feeding, can initiate the inflammatory process by activating the translocation of NF-KB on macrophages and adipose cells. Once activated, NF-KB in macrophages and adipocytes is able to induce secretion of the pro-inflammatory signaling molecules inducible nitric oxide synthase, interleukin 1 and 6 (IL-1, IL-6), tumor necrosis factor- (TNF-), and leptin. These cytokines stimulate the inflammatory process and formation of C-reactive protein (CRP), complement factors, acute phase proteins, and free fatty acids (FFA), triglycerides, and very low density lipoproteins (VLDL). The over-abundance of immune mediators in obesity increases the risk of insulin resistance, hypertension, and cardiovascular disease. The resulting reactive oxygen species, and free fatty acids may create a positive feedback loop, activating additional NF-KB, leading to auto-toxicity, and differentiation of mature adipocytes. Sodium salicylate and aspirin decrease the inflammatory response by preventing activation of NF-KB.

In response to these cytokines, hepatocytes synthesize and secrete acute phase proteins, primarily C-reactive protein (CRP), complement system proteins C3, C9, and Factor B, as well as serum amyloid A, plasminogen activator inhibitor-1, fibrinogen, and angiotensinogen. At the same time, other plasma protein concentrations, including albumin, transferrin, and insulin-like growth factor I decline in inflammation. These important changes in cytokine and plasma protein levels are referred to as acute phase responses, although they also occur in chronic inflammation. Synthesis of acute phase proteins is generally enhanced by glucocorticoids, while insulin usually decreases the effects of cytokines on acute phase protein synthesis.[12] Since the serum levels of acute phase constituents, and their physiologic effects, can vary in different inflammatory conditions, it is possible that various combinations or patterns of cytokine production, rather than the specific products, govern immune responses.[12]

CRP is an especially sensitive, but non-specific, marker of systemic inflammation. Factors known to increase serum CRP concentrations include advancing age, cigarette smoking, estrogen use, and inflammatory diseases, such as rheumatoid arthritis. CRP's functions include acting as an opsonin to target some microbes and damaged cells for phagocytosis, stimulating production of cytokines in monocytes, and activating the complement cascade.[12] Contributing to immunologic defenses are the factors of the complement system, a constellation of proteins that coat microbes and damaged cells, marking them for destruction and clearance by phagocytic cells. The classical pathway (Figure 2) requires the host to have had prior exposure to a specific organism, in order to develop specific antibodies able to initiate the process through the formation of antigen-antibody complexes. Upon repeat exposure to the antigen, antigen-antibody complexes form and bind to complement factor 1 (C1), triggering a series of enzymatic reactions. These reactions culminate in the formation of the terminal components of the complement system, C6, C7, C8, C9, and the membrane attack complex, which is able to pierce the cell walls of microorganisms.[13] The influx of water and ions into the perforations destroys the targeted cell.

The alternative pathway (Figure2) protects against pathogens that are unfamiliar to the host, and which cannot evoke specific antibodies. C3 can be spontaneously activated by water, so that C3 is split to form C3b. Small amounts of C3b are always present in the circulation, able to bind to microbial surface molecules, such as polysaccharides from bacterial cell walls. C3b is itself capable of binding Factor B, forming a new C3 convertase to cleave more C3, rapidly amplifying the amount of C3b adhering to the cell surface, and producing large amounts of C3a (anaphylatoxin) in the process.[13] Anaphylatoxin promotes mast cell destruction, and is a powerful chemotactic factor, recruiting additional neutrophils and monocytes to the area. Among the acute phase markers of inflammation are complement components C3 and Factor B. Like the classical pathway, the alternative pathway ends with the creation of the membrane attack complex, and the lysis and death of the invading microorganism.

In addition to cytokine and acute phase protein mediators, innate immune responses often create oxidative stress by generating nitric oxide (NO) and reactive oxygen species (ROS) to defend the host against injury and infection. Macrophages activated by tissue damage, hypoxia, or infection produce inducible nitric oxide synthase (iNOS or NOS2), leading to the generation of very high concentrations of NO from the amino acid L-arginine.[14-16] The iNOS isoform differs from the neuronal and endothelial NOS isoforms, in that it requires stimulation by cytokine or microbial product to initiate its expression. The resulting abundant supply of long-lasting NO combines with oxygen to form large amounts of perioxynitrite ($ONOO^-$), which is directly cytotoxic to many pathogens and some human tumor cells.[16] Furthermore, iNOS in macrophages produces significant amounts of superoxide (O_2^-•) when L-arginine is depleted, as at inflammatory sites during macrophage infiltration and wound healing. When macrophage L-arginine is low, both NO and O_2^-• are generated, leading to high $ONOO^-$ production.[15] Peroxynitrite degrades into hydroxyl radicals (OH^-•), the most reactive species, and a major cause of cellular and tissue damage in inflammation. ROS aid in destroying microbes, but risk damaging healthy host cells by disrupting lipid membranes, nucleic acids, and proteins, particularly essential enzymes. Oxygen free radicals may themselves initiate the inflammatory process.[13]

In response to infections or tissue injury, the increase in iNOS and NO is usually transient. However, in some conditions, such as chronic infection, iNOS may not be adequately inhibited, and persistent production of NO and ROS may perpetuate a chronic state of inflammation. Overproduction of iNOS-mediated O_2^-• and $ONOO^-$ may lead to unchecked auto-toxicity.

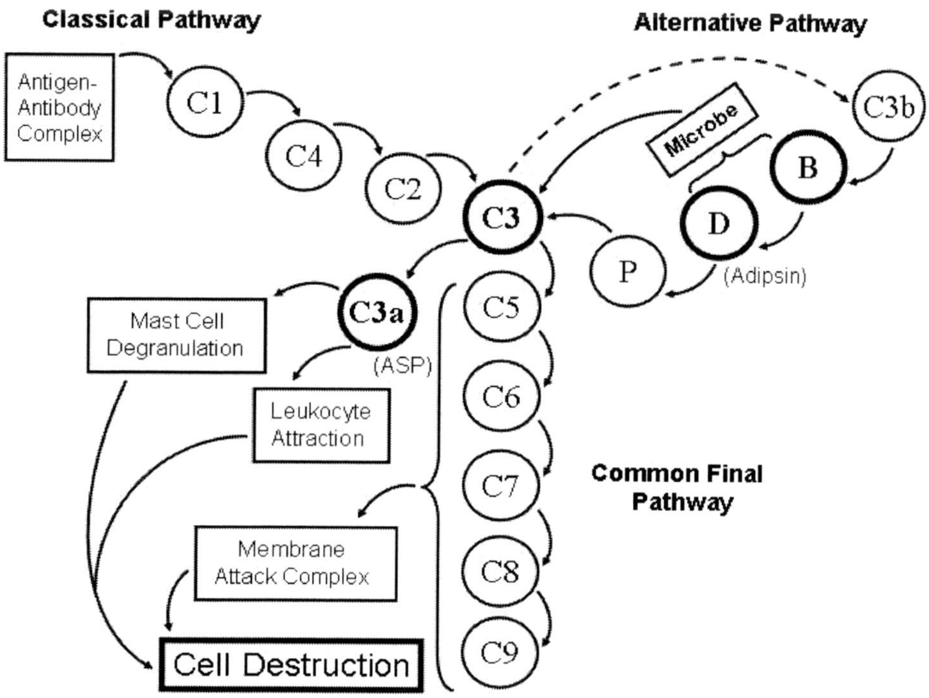

Figure 2. The Complement Cascade. Four molecules of the alternative complement pathway are produced by adipocytes, Factors C3A, C3, and B and D. Initiated by pathogens, the alternative pathway results in the production of large amounts of acylation-stimulating protein (ASP or C3a), an anaphylatoxin, which promotes mast cell destruction and leukocyte chemotaxis to kill the invading microbe. The alternative pathway terminates with the formation of the membrane attack complex that perforates and disrupts the pathogen's cell membrane. Adipsin activates C3 convertase to generate more C3, amplifying the production of anaphylatoxin.

Both NO and peroxynitrite inhibit the activity of cis-aconitase, an enzyme of the tricarboxylic acid cycle, by binding to aconitase's iron-sulfur center. This inhibition of aconitase blocks the first step of the TCA cycle, the conversion of citrate to isocitrate, and prevents the formation of nicotinamide-adenine dinucleotide (NADH), which is necessary to drive oxidative phosphorylation.[15] Blocking aconitase and disrupting the tricarboxylic acid cycle severely impairs a cell's ability to maintain its pool of adenosine triphosphte (ATP); this is an effective defense against microbes, but is potentially harmful to healthy host cells.

Finally, an important response to cellular injury involves fatty infiltration. TNF-α stimulates hepatic and adipose fatty acid synthesis.[17-19] The mobilization of lipids may provide additional substrates to fuel the energy demands of host defenses, or serve as a source of lipoproteins needed for cellular repair; FFAs may also bind to lipopolysaccharides, neutralizing pathogens.[12] Unfortunately, when injured cells are unable to metabolize nutrients properly to produce ATP, or when normal pathways for metabolizing fat are interrupted, fatty infiltrates develop when fatty acids and triglycerides accumulate within the cell cytosol. Excess lipid accumulates in adipocytes, as well as in other cells, such as hepatocytes, skeletal muscles, fibroblasts, and vascular endothelial and smooth muscle cells, compromising normal cell function (lipotoxicity).

Oxidative stress and lipotoxicity have been implicated as critical pathological features of insulin resistance and diabetes. Due to the low availability of oxygen radical metabolizing enzymes in pancreatic islets, the beta cells are more vulnerable to oxidative stress than other tissues.[20] Furthermore, intracellular accumulation of fatty acids or their metabolites in the muscle and liver have been found to play an important role in the pathogenesis of insulin resistance and type 2 diabetes.[21]

Adipose Tissue and Innate Immunity

Researchers have discovered preadipocytes and adipocytes secrete an array of cytokines and acute phase proteins that are essential elements of innate immunity. Secretion of these products is often elicited under circumstances that normally evoke their production by monocytes and macrophages. In addition, preadipocytes exert phagocytic and microbicidal functions in vitro, and these immunologic activities are most pronounced when preadipocytes are proliferating, whether they differentiate to mature adipocytes or not.[22] The enhanced generation of the components of innate immunity by adipose tissue forms key intersections between obesity and atherosclerosis, hypertension, insulin resistance and type 2 diabetes (Figure 2).

The discovery of leptin, the product of the ob/ob gene,[23] abolished the conceptualization of adipocytes as simple energy reservoirs, and gave rise to a new understanding of adipocytes as secretory cells with regulatory influences locally and systemically. Leptin is secreted by adipocytes, and first excited interest because it decreases food intake, increases fatty acid oxidation, elevates energy expenditure, and lowers fat mass.[24] However, leptin's regulatory effects are much broader. Leptin influences normal neuron and glial development,[25] sexual maturation and reproduction, [24,26] and angiogenesis.[27-28] Through its stimulatory effect on the sympathetic nervous system, leptin also increases arterial blood pressure. [29]

During starvation in mice and humans, leptin gene expression drops dramatically, and the higher the initial leptin level, the steeper the decline in gene expression with weight loss.[26,30] It seems, therefore, that leptin's actions are the consequence of relative changes in leptin rather than absolute leptin levels. Furthermore, the reduction in leptin occurs prior to any appreciable loss of adipose tissue, indicating that the changes in leptin and fat mass are independent of each other, and are subject to other, as yet, unidentified signaling pathways.[26]

Leptin also exerts influence over both innate and acquired or adaptive immunity. Production of leptin is increased in response to adipose stimulation by LPS and the proinflammatory cytokines TNF-α and IL-1.[31] Leptin selectively amplifies primary T lymphocytes, but not memory T cells, by stimulating proliferative and anti-apoptosis activities, and is necessary for cell-mediated immunity.[30] CD4+ cells orchestrate two patterns of cytokine production. The Th1 response is proinflammatory, and involves secretion of interleukin-2 (IL-2) and interferon-gamma (IFN-γ), while the Th2 response is generally regulatory and produces IL-4. Leptin promotes Th1 responses with the production of IL-2 and IFN-γ, but completely inhibits regulatory IL-4.[30] Stimulation of monocytes by leptin also activates the production of ROS (oxidative burst), consistent with a proinflammatory role for adipocytes in the immune system.[32] Leptin secretion is usually high in obesity, and it has been suggested the high leptin levels represent a resistance to the adipokine's anorexigenic effect.[24,33,34]

Adiponectin, an anti-inflammatory factor, is a second protein secreted by adipocytes. With a high homology to complement factor C1q, adiponectin suppresses adhesion molecule expression in vascular cells, limiting the ability of macrophages and neutrophils to adhere to the endothelium. By inhibiting induction of NF-KB signaling, adiponectin suppresses cytokine production and phagocytic activity in macrophages, and inhibits vascular inflammation.[35-36] Thus, adipocytes are able to mediate pro-inflammatory or anti-inflammatory processes by adjusting secretion of leptin and adiponectin.

Serum adiponectin levels are reduced in obesity, but increase with weight loss following gastric partition surgery,[37] and by thiazolidinediones (TZDs), known PPAR-γ ligands.[38,39] Researchers have shown an inverse relationship between low adiponectin levels in obesity and serum triglycerides, percentage of body fat, and high plasma IL-6 and CRP levels.[37] TNF-α decreases adiponectin levels in a dose-dependent manner.[34] Low plasma adiponectin levels have been demonstrated in a number of disorders characterized by inflammation, including coronary artery disease, type 2 diabetes, and insulin resistance.[35,36] This suggests that, like serum albumin, adiponectin is a negative acute phase protein.

Adipocytes are also a rich source of acute phase proteins. Many of the cytokines produced by macrophages are also secreted by adipocytes and exert parallel immunologic effects. For example, macrophage colony stimulating factor (MCSF) is a cytokine that is both secreted by macrophages and is capable of inducing the proliferation, differentiation, and maturation of new macrophages. However, MCSF is also secreted by adipose cells, inducing differentiation of preadipocytes to adipocytes and promoting adipose hyperlplasia. [40] Among the conditions known to be capable of stimulating MCSF secretion by adipocytes are overfeeding and acute bowel inflammation.[40]

Properly activated, adipocytes secrete pro-inflammatory TNF-α. This cytokine stimulates lipolysis, increases serum free fatty acid (FFA), triglyceride, and very low density lipoprotein (VLDL) concentrations, and is associated with increased peripheral insulin resistance.[40-43] FFAs, triglycerides, and VLDLs derived from lipolysis support the inflammatory process by activating production of iNOS and NO. Researchers have found that the addition of FFAs to cultured islet cells, upregulated iNOS mRNA, increased NO generation, and reduced insulin secretion; these effects were significantly greater for obese prediabetic fa/fa Zucker rats, and their fa/+ lean littermates, than lean Wistar rats.[44] This work demonstrates that one pathway to insulin resistance and diabetes is the induction of NO by free fatty acids, and this pathway can be a reflection of the interaction between genetic predisposition and dietary factors, as well as between genetic constitution and inflammation, since iNOS is genetically controlled.[14] Dietary intake, however, is not the body's only source of free fatty acids, since TNF-α stimulates hepatic fatty acid synthesis. Troglitazone, a PPAR-γ ligand and a thiazolidinedione derivative (TZD), suppresses iNOS expression in adipocytes.[45] Although TZDs often promote weight gain among patients when used to manage Type 2 diabetes, they improve insulin sensitivity and blood glucose levels, at least in part, by lowering a major mediator of inflammation.

Increased IL-6 production by adipocytes, induced by TNF-α, in turn stimulates the proliferation and maturation of T and B lymphocytes, and promotes B lymphocytes' synthesis of immunoglobulins. IL-6 also leads to increased production of colony stimulating growth factors that promote monocyte, macrophage, and granulocyte development. Increased IL-6 secretion by adipocytes helps to explain why Neiman and others found obese adult subjects had higher total leukocytes, neutrophils, monocytes, and T lymphocytes than nonobese subjects.[46]

IL-6, which is increased in obesity, suppresses thyroid stimulating hormone (TSH) production and reduces triiodothyronine concentrations. Lower thyroid function, even if it remains within a clinically normal range, will result in relative energy conservation, while producing two common symptoms of infection and systemic inflammation, namely excessive fatigue and somnolence.[47] The increased secretion of IL-6 contributes to the excessive fatigue and low physical activity common among those who are obese.

Increased IL-6 in obesity also stimulates the secretion of vasopressin, a potent vasoconstrictor, and can cause the syndrome of inappropriate antidiuretic hormone secretion.[48] The resulting vasoconstriction and sodium and water retention raise the blood pressure, contributing to the risk of hypertension in obesity. Adipose-derived IL-6 also induces production of C-reactive protein and fibrinogen in obesity, and these mediators are known risk factors for vascular disease.[3,4,49]

Normally, glucocorticoids, secreted in response to ACTH, provide a negative feedback signal to dampen the inflammatory response, and decrease IL-6 secretion. In obesity, cortisol levels are elevated, yet the inflammatory responses remain activated, as indicated by the persistently high cytokines, acute phase proteins, complement factors, and leptin concentrations. Furthermore, the increased glucocorticoid concentration may have an unfavorable influence on obesity, since glucocorticoids promote adipogenesis,[50] and favor central adipose development.

Adipose tissue produces several proteins of the alternative complement cascade, including C3, Factor B, and adipsin (Factor D), as shown in Figure 2. From a two-step reaction, these factors can also generate acylation-stimulating protein (ASP), which is homologous with anaphylatoxin, complement factor C3a. ASP promotes triglyceride synthesis in adipocytes.[51,52] Increased ASP has been correlated with coronary artery disease,[52] while complement factor C3, but not ASP, has been linked to hyperinsulinemia.[53]

Dietary indiscretion and sedentary lifestyles are frequently cited as the principal causes of the dramatic rise in obesity in contemporary society.[54,55] Special emphasis is placed on the role of high dietary fats in the development of obesity, and over-feeding laboratory animals a diet high in fat is an established method for investigating the physiology of obesity. Experimental infusion of a lipid emulsion and heparin in healthy non-obese subjects resulted in elevated serum free fatty acids, followed by increased NF-KB binding activity in mononuclear cells, increased ROS production by mononuclear and polymorphonuclear leukocytes, and reduced post-ischemic brachial artery dilation.[56] Similarly, ingestion of a 900-kcal mixed meal was followed by increased NF-KB binding activity in mononuclear cells, increased ROS production by mononuclear cells and polymorphonuclear leukocytes, and increased serum CRP concentrations in normal-weight subjects.[57] These studies demonstrate that dietary factors are capable of eliciting pro-inflammatory responses in normal, healthy individuals.

Free fatty acids can both activate NF-KB and increase TNF-α production, initiating the cascade of inflammatory reactions. Therefore, it is possible for prolonged, high-fat dietary intake to give rise to the chronic state of inflammation seen in obesity. On the other hand, inflammatory processes and elevated TNF-α secretion lead to hepatic synthesis of FFAs, and a state of chronic inflammation resulting from subclinical infection or other pro-inflammatory stimuli may promote adipogenesis. As Figure 1 shows, a positive feedback loop may develop in support of inflammation, with adipose-derived cytokines mediating the process.

High concentrations of salicylates, including aspirin, inhibit the NF-KB activation and nuclear translocation needed to produce TNF-α and IL-6. Blocking NF-KB translocation by high doses (120mg/kg/day) of sodium salicylate or aspirin dramatically lowered insulin resistance in Zucker fatty rats, as well as ob/ob mice, and reduced free fatty acid and triglyceride levels, while nonsteroidal anti-inflammatory agents that target cyclooxygenase (COX1 or COX2) failed to achieve these effects.[58] Similarly, in human adult diabetic patients who were treated with high doses (~7g/d) of enteric-coated aspirin for two weeks, fasting blood glucose, serum cholesterol, and triglyceride levels were significantly reduced, and high density lipoproteins were increased.[59]

A number of studies have attempted to determine the effects of dietary restraint and weight loss on cytokine concentrations, particularly levels of the leptin, TNF-α, IL-6, and CRP. In a study of middle-aged women who lost weight following gastric banding, investigators found that one year after surgery, the women had a significant reduction in leptin and CRP from pre-banding levels; however, they found no significant reduction in TNF-α and IL-6. Concomitantly, fasting serum insulin and triglyceride levels were significantly lower, but total cholesterol, high density lipoproteins, and low density lipoprotein levels were not significantly altered from their pre-surgery concentrations.[60] Other researchers have found reductions in CRP following weight loss resulting from a 12-week, very-low-fat, energy restricted diet,[61] but did not measure TNF-α and IL-6 levels. In another study, insulin sensitivity improved and leptin and IL-6 levels were lowered in obese women after three weeks of a very low calorie diet; however, serum and subcutaneous TNF-α concentrations were unchanged.[62] These studies suggest that CRP and leptin levels may be significantly reduced by weight loss, corresponding to the reduction in fat mass, but that TNF-α levels may continue to be elevated in obesity, despite substantial weight loss.

INNATE IMMUNITY PRECURSORS TO OBESITY

With the relationships between adipose-derived inflammatory markers and the risk of insulin resistance and vascular disease becoming more firmly established, the pathways between dietary-induced obesity and obesity co-morbidities are also becoming clear. At the same time, these findings prompt new questions about the possible roles for non-dietary induced immune responses and inflammation in the pathogenesis of obesity. Early evidence indicates pro-inflammatory molecules are antecedents to weight gain and obesity.

Leptin concentrations tend to be higher in obese individuals than in non-obese people, but its role in regulating adiposity is not yet clear. In a study of Pima Indians, relatively lower leptin levels were related to subsequent increases in body fat,[63] whereas in research involving non-obese Japanese-Americans, increased leptin concentrations were associated with fat accumulation.[64] Among obese Caucasian and African-American children and adolescents, girls with high serum leptin levels for their BMI at baseline gained more weight over 2½ years than girls with lower leptin levels. Plasma leptin levels did not predict BMI changes among boys over the same time period.[65]

In an investigation of smoking cessation among a cohort of the Atherosclerosis Risk in Communities (ARIC) Study, researchers found, as expected, subjects who quit smoking gained more weight than adult subjects who continued to smoke over a period of three years.[66] They also found that among the smokers who quit, those with the highest leukocyte counts and fibrinogen concentrations had approximately 50% greater weight gain than quitters with the lowest levels of leukocytes and fibrinogen. Quitters in the highest quartile of leukocyte and fibrinogen levels also had approximately double the odds of a large weight gain ($\geq 90^{th}$ percentile) than those in the lowest quartile for these factors. Although both the leukocyte and fibrinogen levels were within clinically normal limits, the results support a role for chronic, low-grade systemic inflammation in weight gain.

Given the associations between inflammatory markers and the pathogenesis of metabolic syndrome and cardiovascular disease, a study based on data from the ARIC project was conducted to determine whether inflammation predicted weight gain.[67] Markers of low-grade inflammation were shown to be predictors of weight gain among 13,017 middle aged, white and African American men and women over a three year period. After controlling for age, sex, ethnicity, baseline height and weight, cigarette smoking, presence of diabetes, and fasting insulin, analyses revealed fibrinogen, white blood cell count, Factor VIII, and von Willebrand factor, were predictive of weight gain among the population studied. The researchers concluded that mild, chronic inflammatory processes contribute to the pathogenesis of the metabolic syndrome and cardiovascular disease through stimulation of weight gain.[67]

The C-reactive protein level is related to the body mass index and is considered an important risk factor for insulin resistance and cardiovascular disease. However, CRP is also related to weight fluctuation, that is weight loss followed by weight regain. In a study of Japanese men,[66] researchers found an increased prevalence of elevated CRP levels with increasing body mass index, after adjusting for possible confounders, including age, smoking, stroke, myocardial infarction, diabetes mellitus, and asthma. These findings again support the presence of a state of low-grade systemic inflammation in the overweight and obese. Although the researchers made no attempt to distinguish between intentional and unintentional weight loss, they also evaluated the odds ratio of elevated CRP levels with changes in the subjects' recorded weights at ages 20, 25, 30, and five years prior to the study. They found the OR of elevated CRP levels were associated with increasing weight fluctuation, independent of the subjects' final BMI.[68]

Investigating the possible association of bacterial infection with cardiovascular disease, researchers discovered linkage between bacterial infection and obesity. Dart and others analyzed antibodies to *Chlamydia pneumoniae* (IgG seropositive) and coronary heart disease (CHD) risk factors in 170 subjects, including 43 with angina of recent onset.[69] They found no difference in angina or CHD between seropositive and seronegative subjects. However, subjects who were seropositive for *C. pneumoniae* had significantly higher BMI than subjects who were seronegative.[69]

Other researchers have found linkages between infections and obesity in both animal and human studies. In India, researchers discovered the SMAM-1 virus, an avian adenovirus, produces excessive accumulation of visceral adipose tissue in chickens.[70] A follow-up study involving human subjects showed that 19.2% of obese subjects had antibodies to the SMAM-1 virus, and the antibody-positive subjects had higher body weights than the antibody-negative participants.[71]

Investigators have shown the human adenovirus-36 (Ad-36) causes increased adiposity in chickens, mice, marmosets, and rhesus monkeys inoculated with the virus.[72,73] Animals inoculated with the Ad-36 virus developed greater total body fat and visceral fat when given the same food portions as uninfected control animals. Viral DNA was detected in the blood and adipose tissue, but not the hypothalamus, of the animals at the time of sacrifice. Like the marine obesity caused by canine distemper virus, Ad-36 appears to cause increased adiposity through peripheral, rather than central, processes.

Associations between Ad-36 and human obesity also have been found. One study revealed 32% of obese subjects with a BMI greater than 27, tested positive for antibodies to Ad-36, compared with only 4.3% of non-obese individuals who tested positive for Ad-36 antibodies.[74] To eliminate the possible alternative explanation that these findings were due to higher susceptibility to infection among obese people than those who are not obese, researchers analyzed differences in antibodies to three other adenoviruses, Ad-2, Ad-31, and Ad-37, among the obese and non-obese subjects. They found the two groups of subjects were no different in their antibodies to these three viruses, indicating the differences in Ad-36 antibodies were unlikely to be due to immunocompromise in the obese subjects. Researchers have also tested 86 pairs of identical twins for Ad-36 antibodies. In 26 twin pairs, where one twin was Ad-36 antibody positive and the other twin was antibody negative, the seropositive twins had significantly higher BMIs.[75]

The small body of research focusing on immunologic factors in the development or progression of obesity points to obesity as an inflammatory disorder. The research includes a number of animal models of virus-induced obesity syndromes, and hints at several microorganisms that may be important in the etiology of human obesity. How adipose tissue and energy regulation may be co-regulated with innate immunity is worth considering in order to understand why increased adiposity may result from immunologic challenge.

Immunologic Pathways to Obesity

What is often overlooked in research based on dietary-induced obesity, is the influence immune responses have on energy consumption and energy expenditure, the key determinants of obesity. According to Hennessey, "The immune response is quite energetically expensive,"[26, p. 16] and increases oxygen consumption. Conversely, disruption of cellular energy production can trigger an inflammatory response. Cooperative signaling between the immune cells, adipose tissue, and the neuroendocrine regulatory centers to assure adequate energy to meet basal demands, and to mount immunological defenses against the environment, is crucial for the host's survival. Metabolic and immunologic pathways may converge in at least three ways that will promote adipogenesis.

The first convergence of immunity and metabolism occurs through direct pathogen-associated molecular activation of preadipocyte and adipocyte toll-like receptors (Figure 1). Activation of the lipopolysaccharide (LPS) receptor on macrophages induces translocation of NF-KB to the nucleus where it regulates the expression of proinflammatory cytokines. Mouse adipocytes express Toll-like receptors 2 and 4 (Tlr-2 and Tlr-4), and the Tlr-4 receptor serves as a signaling receptor for bacterial lipopolysaccharide, as it does in macrophages.[76] Additionally, direct stimulation of murine adipocytes with LPS leads to increased IL-6 expression.(76) Investigators also have demonstrated that in primary pig adipocytes, LPS activates NF-KB, leading to increased expression and secretion of interleukin-6 (IL-6).[77] Although Interferon-γ (IFN-γ) failed to increase expression of TNF-α or IL-6 in cultured pig adipocytes, it did stimulate expression of IL-15. IL-15 is a critical growth factor in initiating primary T cell proliferation *in vivo*.[78]

Direct toll-like receptor stimulation may account for the adenovirus induced obesity observed in animal models. Researchers demonstrated that the addition of the Ad-36 virus to preadipocyte cultures leads to increased adipocyte differentiation in laboratory cultures of preadipocytes.[79] Adipocyte differentiation in the presence of AD-36 is likely due to stimulation of toll-like receptors that leads to activation of NF-KB.

Wlodek and Gonzales have postulated a second pathway to immunity-related obesity is through disruption of the tricarboxcylic acid (TCA) cycle production of adenosine triphosphate (ATP).[80] By increasing secretion of TNF-α, and activation of iNOS, immunogenic factors generate high concentrations of NO, leading to oxidative stress in the mitochondria. NO and its reactive oxygen species derivatives suppress enzyme-driven reactions essential to the formation of ATP. Aconitase, the enzyme responsible for the conversion of citrate to isocitrate, is particularly sensitive to nitric oxide inhibition. Moreover, diminished energy production resulting from aconitase inhibition may play a unique role in the development and perpetuation of obesity, because citrate conversion to isocitrate is the pivotal step in which fuel substrates are directed to fat synthesis or ATP production.[80] If ATP levels are adequate, citrate is transported out of the mitochondria and committed to fatty acid synthesis in the cytosol; when ATP is low, citrate is converted to isocitrate and dedicated to ATP formation. Inhibiting aconitase at this crucial step will misdirect citrate to lipogenesis and deplete cellular energy, resulting in two unrelenting features of obesity: fatigue and hunger. Driven by low ATP availability, the subject will seek additional food (particularly energy dense foods) to supply the "missing" energy, and will curb energy expenditure.

Through the activation of pro-inflammatory cytokines, a number of environmental factors may contribute to weight gain by interfering with the action of aconitase in the tricarboxcylic acid cycle. The tricyclic antidepressants amitriptyline and nortriptyline often have weight gain as an undesirable side effect, and this weight gain is preceded by increased TNF-α levels.[81] Continuous, but not intermittent, use of nitroglycerine can result in significant weight gain, with nitric oxide suppression of aconitase the likely pathway.[82] Environmental contaminants such as 2, 3,7, 8-tetrachlorodibenzo-p-dioxin (TCDD) and dichloroacetonitrile (DCAN), a by-product of a common water disinfectant, also elevate TNF-α production.[80]

Since mercury promotes formation of hydrogen peroxide and hydroxyl radicals,[83] chronic mercury exposure derived from fish, dental amalgams, and vaccines may contribute to aconitase inhibition and weight gain. There is tangential evidence to support the role of mercury in adipose accretion and its cardiovascular complications. In a study of eastern Finnish men, those men who had consumed freshwater, nonfatty fish were found to have elevated hair mercury content.[84] Dietary fish intake and hair mercury levels were associated with significantly increased risk of acute myocardial infarction, and death from coronary heart disease and cardiovascular disease, compared to men with lower hair mercury content. In a nested case-control subsample from the study, the 24-hour urinary excretion of mercury was independently associated with the risk of acute MI. Furthermore, both the hair and urinary mercury levels were positively associated with titers of immune complexes containing oxidized LDL.[84] Although findings regarding the subjects' weight or BMI characteristics were not reported, the study does indicate mercury level was associated with dyslipidemia, an important feature of obesity. In a case-control study carried out in eight European countries and Israel, [85] researchers analyzed the association of mercury levels in toenail clippings and the risk of a first myocardial infarction among 684 male patients and 724 male controls. When compared to the controls, the patients had significantly higher body mass index, lower HDLs, and were more likely to have hypertension or diabetes, to smoke, and to have a family history of myocardial infarction. After adjustment for age, research center, and coronary risk factors, the mercury levels in the patients were 15 percent higher than those in controls. The odds ratio for myocardial infarction associated with the highest, compared to the lowest, quintile of mercury was 2.16, after controlling for coronary risk factors.[85] The study provides indirect linkage between mercury level and BMI. However, other researchers [86] found no relationship between mercury and the risk of heart disease among men, and provided insufficient information to evaluate possible relationships between mercury exposure and anthropometric characteristics among the subjects. The relationship between mercury and obesity warrants direct investigation.

Based on available research, chronic exposure to infectious agents, medications, and industrial chemicals can interfere with energy metabolism by stimulating cytokine and nitric oxide activity. Until recently, concern about environmental exposures have focused on their possible relationships to neurological degenerative disorders, congenital anomalies, and cancer risk rates, with little attention given to the role these agents may play in the development of obesity. Whether these exposures may account for the rise in the obesity in industrialized nations deserves more research.

A third way in which acute and chronic infections may promote adipogenesis is through suppression of energy consumption, followed by refeeding. Dulloo, Jacquet, & Montani have explained that adipose tissue mass is regulated by dual feedback mechanisms that assures maximum protection of fat stores in response to dietary restriction.[87]) When fat mass is reduced as a consequence of restricted or inadequate energy intake, cellular energy expenditure (thermogenesis) will be lowered, primarily in skeletal muscles. When dietary intake is restored, it will preferentially replenish fat stores at a disproportionately faster rate than the recovery of lean tissue (muscle) mass. The reduction in energy expenditure persists after adequate nutrition is restored, accelerating the accumulation of adipose stores, even when the diet is low in fat. Suppressed thermogenesis during refeeding also leads to decreased blood flow to muscle tissue, and can produce excess circulating free fatty acids and lipid deposition in skeletal muscle.[87] These compensatory changes, while initially adaptive, increase the person's risk of obesity, hypertension, heart disease, and diabetes.

Although excessive adipose recovery (catch-up fat) is exacerbated by a diet high in fat, it has been documented in a variety of conditions, including "on well-balanced diets, often low in fat content, and independently of the level of dietary energy, protein, mineral, and vitamin supplementation."[87] Disproportionate adipose recovery following energy deprivation is characteristic in low-birth weight infants who "catch-up" in growth, independent of obesity. It has also been described in tuberculosis, septic shock, and human immunodeficiency virus infection. Excess fat recovery has also been studied in otherwise healthy subjects who have undergone modest, temporary energy restriction. According to Dulloo, Jacquet, and Montani, energy restriction, particularly when it occurs during critical periods of growth and development, can lead to lasting alterations in the structure or function of tissues and body systems. Such malnutrition-induced "programming" or "imprinting", although adaptive during the period of limited supply of nutrients, is thought to contribute to increased risks for diseases during improved nutrition later in life.[87]

The regulatory processes underlying preferential adipose recovery have not been clearly established. Interestingly, refed animals respond to environmental cold like control animals, with increased sympathetic nervous system-activated energy expenditure through shivering, while still maintaining overall energy conservation favoring catch-up fat. These findings point toward an adipose-specific control of thermogenesis that is independent of the sympathetic nervous system mediated control elicited in response to cold, hyperphagia, or infections.[87] Adipose specific control of energy metabolism has been hypothesized to consist of a feedback loop between adipose tissue and skeletal muscles, capable of suppressing energy expenditure when fat stores are depleted. The signaling pathways for adipose-specific control of thermogenesis have not been identified, but research so far has not supported a role for leptin, or for uncoupling protein 1 or 2.[87] Inflammatory molecules secreted by adipose tissue, especially TNF-α, or IL-6 may play a role in the adipose-specific control of catch-up fat. TNF-α is a possible signaling candidate because it is secreted in abundance by adipose tissue, and is associated with elevated free fatty acids and insulin resistance in skeletal muscles. IL-6 merits study as a signaling molecule in catch-up fat because it suppresses TSH and lowers thyroid function, which would favor energy conservation and weight gain. Weight loss has been shown to result in a reduction in 24-hour energy expenditure mediated, at least in part, by decreased triiodothyronine secretion, and the reduction in thermogenesis can persist for several years, or until the initial weight is recovered.[88,89] Furthermore, the ability of IL-6 to stimulate vasopressin release, with its resultant vasoconstriction, fits the profile of a signaling molecule for catch-up fat, in which there is an increase in skeletal muscle vascular resistance.[87] Since energy restriction is followed by catch-up fat, the generation of ROS, free fatty acids, or VLDLs may initiate the process of adipose recovery by evoking the release of abundant adipocytokines. Research into the absolute and relative changes in cytokine secretion associated with obesity, weight loss, and weight regain, especially adipose recovery, is needed.

Based on the model of excessive adipose recovery, even small, transient periods of energy restriction could have profound effects on the development of obesity, particularly if these episodes occur vulnerable periods of growth and development, such as pregnancy and early childhood. Short episodes of calorie restriction may accompany many common minor infections, such as influenza, pharyngitis, and gastroenteritis. Thus, frequent or recurrent infections, as well as severe illnesses, have the potential to precipitate excessive adipose recovery.

CONCLUSION

Although the defining feature of obesity is the excess accumulation of adipose tissue, it is increasingly apparent that how regulation of energy consumption becomes dissociated with energy expenditure involves multiple genetic and environmental factors. What consistently links many of these factors to the development and progression of obesity and its comorbidities is the activation of immunologic processes ordinarily initiated to protect the host and enhance survivability. Furthermore, in two of the metabolic pathway models described above, the precipitating event is an energy *deficit* relative to cellular energy needs. In one model, the energy deficit stems from the failure of ATP production;[80] in the model of catch-up fat, restricted nutritional intake is followed by disproportionate adipose recovery when isoenergetic intake is restored.[87] This would suggest that the injury to cells from insufficient energy availability elicits an immunologic response that becomes amplified by adipose tissue. Paradoxically, obesity may ultimately stem from the availability of too little energy, rather than too much energy.

The increased secretion of TNF-α, IL-6, CRP, and other immune mediators indicate obesity is an inflammatory disorder. What is distinctive about obesity as inflammatory disorder, however, is the consistency with which adipose tissue mimics and even magnifies, the functions of macrophages. Adipose tissue is a major source of proinflammatory cytokines, and increased levels of adipose-derived cytokines and acute-phase proteins are associated with insulin resistance, type 2 diabetes, and cardiovascular disease. What is less clear is how regulatory mechanisms tilt in favor of excess adiposity. Feeding laboratory animals a high fat diet is a useful way for studying physiological processes and models of obesity, but may not accurately reflect the genesis of obesity in free-living human beings. Longitudinal studies have revealed that body weight is carefully regulated, body weight tends to be stable over years and decades, and is highly resistant to volitional change.[90-92] Energy excesses or deficits are balanced over weeks and months, and are generally offset by changes in energy expenditure, primarily through semi-automatic activity, rather than through deliberate physical exercise.[93] Although dietary glucose and fats are capable of stimulating the secretion of pro-inflammatory cytokines, especially TNF-α, many other environmental factors may be as important as dietary intake in triggering the inflammation associated with obesity, and may be key stimuli in the development of obesity itself. By activating NF-κB, non-dietary, as well as dietary, factors are able to activate inflammation and stimulate adipogenesis. Microorganisms may stimulate toll-like receptors, leading to adipocyte differentiation. Inhibition of aconitase and disruption of the TCA cycle by reactive oxygen species can result from a variety of drugs and environmental toxins, and acute and chronic infections may produce temporary nutrition deprivation that is followed by excessive adipose recovery upon restoration of adequate dietary intake. Furthermore, an interaction between infections or other macrophage-stimulated inflammatory processes may develop into a positive feedback loop, in which the generation of reactive oxygen species, in turn, activate NF-κB and adipogenesis. These mechanisms would help to explain the general failure of calorie restriction in the long-term control of the obesity problem in clinical practice.

Increased research into the relationships between immunologic factors and obesity is needed. The association between common infections – in both children and adults – and subsequent weight gain should be investigated to identify possible microbial causes of obesity. Likewise, the relationships between exposure to such ubiquitous chemical agents as mercury, fluoride, and chlorine and obesity should be studied. Research is also needed to determine if there are associations between the use of common medications, such as antibiotics, and the development of obesity. The possible role of cytokines and acute phase proteins in the signaling pathways mediating excessive adipose recovery following dietary restraint also needs to be investigated. Finally, studies are needed to map changes in cytokine and acute phase protein production associated with the development of obesity, weight loss, and weight regain.

REFERENCES

[1] Lyons MJ, Faust IM, Hemmes RB, Buskirk DR, Hirsch J, Zabriskie JB. 1982. A virally induced obesity syndrome in mice. *Science*, *216*:82-85.

[2] Vgontzas AN, Bixler, EO, Papanicolaou DA, Chrousos GP. 2000. Chronic systemic inflammation in overweight and obese adults. *Journal of the American Medical Association*, *283*:2235-2236.

[3] Visser M, Bouter LM, McQuillan GM, Wener MH, Harris TB. 1999. Elevated C-Reactive protein levels in overweight and obese adults. *Journal of the American Medical Association*, *282*:2231-2135.

[4] Yudkin JS, Stehouwer CDA, Emeis JJ, Coppack SW. 1999. C-Reactive protein in healthy subjects: Associations with obesity, insulin resistance, and endothelial dysfunction: A potential role for cytokines originating from adipose tissue? *Arteriosclerosis, Thrombosis, and Vascular Biology*, *19*:972-978.

[5] Xu H, Barnes GT, Yang Q, Tan G, Yang D, Chou, CJ, Sole J, Nichols N, Ross JS, Tartglia LA, Chen H. 2003. Chronic inflammation in fat plays a crucial role in the development of obesity-related insulin resistance. *Journal of Clinical Investigation*, *112* (12):1821-1830.

[6] Brightbill HD & Modlin RL 2000. Toll-like receptors: molecular mechanisms of the mammalian immune response. *Immunology*, *101*:1-10.

[7] Medzhitov R. & Janeway C. 2000. Innate immune recognition: mechanisms and pathways. *Immunology Review*, *173*:89-97.

[8] Bays H, Mandarino L, & DeFronzo RA. 2004. Role of the adipocyte, free fatty acids, and ectopic fat in pathogenesis of type 2 diabetes mellitus: Peroxisomal proliferator-activated receptor agonists provide a rational therapeutic approach. *Journal of Clinical Endocrinology and Metabolism*, *89*:463-478.

[9] Fink MP. 2004. Research: Advances in cell biology relevant to critical illness. *Current Opinion in Critical Care*, *10*:279-291.

[10] Seitz, M. 2003. Toll-like receptors: sensors of the innate immune system. *Allergy*, *58*:1247- 1249.

[11] Travis J. 2001. Immunity's Eyes: Biologists reveal the proteins that first see dangerous microbes. *Science News*, *160*:152.

[12] Gabay C & Kushner I. 1999. Acute-phase proteins and other systemic responses to inflammation. *New England Journal of Medicine*, *340*:449-454.

[13] Rote NS. 2002. Inflammation. In K. L. McCance & S. E. Huether (Eds.), *Pathophysiology: The biologic basis for disease in adults and children,* 4th ed. St. Louis: C.V. Mosby.

[14] Burgner D, Rockett K, & Kwiatkowski D. 1999. Nitric oxide and infectious diseases. *Archive of Diseases of Childhood*, *81*:185-188.

[15] Xia Y & Zwier JL. 1997. Superoxide and peroxynitrite generation from inducible oxide synthase in macrophages. *Proceedings of the National Academy of Sciences USA*, *94*:6954-6958.

[16] Droge W. 2002. Free radicals in the physiological control of cell function. *Physiological Reviews*, 82:47-95.

[17] Grunfeld C, Verdier JA, Neese R, Moser AH, & Feingold KR. 1988. Mechanisms by which tumor necrosis factor stimulates hepatic fatty acid synthesis in vivo. *Journal of Lipid Research*, *29*:1327–1335.

[18] Grunfeld C, Soued M, Adi S, Moser AH, Dinarello CA, & Feingold KR. 1990. Evidence for two classes of cytokines that stimulate hepatic lipogenesis: relationships among tumor necrosis factor, interleukin-1 and interferon-alpha. *Endocrinology*, *127*:46–54.

[19] Lopez-Soriano J, Argiles JM, & Lopez-Soriano FJ, 1995. Metabolic effects of tumour necrosis factor-alpha on rat brown adipose tissue. *Molecular and Cellular Biochemistry*, *143*:113-118.

[20] Hohmeier HE, Tran VV, Chen C, Gasa R, & Newgard CB. 2003. Inflammatory mechanisms in diabetes: Lessons from the β-cells. *International Journal of Obesity*, *27*:S12-S16.)

[21] Perseghin G, Peterson K, & Shulman GI. 2003. Cellular mechanism of insulin resistance: Potential links with inflammation. *International Journal of Obesity*, *27*:S6-S11.

[22] Cousin B, Munoz O, Andre M, Fontanilles AM, Dani C, Cousin JL, Laharrague P, Casteilla L, & Penicaud P. 1999. A role for preadipocytes as macrophage-like cells. FASEB Journal, 13:305-312.

[23] Zhang Y, Proecna R, Maffei M., Barone M, Leopold L, & Friedman JM. 1994. Positional cloning of the mouse obese gene and its human homologue. *Nature*, *372*:425-432.

[24] Friedman JM. & Halaas JL, 1998. Leptin and the regulation of body weight in mammals. *Nature*, *395*:763-770.

[25] Ahima RS, Bjorbaek C, Osei S, & Flier JS. 1999. Regulation of neuronal and glial proteins by leptin: implications for brain development. *Endocrinology*, *140*:2755–2762.

[26] Hennessey J. 2003. Leptin As Fat Orchestrator. *Reviews in Undergraduate Research*, *2*:12-18.

[27] Margetic S, Gazzola C, Pegg GG, & Hill RA. 2002. Leptin: A review of its peripheral actions and interactions. *International Journal of Obesity*, *26*:1407-1433.

[28] Bouloumie A, Drexler HCA, Lafontan M, & Busse R. 1998. Leptin, the product of the *ob* gene, promotes angiogenesis. *Circulation Research*, *83*:1059-1066.

[29] Beltowski J, Wojcicka G, & Jamroz A. (2003). Stimulatory effect of leptin on nitric oxide production is impaired in dietary-induced obesity. *Obesity Research*, *11*:1571-1580.

[30] Lord GM, Matarese G, Howard JK, Baker RJ, Bloom SR, & Lechler RI. 1998. Leptin modulates the T-cell immune response and reverses starvation-induced immunosuppression. *Nature, 394:*897-901.

[31] Fantuzzi G & Faggioni R. 2000. Leptin in the regulation of immunity, inflammation, and hematopoiesis. *Journal of Leukocyte Biology*, *68*:437446.

[32] Sanchez-Pozo C, Rodriguez-Bano J, Dominguez-Castellano A, Muniain MA, Goberna R, & Sanchez-Margalet V. 2003. Leptin stimulates the oxidative burst in control monocytes but attenuates the oxidative burst in monocytes from HIV-infected patients. *Clinical and Experimental Immunology*, *134*:464-9.

[33] Considine RV, Sinha MK, Heiman MI, Kriaucinas A, Stephens TW, Nyce MR, Ohannesian, JP, Marco CC, McKee LJ, Bauer TL, & Caro JF. 1995. Serum immunoreactive-leptin concentrations in normal-weight and obese humans. *New England Journal of Medicine*, *334*:292-295.

[34] Rajala MW & Scherer PE. 2003. Minireview: The adipocyte – At the crossroads of hemostasis, inflammation, and atherosclerosis. *Endocrinology*, *144*:3765-3773.

[35] Engeli S, Feldpausch M, Gorzelniak K, Hartwig F, Heintze U, Janke J, Mohlig M, Pfeiffer A, FH, Luft FC, & Sharma AM. 2003. Association between adiponectin and mediators of inflammation in obese women. *Diabetes*, *52*:942-947.

[36] Wiecek A, Kokot F, Chudek J, & Adsmczak, M. 2002. The Adipose Tissue – a novel endocrine organ of interest to the nephrologist. *Nephrology Dialysis Transplantation*, *17*:191- 194.

[37] Yang W-S, Lee W-J, Funashi T, Tanaka S, Matsuzawa Y, Chao C-L, Chen C-L, Tai T-Y, & Chuang L-M. 2001. Weight reduction increases plasma levels of adipose-derived anti- inflammatory protein, adiponectin. *Journal of Clinical Endocrinology and Metabolism*, *86*:3815-3819.

[38] Maeda N, Takahashi M, Funahashi T, Kihara S, Nishizawa H, Kishida K, Nagaretani H, Matsuda M, Komuro R, Ouchi N, Kuriyama H, Hotta K, Nakamura T, Shimomura I, Matsuzawa Y. 2001. PPARγ ligands increase expression and plasma concentrations of adiponectin, an adipose-derived protein. *Diabetes*, *50*:2094-2099.

[39] Yang WS, Jeng CY, Wu TJ, Tanaka S, Funahashi T, Matsuzawa Y, Wang JP, Chen CL, Tai TY, & Chuang LM. 2002. Synthetic peroxisome proliferator-activated receptor-γ agonist, rosiglitazone, increases plasma levels of adiponectin in type 2 diabetic patients. *Diabetes Care*, *25*:376 -380.

[40] Levine JA, Jensen MD, Eberhardt NL, O'Brien T. 1998. Adipocyte macrophage colony- stimulating factor as a mediator of adipose tissue growth. *The Journal of Clinical Investigation*, *101*:1557-1564.

[41] Hotamisligil GS, Shargill NS, Spiegelman BM. 1993. Adipose expression of tumor necrosis factor-alpha: Direct role in obesity-linked insulin resistance. *Science*, *259*:87-91.

[42] Pickup JC, Mattock MB, Chusney GD, & Burt D. 1997. NIDDM as a disease of the innate immune system: Association of acute-phase reactants and interleukin-6 with metabolic syndrome X. *Diabetologia*, *40*:1286-1292.

[43] Bullo M., Garcia-Lorda P, Megias I, & Salas-Salvado J. 2003. Systemic inflammation, adipose tissue tumor necrosis factor, and leptin expression. *Obesity Research*, *11*:525-540.

[44] Shimabukuro, M., Ohneda, M., Lee, Y., & Unger, R. H. (1997). Role of nitric oxide in obesity-induced beta cell disease. *Journal of Clinical Investigation*, *100* (2): 290-295.

[45] Dobashi, K., Asayama, K., Nakane, T., Kodera, K., Hayashibe, H., & Nakazawa, S. (2000). Troglitazone inhibits the expression of inducible nitric oxide synthase in adipocytes in vitro and in vivo study in 3T3-L1 cells and Otsuka Long-Evans Tokushima fatty rats. *Life Sciences*, *67*:2093-2101.

[46] Nieman DC, Nehlsen-Cannarella SL, Henson DA, Butterworth DE, Fagoaga OR, Warren BJ, & Rainwater MK. 1996. Immune response to obesity and moderate weight loss. *International Journal of Obesity*, *20*:353-360.

[47] Vgontzas AN, Papanicolauo DA, Bixler EO, Kales A, Tyson K, & Chrousos GP. 1997. Elevation of cytokines in disorders of excessive daytime sleepiness: Role of sleep disturbance and obesity. *Journal of Clinical Endocrinology and Metabolism*, *82*:1313-6.

[48] Papanicolauo DA, Wilder RL, Manolagas SC, Chrousos GP. 1998. The pathophysiologic roles of interleukin-6 in human disease. *Annals of Internal Medicine*, *128*:127-137.

[49] Tracy RP. 2003. Emerging relationships of inflammation, cardiovascular disease and chronic diseases of aging. *International Journal of Obesity*, *27*:S29-S34.

[50] Ailhaud G. 1997. Obesity and the adipocyte: Molecular mechanisms of adipocyte differentiation. *Journal of Endocrinology*, *155*:201-202.

[51] Baldo A, Sniderman AD, St-Luce S, Avramoglu RK, Maslowska M, Hoang B, Monge JC, Bell A, Mulay S, & Cianflone K. 1993. The adipsin-acylation stimulating protein system and regulation of intracellular triglyceride synthesis. *Journal of Clinical Investigation*, *92*:1543-1547.

[52] Cianflone K, Zhang X-J, Genest J, Sniderman A. 1997. Plasma acylation-stimulating protein in coronary artery disease. *Arteriosclerosis, Thrombosis and Vascular Biology*, *17*:1239- 1244.

[53] Weyer C, Tataranni PA, Pratley RE. 2000. Insulin action and insulinemia are closely related to the fasting complement C3, but not to acylation stimulating protein concentration. *Diabetes Care*, *23*:779-785.

[54] National Heat Lung and Blood Institute Obesity Education Initiative Expert Panel on the Identification, Evaluation and Treatment of Overweight and Obesity in Adults. 1998. *Clinical guidelines on the identification, evaluation and treatment of overweight*

and obesity in adults. Bethesda, MD: National Heart Lung and Blood Institute. NIH Publication No. 98-4083.

[55] U.S. Department of Health and Human Services. (2001). *The Surgeon General's call to action to prevent and decrease overweight and obesity*. Rockville, MD: U.S. Department of Health and Human Services, Public Health Service, Office of the Surgeon General.

[56] Tripathy D, Mahanty P, Dhindsa S, Syed T, Husam G, Aljada A, Dandona, P. 2003. Elevation of free fatty acids induces inflammation and impairs vascular reactivity in healthy subjects. *Diabetes*, 52:2882-2887.

[57] Aljada A, Mohanty P, Husam GH, Abdo T, Tripathy D, Chaudhuri A, & Dandona P. 2004. Increase in intranuclear nuclear factor KB and decrease in inhibitor KB in mononuclear cells after a mixed meal: evidence for a proinflammatory effect. *The American Journal of Clinical Nutrition*, *79*:682-690.

[58] Shoelson SE, Lee J, & Yuan M. 2003. Inflammation and the IKK beta/I kappa B/NF-kappa B axis in obesity- and diet-induced insulin resistance. *International Journal of Obesity & Related Metabolic Disorders: Journal of the International Association for the Study of Obesity*, *27*(Suppl 3):S49-52.

[59] Hundal RS, Petersen KF, Mayerson AB, Randhawa PS, Inzucchi S, Shoelson SE, Shulman GI. 2002. Mechanism by which high-dose aspirin improves glucose metabolism in type 2 diabetes. *Journal of Clinical Investigation*, *109*:1321–1326.

[60] Laimer M, Ebenbichler CF, Kaser S, Sandhofer A, Weiss H, Nehoda H, Aigner F, & Patsch JR. 2002. Markers of chronic inflammation and obesity: A prospective study of the reversibility of this association in middle-aged women undergoing weight loss by surgical intervention. *International Journal of Obesity*, *26*:659-662.

[61] Heilbronn LK, Noakes M, & Clifton PM. 2001. Energy restriction and weight loss on very-low-fat diets reduce c-reactive protein concentrations in obese, healthy women. *Arteriosclerosis, Thrombosis, and Vascular Biology*, *21*:968-970.

[62] Bastard JP, Jardel C, Bruckert E, Blondy P, Capeau J, Laville M, Vidal H, & Hainque B. 2000. Elevated levels of interleukin 6 are reduced in serum and subcutaneous adipose tissue of obese women after weight loss. *Journal of Clinical Endocrinology and Metabolism*, *85*:3338-3342.

[63] Ravussin E, Pratley RE, Maffei M, Wang H, Friedman J, Bennett PH, & Bogardus C. 1997. Relatively low plasma leptin concentrations precede weight gain in Pima Indians. *Nature Medicine*, *3*:238-240.

[64] Chessler SD, Fujimoto WY, Shofer JB, Byoko EJ, Weigle DS. 1998. Increased plasma leptin levels are associated with fat accumulation in Japanese-Americans. *Diabetes*, *47*:239- 243.

[65] Savoye M, Dziura J, Castle J, DiPietro L, Tamborlane WV, & Caprio S. 2002. Importance of plasma leptin in predicting future weight gain in obese children: a two-and-a-half-year longitudinal study. *International Journal of Obesity*, *26* :942-946.

[66] Duncan BB, Schmidt MI, Chambless LE, Folsom AR, & Heiss G. 2003. Inflammation markers predict increased weight gain in smoking quitters. *Obesity Research*, *11*:1339-1344.

[67] Duncan BB, Schmidt MI, Chambless LE, Folsom AR, Carpenter M., & Heiss G. 2000. Fibrinogen, other putative markers of inflammation, and weight gain in middle-aged adults – The ARIC study. *Obesity Research*, *8*:279-286.

[68] Tamakoshi K, Yatsuya H, Kondo T, Ishikawa M, Zhang H, Murata C, Otsuka R, Mabuchi T, Hori Y, Zhu S, Yoshida T, & Toyoshima H. 2003. Long-term body variability is associated with elevated C-reactive protein independent of current body mass index among Japanese men. *International Journal of Obesity*, *27*:1059-1065.

[69] Dart AM, Martin JL, & Kay S. 2002. Association between past infection with *Chlamydia pneumoniae* and body mass index, low density lipoprotein particle size, and fasting insulin. *International Journal of Obesity*, *26*:464-468.

[70] Dhurandhar NV, Kulkarni PR, Ajinkya SM, & Sherikar AA. 1992. Effect of adenovirus infection on adiposity in chickens. *Veterinary Microbiology*, *31*:101-107.

[71] Dhurandhar NV, Kulkarni PR, Ajinkya SM, Sherikar AA, & Atkinson RL. 1997. Association of adenovirus infection with human obesity, *Obesity Research*, 5:464-469.

[72] Dhurandhar NV, Israel BA, Kolesar JM, Mayhew GF, Cook ME, & Atkinson RL. 2000. Increased adiposity in animals due to a human virus. *International Journal of Obesity*, *24*:989-996.

[73] Dhurandhar NV, Whigham LD, Abbott DH, Schultz-Darken NJ, Israel BA, Bradley SM, Kemnitz JW, Allison DB, & Atkinson RL. 2002. Human adenovirus Ad-36 promotes weight gain in male Rhesus and marmoset monkeys. *Journal of Nutrition*, *132*:3155–3160.

[74] Atkinson RL, Dhurandhar NV, Allison DB, Bowen R, & Israel BA. 1998. Evidence for an association of an obesity virus with human obesity at three sites in the United States, *International Journal of Obesity*, *22*:S57.

[75] Holmes B. 2000. The obesity bug. *New Scientist*. Retrieved October 6, 2000 from *http://www.newscientist.com/features/features_225015.html*.

[76] Lin Y, Lee H, Berg AH, Lisanti MP, Shapiro L, & Scherer PE. 2000. The lipopolysaccharide-activated toll-like receptor (TLR)-4 induces synthesis of the closely related receptor TLR-2 in adipocytes. *Journal of Biological Chemistry*, *275*:24255-24263.

[77] Ajuwon KM, Jacobi SK, Kuske JL & Spurlock ME. 2004. Interleukin-6 and interleukin-15 are selectively regulated by lipopolysaccharide and interferon-gamma in primary pig adipocytes. *American Journal of Physiology - Regulatory Integrative & Comparative Physiology, 286:*R547-53.

[78] Li XC, Demirci G, Ferrari-Lacraz S, Groves C, Coyle A, Malek TR, & Strom TB. 2001. IL-15 and IL-2: a matter of life and death for T cells *in vivo*. *Nature Medicine*, *7*:114-118.

[79] Dhurandhar NV, Sheele J, Israel BA, & Atkinson RL. 1999. Adenovirus that induces adiposity in animals also influences in-vitro differentiation of preadipocytes. *Obesity Research*, *7*:39S.

[80] Wlodek, D., & Gonzales, M. (2003). Decreased energy levels can cause and sustain obesity. *Journal of Theoretical Biology*, *225*:33-44.

[81] Hinze-Selch D, Schuld A, Kraus T, Kuhn M, Uhr M, Haack M, & Pollmacher T. 2000. Effects of antidepressants on weight and on the plasma levels of leptin, TNF-alpha and soluble TNF receptors: a longitudinal study in patients treated with amitriptyline or paroxetine. *Neuropsychopharmacology*, *23*:13–19.

[82] Parker JD, Farrell B, Fenton T, Cohanim M, & Parker JO. 1991. Counter-regulatory responses to continuous and intermittent therapy with nitroglycerin. *Circulation*, *84*:2336–2345.

[83] Lindh U, Hudecek R, Danersund AS, & Lindvall A. 2002. Dental amalgam removal improves health. *Neuroendocrinology Letters*, *23*:459-482.

[84] Salonen JT, Seppanen K, Nyyssönen K, Korpela H, Kauhanen J, Kantola M, Tuomilehto J, Esterbauer H, Tatzber F, & Salonen R. 1995. Intake of mercury from fish, lipid peroxidation, and the risk of myocardial infarction and coronary, cardiovascular, and any death in eastern finnish men. *Circulation*, *91*:645-655.

[85] Guallar E, Sanz-Gallardo MI, van't Veer P, Bode P, Aro A, Gomez-Aracena J, Kark JD, Riemersma RA, Martin-Moreno JM, & Kok FJ. 2002. Heavy Metals and Myocardial Infarction Study Group. Mercury, fish oils, and the risk of myocardial infarction. *New England Journal of Medicine*, *347*:1747–1754.

[86] Yoshizawa K, Rimm EB, Morris JS, Spate VL, Hsieh CC, Spiegelman D, Stampfer MJ, & Willett WC. (2002). Mercury and the risk of coronary heart disease in men. *New England Journal of Medicine*, *347*:1755–1760.

[87] Dulloo AG, Jacquet J, & Montani J-P. 2002. Pathways from weight fluctuations to metabolic diseases: focus on maladaptive thermogenesis during catch-up fat. *International Journal of Obesity*, *26* Suppl2:S46-S57.

[88] Rosenbaum M. & Leibel RL. 1998. The physiology of body weight regulation: Relevance to the etiology of obesity in children. *Pediatrics*, 101:525-539.

[89] Rosenbaum M, Leibel RL, & Hirsch J. 1997. Medical progress: Obesity. *New England Journal of Medicine*, *337*:396-407.

[90] Leibel RL, & Hirsch J. 1984. Diminished energy requirements in reduced-obese patients. *Metabolism*, *33*:164-170.

[91] Crawford D, Jeffery RW, & French SA. 2000. Can anyone successfully control their weight? Findings of a three year community-based study of men and women. *International Journal of Obesity*, *24*:1107-1110.

[92] Heo M, Faith MS, & Pietrobelli A. 2002. Resistance to change of adulthood body mass index. *International Journal of Obesity*, *26*:1404-1405.

[93] Levine JA, Eberhardt NL, & Jensen MD. 1999. Role of nonexercise activity thermogenesis in resistance to fat gain in humans. *Science*, *283*:212-214.)

In: New Developments in Obesity Research
Editor: Lawrence F. Ditmier, pp. 175-201

ISBN 1-60021-296-4

Chapter 7

ALSTRÖM SYNDROME

Pietro Maffei[1*], Jan D. Marshall[2], Roberto Vettor[1], Gayle B. Collin[2], Nicola Sicolo[1] and Jürgen K. Naggert[2]

[1]Department of Medical and Surgical Sciences, Clinica Medica 3^,
University School of Medicine, Padua, Italy
[2]The Jackson Laboratory, Bar Harbor, ME, USA

ABSTRACT

Alström syndrome (ALMS) (OMIM 203800) is a rare autosomal recessive genetic disorder with approximately 350 cases known worldwide. ALMS patients typically present in the first months of life with an ocular disorder characterized by progressive retinal dystrophy that leads to blindness by the end of the second decade. Other early features of ALMS include obesity, beginning in early childhood, hyperinsulinemia or insulin-resistance, type 2 diabetes, and hyperlipidemia. Bilateral neurosensory hearing loss develops in childhood and usually progressively worsens. Impairment of heart, kidney, or liver function at different stages of life may severely affect patients' prognosis and survival. Other less frequently observed features involve the pulmonary, bone-skeletal, neuromuscular, hemopoietic and endocrine systems. In most of the patients there is normal intelligence, although there are some reports that showed a possible delayed intellectual development.

In 2002, the gene responsible for ALMS, *ALMS1*, on chromosome 2p13 was identified. *ALMS1*, a novel gene of unknown function, is comprised of 23 exons and encodes a protein of 4,169 amino acids. *ALMS1* is ubiquitously expressed at low levels in all tissues reported to be pathologically affected in ALMS patients and thus far does not share significant sequence homology with other known genes. The diagnosis of Alström syndrome remains a difficult task both for clinicians and geneticists, considering the overlap between ALMS and other genetic disorders with retinopathy, obesity, or diabetes.

[*] Correspondence concerning this article should be addressed to Dr. Pietro Maffei, Dipartimento di Scienze Mediche e Chirurgiche, Clinica Medica 3ª, Azienda Ospedaliera di Padova, Via Giustiniani 2, 35100 Padova, Italy. Tel.: +39 049 8218745; Fax: +39 049 8218744; e-mail: pietromaffei@libero.it.

Initially, many patients are diagnosed clinically without genetic confirmation, which often results in multiple misdiagnoses and delayed diagnosis. No specific treatment of ALMS currently exists. The ALMS syndrome deeply impacts the life of patients and families. The formation of various pro-active groups, including Alström Syndrome International, Alström Syndrome-Canada, and Alström Syndrome-UK have had a fundamental role, in raising interest in the syndrome, organizing scientific meetings, and giving support to patients and their families. Moreover, an International Scientific Advisory Board on ALMS has been established which has the role of translating scientific achievements into everyday practice.

Keywords: Alström Syndrome, ALMS1, obesity, retinopathy, diabetes, cardiomyopathy

INTRODUCTION

Alström syndrome (ALMS) (OMIM 203800) is a rare recessively inherited disorder described initially in 1959 by Carl-Henry Alström in three Swedish patients from a large Scandinavian pedigree (Alström et al, 1959). Alström's early description of a unique combination of retinal degeneration, hearing impairment, obesity, and type 2 diabetes, with clinical features distinct from Laurence-Moon and Bardet-Biedl syndromes, has remained essentially unchanged. ALMS patients typically present in the first months of life with an ocular disorder characterized by nystagmus, photophobia, and progressive retinal dystrophy that leads to blindness by the end of the second decade (Alström et al, 1959; Millay et al, 1986; Tremblay et al, 1993; Russell-Eggitt et al, 1998). Other early features of ALMS include obesity, beginning in early childhood, hyperinsulinemia or insulin-resistance, type 2 diabetes, and hyperlipidemia. Bilateral neurosensory hearing loss develops in early childhood and usually progressively worsens (Marshall et al, 1997). Impairment of heart, kidney, or liver function at different stages of life may severely affect patients' prognosis and survival. Other less frequently observed features involve the pulmonary, bone-skeletal, neuromuscular, hemopoietic and endocrine systems. In most of the patients there is normal intelligence, although there are some reports that have suggested delayed intellectual development (Marshall et al, 2005).

The syndrome is characterized by symptoms involving many different organ systems. Different phenotypic expression has been reported in patients within the same family or in patients from different families with identical mutations. Genetic modifiers and environmental interactions are likely to explain some phenotypic differences among sibs with the same genetic mutation. Because of the progressive nature of the disorder, Alström syndrome is difficult to diagnose, and patients often receive multiple misdiagnoses or never receive a diagnosis in the course of their lifetime.

Collin et al (1997) localized the gene responsible for ALMS to the short arm of chromosome 2 (2p13), resulting in an increased interest in the disorder (Maffei et al 2002a, Hopkinson et al, 2003, Marshall et al, 2005).

Five years later, the *ALMS1* gene was identified by two independent groups of investigators, one from The Jackson Laboratory (Bar Harbor, ME, USA) and the other from University of Southampton, UK.(Collin et al 2002; Hearn et al 2002). Notwithstanding these important achievements, the etiopathogenetic mechanisms responsible for the multiorgan involvement of ALMS remain completely unknown. Eventual understanding of the function of *ALMS1* will likely add relevant pathophysiological interpretations of more common disorders in the general population.

In this chapter, we give a complete overview of the many features of the syndrome. A particular emphasis will be placed on the metabolic syndrome, obesity, and endocrine system alterations and interactions.

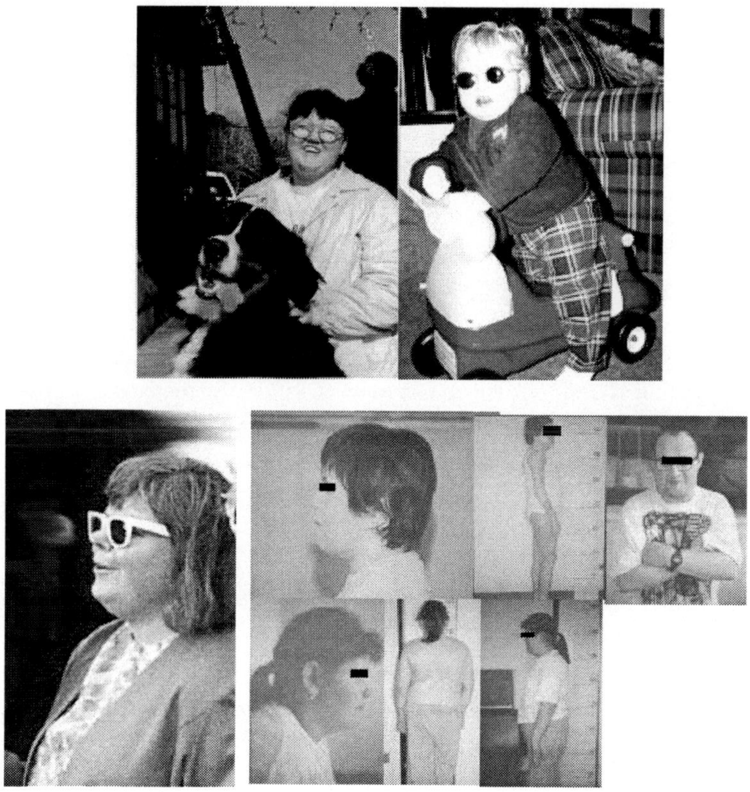

Figure: patients with Alström Syndrome

Epidemiology

ALMS is caused by mutations in a single gene transmitted in an autosomal recessive manner. Therefore, males and females share the same probability (M:F=1:1) of inheriting or transmitting the disorder. However, close communities and consanguinity may increase the frequency of homozygosity for mutations causing the syndrome. The autosomal recessive transmission of ALMS was first hypothesized in a Swedish family (C.H. Alström et al, 1959) and confirmed, several years later, in another kindred with three female subjects (Goldstein and Fialkow, 1973). The syndrome has been reported in different ethnic populations and many geographic areas including Europe, Asia, Africa, and North America (Marshall JD, personal communication). The total number of ALMS cases known, including the deceased patients, is approximately 350. The highest incidence of the syndrome occurs in individuals of Northern and Central European descent (Marshall JD, 2003). The syndrome has been well reported in the Acadian population with a common founder originating from Nova Scotia. (Marshall et al, 1997, Collin et al, 1997).

VISUAL SYSTEM

The probability that the *ALMS1* gene is directly involved in retinal disorders of ALMS patients is very high because retinal dystrophy appears early in all the individuals, usually during the first year of life. According to Russell-Eggitt, et al (2003), there is little variability in the expression of the retinal dystrophy within or between sibships. All of the cases observed had severe vision handicap beginning from the first months after birth. However, some mutation-confirmed ALMS individuals, reported first vision problems as late as 40 weeks of age (Marshall et al, 2005).

Millay et al (1986) first focused on the ophthalmologic manifestations in ALMS, providing follow-up information on three patients reported by Goldstein's et al (1973), and reviewed a total of 15 cases from past literature, including two new cases. They observed that pigmentary retinopathy with absence of bone spicules was a consistent finding in ALMS and could be a distinctive feature in comparison with other forms of retinopathy. In all of these patients, vision impairment manifested itself in the first year of life and visual symptoms were mostly featured by nystagmus, blepharospasm due to light sensitivity, and apparent decrease of vision. The authors reported that severe visual loss progresses during the first decade of life and leads to blindness with no light perception by the second decade. Another typical aspect of the eye phenotype in ALMS is posterior subcapsular cataract or diffuse or mature cataract. In some patients they observed iris atrophy, bone spicules, and optic disk drusen. Millay et al (1986) also reported pale and waxy optic discs and ghosting of vessels, loss of foveal reflex, cellophane-like appearance of the macula, and a prominent salt-and-pepper appearance involving the periphery and posterior pole. The fluorescein angiogram showed diffuse, severe, peppery, and confluent retinal pigment epithelial atrophy.

Further literature confirmed that nystagmus is among the first symptoms related to the retinal disorder in ALMS and it usually appears in the first months of life (Marshall et al, 1997). The nystagmus tends to diminish slightly as the patients age, and sometimes it disappears completely. Retinal degeneration is characterized by early loss of central vision. The photoreceptors have been described as abnormal in structure and function with some residual function of rod receptors (Russell-Eggitt, 2003). The typical signs of strabismus, ocular tics, and photophobia that often appear in the first year of life are thought to be stimulated by the retina responding to excessive light exposure. Extreme photophobia develops due to the total loss of the cones resulting in rod oversaturation, especially under conditions of bright light. With the progression of retinal dystrophy, rod receptor degeneration is usually associated with a reduction of photophobia (Tremblay et al, 1993; Van den Abeele et al, 2001).

Sebag et al. (1984) described quite extensively the ophthalmic histopathology and retinal ultrastructure from a single ALMS patient. The patient had small ocular globes, bilateral posterior subcapsular cataracts, lacy vacuolation of the iris, ciliary process hyalinisation, unilateral asteroid hyalosis, total absence of rods and cones, intraretinal melanin pigment, retinal pigment epithelium atrophy, focal chorioretinal fusion, preretinal fibrosis, bilateral giant optic disc drusen, and optic nerve atrophy. Moreover, by electron microscopy of the retinal field, they demonstrated the presence of large numbers of melanolysosomes, numerous folds of basement membrane material, disruption of Bruch's membrane, and numerous bundles of extracellular collagen fibrils in the retinal pigment epithelium. Deposits of melanin pigment, scattered in the outer and inner nuclear layers, were also described in a preliminary report from Bronson et al (2003) in two different ALMS patients.

In a further description of eight patients, (Michaud et al, 1996), photophobia and nystagmus could be first documented between the ages of 5 months and 15 months. In all the patients, electroretinography initially showed a severe cone impairment with mild (2/8) or no (6/8) rod involvement. Electroretinograms, obtained again at ages 9 to 22 years in four patients, revealed extinguished rod-and-cone responses.

In most patients, total blindness occurs before the second decade of life. The course of retinal dystrophy is progressive with the patient's visual acuity of 6/60 or less by 10 years of age and no light perception by 20 years of age (Russell-Eggitt et al, 1998). The electroretinogram usually confirms the retinal disorder and its repetition over time is helpful to exclude rod monochromatism, which is a non-progressive form of retinal degeneration. In ALMS patients, the electroretinogram reflexes are usually completely extinguished at 7-15 years of age.

Fundoscopic examination is not useful for the early diagnosis of ALMS because it can be completely normal, especially in the first years of life. When the patients age, the funduscopic examination becomes more useful and the typical reduction of vascular diameters, the pallor in the optical disk area, a poor definition of the macular region and tapeto-retinic degeneration can be detected. However, funduscopic examination may be impeded due to the sub-capsular cataracts that often occur in individuals with ALMS. The development of cataracts is thought to be associated with unknown substances released from damaged retinal tissue (Millay et al, 1986).

The earliest retinal alterations are not unique to ALMS and do not permit a differential diagnosis with other retinal disorders. For example, the retinal dystrophy in ALMS cannot be differentiated from Leber's congenital amaurosis, another autosomal recessive disorder characterized by a precocious retinal degeneration. In fact, only the further clinical course of ALMS will permit a clear distinction between the two disorders (Russell-Eggitt et al, 1989).

Auditory System

Progressive, bilateral sensorineural hearing impairment may be a typical feature of ALMS. There is some evidence that the vestibular function is also compromised in ALMS patients (C. Moller, personal communication). Middle ear infections manifesting very early in childhood seem to occur more frequently in ALMS patients in comparison with the general population and may be exacerbated by the presence of hearing aids. In the series described by Marshall et al (1997), a slowly progressive bilateral sensorineural hearing loss was first noted between 1-5 years of age and most of the individuals, with the exception of the youngest subjects, required hearing aids. Also in the large series described by Russell-Eggitt et al (1998), 14 out of 22 ALMS patients suffered from sensorineural hearing loss, which was usually detected at approximately 7 years of age. However, the authors reported on two older patients, at 12 and 19 years of age respectively, with normal audiograms. Therefore, hearing loss may not occur in all patients.

Because the hearing impairment progresses slowly over time and begins after the acquisition of verbal skills, language skills are usually not compromised in ALMS. The first symptoms related to auditory system disease may start with tinnitus, and the patients lose the ability to comprehend words or follow low-pitched conversations (Johnson, 1961). The audiograms usually show a progressive and symmetrical worsening both for low and high frequencies.

Although the hearing loss for most patients with ALMS is usually correctable through hearing aids, there are a small number of ALMS patients who are profoundly deaf.

Metabolism and Endocrinology

Alström syndrome severely affects the endocrine system and metabolism. Every patient presents with impairment of both these systems at some point along the clinical course of the disease. Different levels of insulin resistance ranging from simple basal hyperinsulinism to overt type 2 diabetes, obesity, and hyperlipidemia are featured in the metabolic disorders. Among the endocrine alterations, a number of hormonal dysfunctions have been described which include hypogonadism, hypothyroidism and growth hormone deficiency (Maffei et al, 2002; Hopkinson et al, 2003).

Diabetes

The molecular basis of insulin resistance in ALMS is not yet understood. In the literature, there are only two contrasting studies that specifically address the basic mechanisms of this important clinical issue (Rudiger et al 1985; Boor et al, 1993). In the first study, Rudiger et al (1985), investigated the cellular basis of insulin resistance in ALMS by means of skin fibroblast cultures incubated under different insulin concentrations followed by the determination of insulin binding activity, glucose uptake and RNA-synthesis. In summary, the fibroblasts from two different ALMS patients showed a reduction of insulin binding activity under incubation at low insulin concentrations in comparison to controls. However, the defect disappeared during incubation at a very high insulin concentration, in the range of 100 ug/ml. The insulin-mediated uptake of 2-desoxyglucose was found to be normal in only one of the two patients studied. Moreover, insulin-mediated stimulation of RNA synthesis in both the patients was not different from controls.

Boor et al (1993) evaluated the insulin receptor binding activity in two young-adult individuals diagnosed with ALMS having atypical features such as mental retardation and absence of renal insufficiency. A maximum insulin binding activity in the range of 13-16% was evidenced in patients compared to a 30% activity in healthy controls. Interestingly, both basal hyperinsulinemia and hyperproinsulinemia in the presence of normal fasting glucose were described in a single case (Charles et al, 1990). The same authors reported the absence of antibodies against insulin receptor in ALMS. Zumsteg et al (2000) found islet cell antibodies to be absent in three ALMS individuals; other reports further confirmed both findings about diabetes and autoimmunity in Alström syndrome (Satman et al, 2002).

In an analysis of previously published cases, Millay et al (1986) documented that diabetes mellitus was found in 12 out of 15 ALMS patients. Since this first overview, it became clear that both hyperinsulinemia in childhood and non-insulin dependent (type 2) diabetes mellitus were constant features of the syndrome. Interestingly, they observed in one of their new cases that, in spite of subnormal insulin response during OGTT, there was a good control of diabetes with oral hypoglycemic agents alone. According to further observations (Holder et al, 1995; Zumsteg et al, 2000; Worthley et al, 2001; Tai et al, 2002) and our personal experience, ALMS individuals with diabetes tend to be more responsive to oral hypoglycemic agents, diet, and exercise than insulin. The latter, when necessary, usually requires very high dosages to reach a good metabolic control.

In the series of 8 patients described by Michaud et al (1996), the five oldest individuals, two of whom were diabetic and two glucose intolerant, showed the highest baseline insulin levels. Elevated fasting insulin levels were also observed in all the individuals reported by Marshall et al (1997) and hyperinsulinemia persisted in these subjects even after their obesity moderated as they became young adults. In a further study, Marshall et al (2005) reported 82% of patients over the age of 16 years had developed clinical type 2 diabetes. Although only 5 out of 22 patients described by Russell-Eggitt et al (1998) had clinical diabetes mellitus all but two of the patients without diabetes were younger than 15 years of age.

Recently, Satman et al (2002) reported on type 2 diabetes in ALMS and gave long-term prospective follow-up of three Turkish siblings. The authors ruled out the possibility of type 1 diabetes and, in contrast with literature, some of their cases became insulin-requiring with time and, despite additional Metformin (850 mg b.i.d.), eventually needed high insulin doses (1.7 U/Kg/day) to reach adequate metabolic control. Interestingly, one of the patients was admitted to a hospital at 14-years of age for diabetic ketoacidosis which abated after diet treatment and insulin infusion. The insulin resistant state of the patient was further investigated by means of the euglycemic-hyperinsulinemic clamp after overnight normoglycemia intravenous insulin infusion. The authors showed a reduction of glucose disposal rate in the range of 1.83 mg/Kg/m (normal range: 6.0-9.0 mg/Kg/m). The two other patients were tested by means of euglycemic-hyperinsulinemic clamp, which confirmed, once again, marked insulin resistance in ALMS.

Medical treatment of diabetes in ALMS remains one of the major clinical issues due to the marked insulin resistant state that has been well documented in most of the patients. Ten et al (2004) recently described, in abstract form, an 8-year-old boy with ALMS who developed progressive weight gain beginning at 9 months of age. Acanthosis nigricans, impaired glucose tolerance, and hyperinsulinemia were seen at 6 years of age. The patient was given Metformin to 1000 mg b.i.d. and 4 mg rosiglitazone was added for the subsequent 12 months. After 20 months of treatment, triglycerides remained elevated and metabolism deteriorated to the level of overt diabetes. However, after another 12 months of treatment, OGTT remained stable, and there was no further decline in insulin secretion.

Hyperlipidemia

The Alström syndrome phenotype includes a metabolic syndrome which severely impacts both carbohydrate and lipid balance. Recent data from a large number of ALMS patients suggest that increased levels of triglycerides largely driven by the insulin resistant state or diabetes (Paisey et al, 2004).

In the review from Millay et al (1986), five of six patients had increased triglycerides levels ranging from 300 to 1625 mg/dl in association with only mild hypercholesterolemia. The authors postulated that hypertriglyceridemia represented an intrinsic finding in ALMS independent from diabetes or obesity.

In the reviews from Marshall et al (1997, 2005) hyperlipidemia was reported as a frequent laboratory alteration both in previous series and their new reports. Detailed chemistry measurements from nine individuals with ALMS documented an atherogenic lipid profile with increased levels of triglycerides (6/9) and total cholesterol (2/9) in most of the subjects. Moreover, a significant reduction of HDL cholesterol (7/9) and apolipoprotein B (2/9) was observed. Low HDL cholesterol levels in ALMS have also been documented in another series (Maffei et al, 2002a).

In three sibs from a consanguineous kindred of Turkish origin, Zumsteg et al (2000) explored the apolipoprotein E amino-acid polymorphism including the □4 and □2 alleles as the possible cause of dyslipidemia in ALMS. However, the presence of the □2 allele mutation did not co-segregate with the clinical feature of hypercholesterolemia. In this series they observed a patient with very high levels of serum cholesterol (10.3 mmol/L) and triglycerides (78.4 mmol/L), however, she was the only patient with marked fasting hyperglycemia (23.0 mmol/L) and elevation of HbA1c (11.3 %). This observation seems to point out that uncontrolled diabetes is a possible cause of hyperlipidemia because in this cohort the other two subjects presented normal fasting glucose and no alteration of lipid parameters.

Paisey et al (2004) specifically addressed causes and associations of hypertriglyceridemia in a series of 37 ALMS patients. It was documented that although many patients showed moderately (4.01-8.0 mmol/L) or grossly elevated (> 8 mmol/L) serum triglyceride levels there was a subgroup of patients without significant hypertriglyceridemia (0-2 mmol/L). No relationship could be found between serum triglyceride levels and age, BMI, HOMA index, renal or hepatic function in the patients. Furthermore, they found no significant correlation between triglycerides and insulin levels. However, venn diagrams showed that all 23 patients with hypertriglyceridemia were also hyperinsulinemic.

Thus far, there are no specific recommendations about the treatment of lipid disorders in ALMS. However, it is useful to mention that Paisey et al (2004) reported that two of the most severe hypertriglyceridemic patients responded to treatment with atorvastatin 80 mg plus nicotinic acid 500 mg tid, low fat diet and weight loss favored by daily exercise.

Growth Hormone Deficiency

Short stature is a typical clinical feature of adult ALMS individuals (Goldstein et al, 1973), while younger patients often present with normal to slightly increased height. In addition, advanced bone age is a typical feature of ALMS in childhood. In the large series described by Russell-Eggitt et al (1998), the height of 7 out of 22 patients was below the 25th percentile. Although pituitary dysfunction might explain, at least in part, short stature in ALMS, very few studies specifically addressed this topic and provide information about the possible effects of growth hormone replacement therapy (Alter et al, 1993; Zumsteg et al, 2000; Tai et al, 2002).

Alter et al (1993) gave a detailed endocrinological investigation of two siblings with typical clinical features of ALMS and normal growth early in life. One of the patients presented in infancy an under-average height and, at 11 years old, was in the 25th percentile, whereas the height of his 10-year-old sister was below the 5th percentile. Although both patients showed advanced bone age, they reported consistent findings of GH deficiency by documentation of low GH concentrations after arginine and clonidine administration. Moreover, they observed low serum GH concentrations in the pooled 12-hour overnight samples. Insulin-like growth factor I (IGF-I) was in the low-normal range in both the individuals.

Zumsteg et al (2000) observed biochemical growth hormone deficiency with blunted GH concentrations following arginine stimulation in a group of three affected siblings with typical ALMS phenotype and genetic linkage to chromosome 2p13. In these sibs there was a normal growth velocity in early childhood, advanced bone age at puberty, and a marked reduction in final height at the age of 15. However IGF-I and IGFBP-3 were measured and found to be normal in all three patients.

Data reported by Maffei et al (2002b) confirms growth hormone deficiency in ALMS. In summary, the authors investigated GH secretion through well-standardized and accepted protocols (GHRH + Arginine test) in five Italian individuals with ALMS from four different families. They documented a severe reduction of GH response in two cases, partial deficiency in one case, and a low-normal GH response in the remaining two patients. They found low IGF-I values in two cases. The IGF system has been further investigated in a cohort of 26 ALMS patients (Maffei et al, 2002c). Preliminary results from this series point to a major impairment of the IGF system (IGF-I; IGFBP3, ALS) that was unrelated to liver and kidney function in adult patients in comparison with prepubertal ALMS patients.

Tai et al (2003) described detailed information on the effects of growth hormone replacement therapy on some endocrinological parameters, metabolism and body composition in a 15-year-old Taiwanese boy who presented with clinical features consistent with ALMS. The patient's height was reported to be below the 3rd percentile, while his bone age was similar to an 11-year-old boy. Growth hormone deficiency was evidenced through insulin-induced hypoglycemic stimulation and clonidine test which both showed that the GH response was inadequate and IGF-I values were below the normal range. Notwithstanding the possible negative effects on glucose metabolism, considering the elevated post-prandial glucose levels and marked hyperinsulinemia after glucose load, the patient was eventually placed under treatment with recombinant GH (rGH) at a dose of 0.10 IU/Kg/day s.c. Interestingly, the 1-year follow-up evaluation documented normal IGF-I and CT imaging disclosed a significant reduction of liver fat content and visceral fat mass. There was a general amelioration of lipid parameters including total cholesterol, LDL and HDL cholesterol, and triglycerides. At the beginning of rGH treatment, a transient increase of blood glucose was observed. However, after one year of treatment, there was a consistent amelioration of several metabolic parameters including baseline HbA1c, C-peptide, and insulin, thus, suggesting a GH mediated general improvement of insulin sensitivity. Finally, during rGH treatment there were no significant side effects observed.

Hypogonadism

Male hypergonadotrophic hypogonadism, with baseline elevation of LH-FSH and testosterone reduction, has been documented by Millay et al (1986) in a 23year old ALMS patient. In this case the testicles were small, measuring approximately 1.5 cm in length. Interestingly, in this paper the authors reviewed previous literature reports of 13 ALMS patients and found that six out of seven males had small soft testicles with small or normally sized phalluses, hypergonadotrophinuria or hypergonadotrophinemia, and decreased or borderline normal serum testosterone (Alström et al, 1959; Klein et al, 1969; Weinstein et al, 1969; Goldstein et al, 1973; Kopecky et al, 1978; Sebag et al, 1984). However, secondary sex characteristics were normal. Autopsy findings showed thickening and hyalinization of seminiferous tubules with occasional spermatogonia. Gynecomastia has been reported in ALMS and it may be due to the condition of hypogonadism with imbalance in testosterone and estrogen ratios (Russell-Eggitt et al, 1998).

Marshall et al (1997) further confirmed the presence of low sex steroid levels in almost all pubertal subjects. Moreover, low ratios of testosterone to estradiol and normal or elevated gonadotropins suggested a primary gonadal failure in ALMS males.

Females suffering from ALMS may present regular or irregular menses and hyperandrogenism, although hypogonadism has never been reported.

As far as we know no ALMS individual has reproduced.

Other Endocrine Disorders

In addition to growth hormone deficiency and hypogonadism, a few other less frequent endocrinological alterations are peculiar in ALMS. Pituitary defects involving TSH and adiuretine have been sporadically reported (Charles et al, 1990; Aynaci et al, 1995; Marshall et al, 1997, 2005).

Some authors have observed primary hypothyroidism, although there is only one paper, which reported a condition of clinical hypothyroidism with low fT4 values, normal baseline TSH, and exaggerated TSH response after TRH stimulation in a 23-year-old girl (Charles et al, 1990). These results were compatible with secondary hypothyroidism resulting from a central pituitary or hypothalamic defect, thus postulating a widespread abnormality of hormone receptors in ALMS. Several patients affected from primary hypothyroidism and treated with L-thyroxin have been confirmed in another series (Marshall et al, 1997, 2005).

Aynaci et al (1995) described a 16-year-old male patient who presented a mild form of diabetes insipidus. The diagnosis was based on the findings of the water deprivation test, the absence of normal signal in the posterior pituitary gland, and decreased daily urinary output on administration of DDVAP.

OBESITY

Since the first descriptions of ALMS, obesity has been a constant clinical feature observed in nearly all the individuals (Alström et al, 1959; Goldstein et al, 1973; Sebag et al,

1984; Millay et al, 1986; Marshall et al, 2005). According to Millay et al (1986), the cause of obesity is unclear and can vary, with ideal body weight ranging from 71% to 230%. Moreover, it became clear from reports in the early literature that obesity was a feature of early infancy and childhood, while the obese phenotype was minimal or absent in the third decade of life. The pathophysiology, which stands behind the unusual clinical picture of early obesity in childhood and progressive weight normalization as the patients age, still remains among the unexplained features of the syndrome. Notably, the weight, the height, and BMI at birth are reported to be normal in ALMS (Marshall et al, 2005).

In the series of 8 patients described by Michaud et al (1996), obesity was documented in seven individuals. In the first few months of life, BMIs were in the normal range but began to increase during early childhood (11 months-2 years). The only non-obese patient in this series suffered from heart failure in childhood and this major clinical problem could have influenced his obese phenotype. In fact, this patient progressed from underweight below 3rd percentile immediately after recovering from heart failure to a progressively normal weight in the following years until he reached the 50th percentile. Weight loss and underweight has been also reported by Millay et al (1986) in a patient with significant kidney failure.

In the large series of Marshall et al (1997), which also included a summary of 49 ALMS cases from the literature, obesity was reported as a constant phenotype. In a large cohort of 12 individuals they specifically reported that BMI ranged from 23.8 to 36.5 and % IBW of all subjects was greater that 100%. The general impression of Marshall et al (1997) about obesity's appearance was that of a normal weight at birth, which could increase rapidly in childhood. The authors reported infantile truncal obesity with hyperphagia that persisted in adolescence, apparent normal caloric intake, and difficulty in maintaining static weight despite caloric restriction. Compulsive hyperphagia has also been reported in other series (Maffei et al, 2002a).

Obesity is well represented in the large cohort from Russell-Eggitt et al (1998). The authors evidenced a weight > 90th percentile in 20 out of 22 individuals, and the only exceptions were two patients, both suffering from congestive heart failure.

The 100% prevalence of obesity in ALMS patients has been further confirmed in a genetic study by Collin et al (1999a) in a series of 21 subjects, including 9 individuals previously reported and 12 new cases. In this cohort, ranging from 5 to 40 years of age, obesity was considered to be present when the weight of patients was above the 95th percentile.

Hung et al (2001) report on two siblings, a 12-year old girl and her 10-year-old brother, with typical signs and symptoms of ALMS. Interestingly, leptin concentrations were found to be elevated in both siblings.

Cardiology

Dilated cardiomyopathy (DCM) usually manifests itself with congestive heart failure, progressing to affect both the left and right ventricles. DCM in 63% of patients seems to fall into two distinct groups: infantile onset before vision disturbances are noted, and adolescent/adult onset (Marshall et al, 2002; Marshall, 2003). However, there is also a persistent risk for adolescent/adult individuals who passed infancy without heart disturbances to develop DCM. Preliminary data shows that, for about 45% of ALMS patients, the onset of sudden heart failure at around two months of age is the first manifestation of the syndrome (Carey, 2003). Over 80% of these infants survive the acute episode although, unfortunately, 20% of survivors will eventually experience a recurrence of DCM and develop progressive congestive heart failure in adolescence or adulthood (Carey, 2003). However, some patients with infantile DCM do not recover. In fact, in an Italian patient with a confirmed mutation in *ALMS1,* heart failure started in the first months of life and continued on a progressive course until death at age 14 due to pulmonary edema and severe secondary liver failure. (Maffei et al, 2002a; Collin et al, 2002).

Several authors report that adult ALMS patients without cardiological dysfunction still maintain a risk for developing a dilated cardiomyopathy when they age (Warren et al, 1987; Worthley et al, 2001). Two brothers described by Weinstein et al (1969) were reviewed in a follow-up by Warren et al (1987). This study has been particularly useful because the authors noticed during the follow-up, at the age of 36 and 37, that the two brothers had developed DCM. Unfortunately, the two brothers died, but results of heart pathology were available by means of myocardial biopsy and autopsy investigation. The most important pathological findings were extreme myocardial fibrosis, which was certainly related to impairment of heart function. Moreover, the authors concluded that coronary atherosclerosis could not explain, by itself, the degree of heart insufficiency (Warren et al, 1987). In the 26-year-old patient described by Worthley et al (2001), symptoms consistent with DCM developed at age 24; however, after 6 months of therapy the functional status of the patient improved from class III New York Heart Association (NYHA) heart failure to class II.

The report from Michaud et al (1987) on 8 ALMS patients pointed out that in five children the cardiomyopathy was seen for the first time very early, between 3 weeks and 4 months of age. Several of their patients had normal blood levels of carnitine, lactate, and urinary organic acids, which seemed to rule out a respiratory chain defect to be the cause of their cardiomyopathy. Moreover histologic and ultrastructural studies revealed no specific abnormality. Endomyocardial biopsies were performed in two patients and showed marked focal myocyte hypertrophy, vacuolar degeneration, and interstitial fibrosis. There was no significant lipid accumulation or inflammatory infiltrate in the myocardial tissue. Myocyte mitochondria had normal morphology and number when studied by electron microscopy. The clinical course of the five patients was characterized by progressive improvement with minimal but persistent ventricular dilatation on follow-up (Michaud et al, 1987).

Marshall et al (1997) described three out of 12 individuals who were diagnosed with infantile cardiomyopathy; however, another patient developed DCM at age 14 and there was a patient described with mild congestive heart failure at age 23. The clinical course of the infantile cardiomyopathy showed a progressive improvement with eventual resolution of symptoms in all the subjects.

In a study from the UK, 18 out of 22 ALMS patients investigated were affected with DCM presenting as cardiac failure between two weeks and 7 months of age (Russell-Eggitt et al, 1998). Also in this series, there was a patient with late presentation of the cardiomyopathy at 16 years of age. Thirteen patients showed a consistent improvement of heart function so that at some point in their clinical history therapeutic intervention was stopped. During the study five deaths occurred and all were attributed to heart failure. Post-mortem structural abnormalities were reported in only one single case featured by pulmonary valve stenosis, left ventricular subendocardial fibrosis, patent ductus arteriosus, and anomalous venous drainage.

A case of ALMS presenting with cardiomegaly, acute respiratory distress, and pulmonary edema very early, at seven months after birth, has been recently documented by Makaryus et al (2003). Transthoracic echocardiogram demonstrated a 4-chamber dilatation and severely depressed systolic function with a calculated left-ventricular ejection fraction less than 10%. The remainder of the intracardiac anatomy, including valves, was reported to be normal apart from mild mitral insufficiency. Unfortunately, the patient underwent numerous hospitalizations due to heart failure, and eventually died at less than 2 years of age.

In ALMS both hypertension and precocious atherosclerotic lesions might appear, but to date no systematic investigations have been done. Curiously, although myocardial fibrosis is a histopathological feature of the syndrome there are no reports on arrhythmogenic abnormalities or heart conduction disorders in ALMS.

Urology and Renal Function

Urinary tract disorders are present in ALMS patients with higher frequency in comparison to the general population. Data from a large series of 182 cases documented mild to severe urological dysfunction in 48% of the patients (Marshall et al, 2005). Charles et al (1990) reported on a 24-year-old woman who presented with recurrent urinary tract infections associated with a tight urethral stenosis. The patient later developed incontinence necessitating permanent catheterization. Further urodynamic studies showed a large capacity bladder with diminished sensation and detrusor instability of unknown origin without features suggestive of diabetic neuropathy. In some preliminary reports on 21 ALMS patients, it has been confirmed that urinary tract problems affect a high proportion of individuals (MacDermott, 2001 and 2003). According to this author, minor symptoms included long intervals between voiding, urgency to void, whereas others had difficulty starting to void plus poor flow and lower abdominal pain. Moderate to severe symptoms were observed in one third of the patients and included marked urinary frequency, often with urgency to void and even extending to incontinence. Abdominal and perineal pain may be a clinical symptom related to urinary disorders in ALMS individuals. In two patients, major surgery for urinary diversion has been done to resolve a highly overactive bladder, although, sometimes in other cases, self-catheterization is enough to manage voiding difficulties. Recurring urinary infections are also increased in frequency, especially in diabetic patients.

Chronic renal failure is considered a cardinal clinical feature of ALMS, especially in older patients. Histopathological sections of the kidney showed thickening of glomerular and tubular membranes, and some degree of glomerular hyalinization (Goldstein et al, 1973). Progressive glomerular sclerosis eventually leads to progressive kidney disease, electrolytic imbalance, and death (Sebag et al, 1984). In an autopsy study from Awazu et al (1997), it has been observed by light and electron microscopy that glomerular basement membranes were thickened, layered, wrinkled, and widened.

End-stage renal failure was the main cause of death in three out of 12 individuals described by Marshall et al (1997) and the severity of nephropathy was once again highly variable at the age of presentation.

In a recent report from a Turkish family, calyceal deformities, narrowed uretero-pelvic angles and position of kidneys parallel to vertebral column were described in three sisters aged 7 to 20 years and a dilated ureter occurred in one of the cases (Satman et al, 2002). A renal biopsy in one patient revealed mesangial proliferative type glomerulopathy with hyaline sclerosis, mild interstitial fibrosis, and no sign of amyloidosis. Despite renal histopathological abnormalities, hypertension and proteinuria, the authors considered the term of "slowly progressive nephropathy" appropriate because the renal involvement did not progress any more. Both proteinuria and hypertension were reported in the other two siblings, as well.

Successful kidney transplantation has been recently reported in a 13-year-old ALMS patient (Porcellini et al, 2003). Three years after the transplantation, the graft was functioning well and the patient was under triple anti-rejection therapy with cyclosporin, steroids and mycophenolate. Glucose tolerance was normal before and during steroid treatment.

HEPATOLOGY

Individuals with ALMS frequently present with some signs, symptoms or laboratory abnormalities, which indicate some degree of liver dysfunction. The hepatopathy can present very early in the first decade of life and severely affect the quality of life of ALMS patients (Connolly et al, 1991; Quiros-Tejeira et al, 2001). Acute liver failure has been reported as the primary cause of death in ALMS (Awazu et al, 1997; Quiros-Tejeira et al, 2001). It has been also hypothesized that a mitochondrial functional abnormality could explain the process of acute liver failure (Quiros-Tejeira et al, 2001).

The mechanisms behind the liver disorder are still completely unknown and span from steatosis or isolated liver enlargement to cirrhosis in the final stages (Ikeda et al, 1974; Horiuchi et al, 1976). In particular, the patient reported by Ikeda et al. (1974) presented with situs inversus and liver biopsy showed biliary cirrhosis. The patient described by Horiuchi et al. (1976) died of hemorrhage from an esophageal varix at 25 years of age.

Extensive investigation by Connolly et al (1991) failed to elucidate a viral, autoimmune, or metabolic etiology of liver failure. Connolly et al (1991) reported on an 11-year-old girl in whom liver dysfunction became evident at 8 years of age, although minor impairment was first noted at 5 years by the increase of plasma alkaline phosphatase. A detailed histological examination was done through a liver biopsy, which showed portal septal fibrosis, portal inflammation, and mild piecemeal necrosis of the periportal hepatocytes. The bile ducts were described to be normal. However, there was minimal cholestasis with hepatocellular swelling and moderate macrovescicular steatosis. Ultrastructurally, mild steatosis was observed in hepatocytes, viral inclusions were excluded, organelles were normal, and mitochondria showed mild pleomorphism. Connolly et al (1991) treated this patient with prednisone and azathioprine with some improvement of serum AST and GGT. However, a repeated liver biopsy demonstrated persistent histopathological findings.

Chronic elevation of liver enzymes are often observed for some years before the full clinical picture of liver failure (Awazu et al, 1997). Awazu et al (1997) reports results from both liver biopsy and autopsy on two sibs with ALMS. Histopathological examination confirmed the presence of hepatocytes with fatty vacuoles and inflammatory infiltrates, perivenular fibrosis, and eroded limiting plate. Liver autopsy showed nodules of varying size and fibrous septa, piecemeal necrosis, marked lymphocyte infiltration and regenerative nodules. Electron microscopic examination showed no specific changes in basement membrane or organelles.

In the cohort from Marshall et al (2005), 92% of individuals with ALMS exhibited elevation of liver transaminases providing some evidence of liver damage. A liver biopsy from one of the individuals showed bridging portal fibrosis, hepatocellular steatosis, and mild active hepatitis, which was later confirmed by post-mortem examination.

Steatosis is unlikely to be the primary disorder of the liver in ALMS because it has been reported to be associated with both obesity and type 2 diabetes (Awazu et al, 1995). Moreover, Tai et al (2003) observed in a 15-year-old diabetic ALMS patient a marked decrease of hepatic fat content after growth hormone replacement therapy. The authors speculate that the decreased body fat mass with concomitantly lowered circulating free fatty acids might have induced a stimulation of triglyceride synthesis in hepatocytes.

MISCELLANEOUS FINDINGS

Cognition

Since the first observations, normal intelligence has been considered a constant and typical clinical feature of ALMS and helpful in the differential diagnosis to distinguish ALMS individuals from those suffering from other genetic disorders (Alström et al, 1959; Goldstein et al, 1973; Sebag et al, 1984; Millay et al, 1986). However, formal testing by psychometric studies are rather difficult to be done in ALMS, considering both visual impairment and neurosensorial deafness. We suggest that psychometric investigations in ALMS should be done by psychologists trained to test impaired children. Millay et al (1986) described a patient with ALMS who was an award-winning musician. Moreover, in their review of 15 patients they observed that most attended school, learned Braille, and were good at manual dexterity tasks. We have also seen an Italian ALMS patient, with confirmed ALMS1 mutations, who have graduated from a high-school level education (data not published).

A contrasting report by Connolly et al (1991) described an 11-year-old girl affected with ALMS who was found to learn at a slower rate with specific problems in abstractation and generalization. Also, Alter et al (1993) reported on a 2-year delayed reading comprehension in an 11-year-old boy. It has been also suggested that mental retardation in ALMS could be the result of educational deprivation (Aynaci et al, 1995). Michaud et al (1996) documented a mild to moderate developmental delay in two out of eight ALMS patients when they were formally assessed in early childhood. Although no formal assessment was done in the six other patients, they were considered by their physicians to have developmental delay. Marshall et al (1997; 2002; 2005) observed in their 135 cohorts a great variability in cognitive development reporting both normal development and a number of cognitive disorders including mental retardation, delayed expressive language skills in the first years of life, mild generalized cognitive impairment at formal testing, and difficulty in initiation and problem solving. Borderline normal intelligence (1/3) and mildly retarded psychomotor activity (2/3) were once again recently documented (Satman et al, 2002).

In preliminary data from a large series of 134 ALMS individuals, Marshall et al (2002) reported neurological symptoms in 39% including absence seizures, Asperger's-like behaviors, clonic tics, or motor/language delays.

Mental retardation with coarse face in ALMS has been reported by Edwards et al (1976) and Boor et al (1993), the latter in two siblings, although ALMS was not genetically proved at that time and the clinical features of these patients do not fit completely with ALMS.

Oral Disorders

Among the clinical manifestations of ALMS, the presence of some oral disorders featured by enamel opacities, discolored enamel bands, and enamel hypoplasia has been reported in a single Turkish kindred (Koray et al, 2001). Moreover, it was documented by light and transmission electron microscopy of the gingival field an irregular thickness of the basal lamina associated with delamination of the myelin sheath. Although the cause of such alterations remains unknown and unexplained, the authors suggested that a systemic disorder in connection with ALMS might have been the etiological factor, which influenced tooth formation for a limited time.

Others

Pancreatitis related to dyslipidemia has been reported in ALMS individuals and the cause of it seems to be hypertriglyceridemia. In particular, a link between hypertriglyceridemia and pancreatitis has been recently documented in a 21-year-old female individual with a diagnosis of ALMS based on clinical picture (Wu et al, 2003). Furthermore, in the large series investigated by Paisey et al (2004) 4 out of 37 patients suffered from severe hypertriglyceridemia (> 8 mmol/L) resulting in pancreatitis and in absence of severe renal or hepatic dysfunctions. Notably, the patients with hypertriglyceridemia and pancreatitis were all affected from type 2 diabetes (Wu et al, 2003; Paisey et al, 2004). Pancreatitis has also been recently reported in another ALMS patient (Beck, 2003).

In ALMS, splenomegaly has been observed, but secondary to liver disorders which could affect the portal blood flow. Furthermore, spleen enlargement could explain the low platelet values observed in some individuals with ALMS. Portal hypertension is a complication of liver cirrhosis and often manifests with encephalopathy and esophageal varices. Gastro-esophageal-reflux-disorders (GERD) are a newly recognized frequent feature reported in 35% of patients from a large series of 182 (Marshall et al, 2005).

In the review of literature by Millay et al (1986), a number of variable features not observed in all the patients were summarized, including acanthosis nigricans (9/15), baldness (8/11), hyperuricemia (5/7), scoliosis (6/14), and hyperostosis frontalis interna (3/10). These features have been further confirmed by many other authors (Collin et al, 1999a; Marshall et al 2005). Squared iliac bones and increased angulation of both femoral heads have been observed in a single report (Charles et al, 1990). Advanced bone age and several skeletal abnormalities including flat feet have been confirmed (Maffei et al, 2002a). Coxa valga, severe genus varus, tibial metaphyseal dysplasia, and narrowing of the iliac wing in two patients have also been documented (Quiros-Tejeira et al, 2001).

Pulmonary disorders have been reported in ALMS, including bronchial non-atopic asthma by the age of three months (Marshall et al, 1997; Titomanlio et al, 2004). However, dyspnea in ALMS might not be a primary disorder of the lung, therefore dilated cardiomyopathy and secondary pulmonary involvement should be excluded. In a large cohort of 182 individuals, Marshall et al (2005) documented persistent pulmonary symptoms in 53% of the patients, including bronchitis, asthma, chronic obstructive pulmonary disease, and acute respiratory distress syndrome. Severe pulmonary fibrosis was documented histopathologically (Marshall et al, 2005).

Finally, there have been some isolated case reports that showed an association between ALMS and congenital dyserythropoietic type III anemia (Cagianut et al, 1977), situs inversus totalis (Ikeda et al, 1974), central diabetes insipidus (Aynaci et al, 1995), acute lymphoblastic leukemia (Chen et al, 2000), and Klinefelter's karyotype (Evans et al, 2000).

Pathology

Histopathological findings from post-mortem evaluation of different tissues including the kidney, liver, heart, pancreas, testis, ovaries, urinary bladder, lungs, and eyes have been systematically described (Bronson et al, 2003, Marshall et al, 2005). The most consistent findings were interstitial fibrosis, sometimes scattered, in all the tissues examined. However, it is not clear whether the fibrosis is the primary disturbance of the syndrome, i.e. the primum movens, or simply the epiphenomenon of another disorder that results in end stage organ failure and fibrosis.

Genetics

ALMS, caused by mutations within a single gene, is eminently suitable for molecular genetic analysis. Understanding the genetic and molecular basis of Alström syndrome will be important not only from the viewpoint of understanding how a mutation in particular gene causes its pleiotrophic effects, but it also may provide access to novel metabolic and regulatory pathways involved in the etiology of neurosensory disease, obesity, type 2 diabetes and related disorders.

In 1992, researchers at The Jackson Laboratory began to study the phenotypic and genetic characteristics of a large kindred living in the Acadian region of Nova Scotia (Marshall et al, 1997). The Acadians migrated from Northern France to Acadia (Canada) in the first decades of the 17th century and lived for many years in relative isolation. The "Grand Deportation" of the Acadians from their original settlements occurred in 1755 as a result of the Seven Year's War in Europe. Following the Deportation, Acadians established settlements in France, Quebec, Louisiana, South Carolina, Massachusetts, and other locations in the American Colonies (Marshall et al, 1997). Most Acadians returned to their homeland and there remained largely isolated in small communities in the coastal regions of what is now called Southwest Nova Scotia and Northern New Brunswick (Canada). The extended Acadian pedigree studied consisted of 5 affected individuals descended through thirteen generations from a common ancestral couple.

In a genome-wide screen and homozygosity mapping using this pedigree, haplotype sharing for a region on chromosome 2p13 was observed in all affected individuals. By two-point linkage analysis, the disease gene was localized to a 14.9-cM region on 2p14-p13 (Collin et al, 1997). Later, Macari et al (1998) confirmed the ALMS locus in a large, consanguineous Algerian family. Genetic analysis of meiotic recombination events in 12 additional unrelated ALMS families localized the critical region to a 6.1-cM interval with a cumulative multipoint LOD score of 9.16 (Collin at al 1999a).

Several attractive candidate genes were found to lie within the critical region. Dynactin, DCTN1 (Collin et al 1998), transforming growth factor-alpha gene, TGFA (Collin et al 1999b), and the human homolog of Lbx2 (Chen et al 1999), were evaluated and eliminated as candidates for *ALMS1*.

In 2002, two independent research groups from The Jackson Laboratory (USA) and Southampton University (UK) cloned the *ALMS1* gene and detected mutations resulting in Alström Syndrome (Collin et all 2002; Hearn et al 2002). Collin et al assembled the sequences throughout the ALMS critical region and identified 16 transcripts. Thirteen of the 16 transcripts were tested by direct sequencing of coding exons. They eventually identified an uncharacterized transcript, KIAA0328, in which affected individuals from the Acadian pedigree had an insertion of 19 bp in exon 16, causing a frameshift resulting in early termination at codon 3530. All 5 affected patients from the pedigree were homozygous for the insertion. Transmission of the insertion allele in unaffected carriers was consistent with previously reported haplotypes (Collin et al 1997). In addition, in another consanguineous Italian family with ALMS, a homozygous 8383C-T transition in the same gene caused a nonsense change, glu2795 to ter (G2795X). Four frameshift mutations and 2 nonsense mutations in KIAA0328, segregated with Alström syndrome in 5 additional unrelated families (Collin et al 2002). These findings provided strong evidence that *KIAA0328* was the Alström gene, *ALMS1*.

At the same time, Hearn et al (2002) studied a patient with the clinical features of ALMS whose karyotype showed a familial balanced reciprocal translocation with a 2p13 breakpoint. The translocation break on the maternal allele involved a genomic fragment of 1.7 kb which contained exon 4 and the start of exon 5 of *KIAA0328* making it a strong candidate for *ALMS1*. The paternal allele was disrupted by a deletion of 2 bp in exon 8, 2141delC-T, a frameshift mutation that was predicted to cause premature termination 5 codons downstream of the deletion. Additionally, in 2 sibs, they found compound heterozygosity for a 10775delC mutation and a trp3664-to-ter mutation in *KIAA0328* gene. They found the 10775delC mutation in 3 additional families not known to be related to the original family.

The ALMS1 Gene

ALMS1 is a novel gene, of unknown function, encoding a large cDNA product comprised of 23 exons. The 12.9-kb transcript of *ALMS1* encodes a protein of 4,169 amino acids. The protein contains a large tandem-repeat domain comprised of 34 imperfect repetitions of 47 amino acids. At least 44 *ALMS1* mutations have been identified thus far (unpublished results). These mutations have been nonsense or frameshift mutations found on exons 2, 8, 10, and 16 (unpublished results; Collin et al 2002; Hearn et al 2002).

ALMS1 is ubiquitously expressed at low levels in all tissues reported to be pathologically affected in ALMS patients and thus far does not share significant sequence homology with other known genes. Several splice variants have been reported (Collin et al 2002).

Recent proteomic and antibody studies have shown ALMS1 to localize to centrosomes and basal bodies of ciliated cells (Andersen et al, 2002; Hearn et al, 2005). Centrosomal function defects have been implicated in Bardet-Biedl syndrome, a related disorder with significant phenotypic overlap. However, the mechanism of centrosomal dysfunction in Bardet-Biedl and Alström syndrome remains unclear.

Genetic Variability

Variable presentation of clinical features is common in ALMS. Subsets of affected individuals can present with differing combinations of features such as onset of hearing loss, dilated cardiomyopathy, hepatic dysfunction, hypothyroidism, male hypogonadism, and mild to moderate developmental delay.

Collin (Collin et al 2002) described two unrelated male young adults, each having the identical 10775delC mutation in *ALMS1*. Both presented with congenital retinal dystrophy and infantile cardiomyopathy within the first 2 months of life. Both were obese by 2 years of age and developed type 2 diabetes, and renal insufficiency in their teenage years. However, the first experienced a sudden recurrence of dilated cardiomyopathy at age 18 and had no evidence of hepatic dysfunction, whereas the second presented with severe hepatic failure at age 20 and did not have a recurrence of cardiomyopathy. This difference in disease progression in individuals carrying the same mutation suggests that the phenotypic variability observed in many individuals with ALMS may be the result of genetic or environmental modifiers interacting with the *ALMS1* locus.

Because ALMS patients share some metabolic and clinical features of type 2 diabetes mellitus, it has been recently investigated whether gene variants of *ALMS1* could be found in general population as well ('t Hart et al, 2003). This is the first association study carried out in a large population with type 2 diabetes in age-matched normoglycemic subjects, specifically investigating three different gene variants of *ALMS1*. Genotype and allele frequencies were not found to be significantly different between diabetic population and control subjects. In addition, there was no evidence of any relationship between these specific *ALMS1* gene variants and BMI, glucose or insulin concentrations during OGTT, and insulin resistance as calculated by the HOMA index. Although these results have been negative, only further studies in larger cohorts with additional *ALMS1* gene variants will clarify the role of ALMS1 in the pathogenesis of type 2 diabetes in the general population.

Genetic Testing

The diagnosis of ALMS is still largely based on clinical findings. Up to now, research groups have detected mutations in the following regions of *ALMS1*: exon 2, exon 8, exon 10, and exon 16. Diagnostic screening of *ALMS1* is usually restricted to these regions. Only rough estimates can be made about the percentage of successful detection of mutations with this approach (about 30%). Carrier testing and prenatal testing is available for families with known mutations (S. Beck, personal communication).

The current methods for genetic testing will not detect mutations located in regions of *ALMS1* not analysed (including intron sequences other than the splice junctions and upstream and downstream sequences). Some sequence alterations that may be detected (such as those causing missense changes) will be of unknown clinical significance. Interpretation of test results should be in the context of the patient's clinical and family histories, and other clinical and laboratory test results.

Mouse Model

To obtain a mouse model for ALMS in which the mechanism and etiology can be further studied, a gene-trapped ES cell line was used to generate mice (Collin et al, 2005). Phenotypical characterization of the null mutant of the mouse *Alms1* gene, B6.129-*Alms1*$^{GT(pGT1Lxf)/Pjn}$ (*Alms1-/-*) has shown that the model faithfully replicates the human disease, with some differences in the onset and disease progression. *Alms1-/-* mice develop obesity, hyperinsulinemia, hypogonadism, retinal degeneration, and hearing loss. Immunohistochemistry and electron microscopy of retinal tissues has revealed mislocalization of rhodopsin in the outer nuclear layer and accumulation of vesicles in the inner segments of the photoreceptors, respectively, suggesting a disruption in intracellular trafficking (Collin et al, 2005). A similar retinal phenotype has been demonstrated in Bardet-Biedl mouse model, BBS4-null mice, implying a defect in the formation and function of ciliated cells (Mykytyn et al, 2004). However, Collin et al (2005) showed by scanning and transmission electron microscopy that *Alms1-/-* mice assemble normal primary cilia.

Diagnosis

The diagnosis of Alström syndrome remains a difficult task both for clinicians and geneticists because of the considerable overlap between ALMS and other genetic disorders with retinopathy, obesity, and diabetes and the difficulty of obtaining a genetic diagnosis. From the clinical point of view there are many problems in recognizing ALMS because the syndrome is rare and largely unknown by specialists. Moreover, although ALMS patients tend to share a similar phenotype (personal observations), there has been no clear codification of somatic alterations, which could reliably aid the diagnostic process beginning at the first clinical examination (see figures of patients). All these aspects could explain why the syndrome can still be overlooked in the diagnostic process. Sometimes the clinical picture is not clear and individuals are classified as ALMS even when they present only a few features of the syndrome or a mild typical phenotype (Iannello et al, 2004). Moreover, there could be an overlap of signs and symptoms with different disorders, including Bardet-Biedl (Edwards et al, 1976; Hauser et al, 1990; Boor et al 1993; Cohen et al, 1994). Modifier genes, genetic heterogeneity or new variants could explain some of the cases that have been difficult to classify. As a general rule, we believe that all unusual, mild, or atypical ALMS individuals should be diagnosed only when they are associated with *ALMS1* mutations.

It has been frequently documented that ALMS patients received both multiple diagnoses or significant diagnostic delay. Misdiagnosis has been frequently reported in the literature and especially involves Bardet-Biedl syndrome (Millay et al, 1986; Dyer et al, 1994; Deeble et al 2000; Beck, 2003), Leber's congenital amaurosis (Russell-Eggitt et al, 1998; Deeble et al, 2000), sporadic dilated cardiomyopathy (Deeble et al, 2000), unidentified mitochondrial disorders (Deeble et al, 2000), and Usher syndrome (Deeble et al, 2000).

Diagnostic delay has been also pointed out by Holder et al (1995) when a 15 year-old was diagnosed for the first time after hospital admission because of poor wound healing and hyperglycemia. Richardson et al (2001) have been able to diagnose ALMS in a 28-year-old woman and her 24-year-old brother when she was admitted to hospital complaining of symptoms related to kidney failure. Psychological consequences of the diagnostic delay at age 25 in an ALMS patient from Portugal have been recently reported by Beck (2003) which focused on the patient's grief, skepticism of medical care, and refusal of further medical help.

Treatment

No specific treatment of ALMS currently exists and it is beyond the scope of this chapter to cover all the many therapeutical options related to the treatment of single complications. Some of the specific therapeutic choices have been addressed in the previous paragraphs. Ideally, ALMS patients should be followed using a multidisciplinary approach and by a team of specialists, not only physicians, who also have expertise in ALMS covering ophthalmology, ear-nose-throat disorders, endocrinology and diabetology, cardiology, nephrology and urology, and gastroenterology. The ALMS syndrome deeply affects the life of patients and their family so that general practitioners should be encouraged to carefully follow their patients in every day problems. Considering the complexity of the syndrome and the multiorgan involvement, both pediatricians in childhood, and internal medicine specialists in adulthood, should take care of patients and possibly coordinate therapeutical choices, specialist follow-up, and prevention of long-term complications. A number of other experts, including psychologists, teachers for disabled people, and dieticians, should be involved to support the difficult everyday life of patients.

Special Issues and Conclusion

The ALMS syndrome is a rare disorder, which deeply impacts the life of patients and families. Fortunately, the knowledge on pathophysiological processes behind the *ALMS1* function is growing everyday and now more researchers than in the past are dedicated to this area.

Of interest, web sites that could be useful to obtain further scientific information on the syndrome include: OMIM (Online Mendelian Inheritance in Man; www.ncbi.nlm.nih.gov), a database of the National Institutes of Health on Mendelian genetic diseases; The web-site sponsored by The Jackson Laboratory (www.jax.org/alstrom) is dedicated to Alström syndrome; Smart Group is a web discussion forum composed of people interested in ALMS, including many individuals with Alström syndrome and their physicians (free registration) (www.smartgroups.com).

The web site, www.jax.org/alstrom is especially useful because it covers the problems both of physicians and patients and is regularly updated. Within the site one can find other useful information on the "world" of Alström syndrome and be updated on further International appointments, associations of patients, and physicians. We commend the work of Alström Syndrome International (ASI), which is a world based association that stimulates international meetings of families and physicians. ASI has contributed significantly to progress in research on ALMS. This foundation offers many services, and among its priorities are research and information sharing, identification of new patients, and support for an international network of physicians. Other critical areas covered by ASI include expertise for newly identified individuals, facilitation of genetic counseling and testing, information, and emergency needs. ASI is committed to services which support diagnosis and appropriate treatments, expansion of an internationally accessible database, organization of regular Medical Workshops, and formulation of guidelines for professionals. ASI functions as a clearing house for patient information and referrals. Moreover, there is an International Scientific Advisory Board on ALMS, which has the role of translating scientific achievements into everyday practice.

We would also mention the formation of local associations of ALMS families, including UK, Canada, and France among the most active groups. These associations have had a fundamental role, along with Alström Syndrome International, in raising interest in the syndrome, raising money for research and meetings, and providing substantial help to patients and families.

We wish to suggest that physicians, patients, families, and all who are interested to ALMS join the Smart Group list-serve, which was recently established by S. Beck, a geneticist from Portugal. This is a special discussion forum dedicated to ALMS, with many participants, and it is the "beating heart" of the ALMS community and it testifies to the emotions and everyday problems related to the syndrome. Through the site, through the years, good news and, unfortunately, also bad news has been shared by this community. We think, that sometimes help for families and professionals dealing with other rare diseases like ALMS could be addressed this way. The Smart Group works independently with no moderator. Its success has greatly contributed to helping people who write from every corner of the world to feel less alone. In the best of times, and sometimes even in the worst of what this syndrome has to offer, patients and families just look for someone willing to laugh with them, cry with them, and to share their problems. It has been a great lesson for everyone.

References

Alström CH, Hallgren B, Nilsson LB, Asander H. Retinal degeneration combined with obesity, diabetes mellitus and neurogenous deafness. A specific syndrome (not hitherto described) distinct from Laurence-Moon-Biedl syndrome. A clinical endocrinological and genetic examination based on a large pedigree. *Acta Psychiatr Neurol Scand* 1959; 34 (suppl. 129):1-35.

Alter CA, Moshang TJ. Growth hormone deficiency in two siblings with Alström syndrome. *Am J Dis Child* 1993; 147(1):97-99.

Andersen JS, Wilkinson CJ, Maylor T, Mortensen P, Nigg EA, Mann M. Proteomic characterization of the human centrosome by protein correlation profiling. *Nature* 2003; 426(6966):570-574.

Aynaci FM, Ökten A, Mocan H, Gedik Y, Sarpkaya AÖ. A case of Alström syndrome associated with diabetes insipidus. *Clin Genet* 1995; 48(3):164-166.

Awazu M, Tanaka T, Yamazaki K, Kato S, Higuchi M, Matsuo N. A 27-year-old woman with Alström syndrome who had liver cirrhosis. *Keio J Med* 1995; 44(2):67-73.

Awazu M, Tanaka T, Sato S, Anzo M, Higuchi M, Yamazaki K, Matsuo N. Hepatic dysfunction in two sibs with Alström syndrome: Case report and review of the literature. *Am J Med Genet* 1997; 69(1):13-16.

Beck S. A case of late diagnosis of Alström syndrome. 2003 (Accessed August 4, 2004, at *http://www.jax.org/alstrom/documents/padua_congress.html).*

Boor R, Herwig J, Schrezenmeir J, Pontz BF, SchÖnberger W. Familial insulin resistant diabetes associated with acanthosis nigricans, polycystic ovaries, hypogonadism, pigmentary retinopathy, labyrinthine deafness, and mental retardation. *Am J Med Genet* 1993; 45(5):649-653.

Bronson R, Marshall JD. Pathology of five Alström patients. 2003 (Accessed August 4, 2004, at *http://www.jax.org/alstrom/documents/padua_congress.html).*

Cagianut B, Hochstrasser P, Rhyner K. Juvenile diabetes, tapeto-retinal degeneration, neurogenic hearing disorder (Alström syndrome), associated with congenital dyserythropoietic type III anemia. *Schweiz Med Wochenschr* 1977; 107(13):446-450.

Carey C. Alström's syndrome and the heart. 2003 (Accessed August 4, 2004, at *http://www.jax.org/alstrom/documents/padua_congress.html).*

Charles SJ, Moore AT, Yates JRW, Green T, Clark P. Alström's syndrome: further evidence of autosomal recessive inheritance and endocrinological dysfunction. *J Med Genet* 1990; 27(9):590-592.

Chen BH, Chiou SS, Tsai RK, Lin YF, Wu JR. Acute lymphoblastic leukemia in one of two sibling with Alström syndrome. *J Formos Med Assoc* 2000; 99(10):792-795.

Chen F, Collin GB, Liu KC, Beier DR, Eccles M, Nishina PM, Moshang T, Epstein JA (2001): Characterization of the murine Lbx2 promoter, identification of the human homologue, and evaluation as a candidate for Alström syndrome. *Genomics 74(2)*: 219-227.

Cohen J, Kisch ES. The Alström syndrome: a new variant? *Isr J Med Sci* 1994; 30(3):234-236.

Collin GB, Marshall JD, Cardon LR, Nishina PM. Homozygosity mapping of Alström syndrome to chromosome 2p. *Hum Mol Genet* 1997; 6(2):213-219.

Collin GB, Nishina PM, Marshall JD, Naggert JK (1998): Human DCTN1: Genomic structure and evaluation as a candidate for Alström syndrome. *Genomics 53*: 359-364.

Collin GB, Marshall JD, Boerkoel CF, Levin AV, Weksberg R, Greenberg J, Michaud JL, Naggert JK, Nishina PM. Alström syndrome: further evidence for linkage to human chromosome 2p13. *Hum Genet* 1999; 105(5):474-479 (a).

Collin GB, Marshall JD, Naggert JK, Nishina PM. TGFA: exon-intron structure and evaluation as a candidate gene for Alström syndrome. *Clin Genet* 1999; 55(1):61-62 (b).

Collin GB, Marshall JD, Ikeda A, SoWV, Russell-Eggitt I, Maffei P, Beck S, Boerkoel CF, Sicolo N, Martin M, Nishina PM, Naggert JK. Mutations in ALMS1 cause obesity, type 2 diabetes and neurosensory degeneration in Alström syndrome. *Nature Genetics* 2002; 31(1):74-78.

Collin GB, Cyr E, Bronson R, Marshall JD, Gifford EJ, Hicks W, Murray SA, Zheng QY, Smith RS, Nishina PM, Naggert JK. Alms1-disrupted mice recapitulate human Alström syndrome. *Hum Mol Genet.* 2005;14(16):2323-33.

Connolly MB, Jan IE, Couch RM, Wong LTK, Dimmick JE, Rigg JM. Hepatic dysfunction in Alström disease. *Am J Med Genet* 1991; 40(4):421-424.

Deeble VJ, Roberts E, Jackson A, Lench N, Karbani G, Woods CG. The continuing failure to recognise Alström syndrome and further evidence of genetic homogeneity. *J Med Genet* 2000; 37(3):219.

Dyer DS, Wilson ME, Small KW, Pai GS. Alström syndrome: a case misdiagnosed as Bardet-Biedl syndrome. *J Pediatr Ophthalmol Strabismus* 1994; 31(4):272-274.

Edwards JA, Sethi PK, Scoma AJ, Bannerman RM, Frohman LA. A new familial syndrome characterized by pigmentary retinopathy, hypogonadism, mental retardation, nerve deafness and glucose intolerance. *Am J Med* 1976; 60(1):23-32.

Evans PMS, Ansari BM, Page MD. Alström's syndrome with Kleinfelter's karyotype. *J Inter Med* 2000; 248(1):89-92.

Goldstein JL, Fialkow PJ. The Alström syndrome. Report of three cases with further delineation of the clinical, pathophysiological, and genetic aspects of the disorder. *Medicine (Baltimore*) 1973;52(1):53-71.

Hauser C, Rojas C, Roth A, Schmied E, Saurat JH. A patient with features of both Bardet-Biedl and Alström syndromes. *Eur J Pediatr* 1990; 149(11):783-785.

Hearn T, Renforth GL, Spalluto C, Hanley NA, Piper K, Brickwood S, White C, Connolly V, Taylor JF, Russell-Eggitt I, Bonneau D, Walker M, Wilson DI. Mutations of ALMS1, a large gene with tandem repeat encoding 47 amino acids, causes Alström syndrome. *Nature Genetics* 2002; 31(1):79-83.

Hearn T, Spalluto C, Phillips VJ, Renforth GL, Copin N, Hanley NA, Wilson DI. Subcellular localization of ALMS1 supports involvement of centrosome and basal body dysfunction in the pathogenesis of obesity, insulin resistance, and type 2 diabetes. Diabetes. 2005; 54(5):1581-7.

Holder M, Hecker W, Gilli G. Impaired glucose tolerance leads to delayed diagnosis of Alström syndrome. *Diabetes Care* 1995; 18(5):698-700.

Hopkinson I, Marshall JD, Paisey RB, Carey C, Mac Dermott S. Alström syndrome 2003 (Accessed August 4, 2004, at *http://www.geneclinics.org*)

Horiuchi N, Sasaki A, Suzuki T, Endo Y, Kiyonaga G. A case of Alström syndrome with cirrhosis of the liver. *J Jpn Diab Soc* 1976; 19:353-359.

Hung YJ, Jeng C, Pei D, Chou PI, Wu DA. Alström syndrome in two siblings. *J Formos Med Assoc* 2001; 100(1):45-49.

Iannello S, Bosco P, Camuto M, Cavaleri A, Milazzo P, Belfiore F. A Mild form of Alström disease associated with metabolic syndrome and very high fasting serum free fatty acids: Two cases diagnosed in adult age. *Am J Med Sci* 2004; 327(5) 284-288.

Ikeda Y, Morita Y, Matsuo Y, Akanuma Y, Itakura H, Yamaji T, Kosaka K. A case of Alström syndrome associated with situs inversus totalis and characteristic liver cirrhosis. *Nippon Naika Gakkai Zasshi* 1974; 63(11):1303-1311.

Johnson J. Diabetes, neurogenous deafness, and retinal degeneration. *Brit Med J* 1961; 2:646.

Klein D, Ammann F. The syndrome of Laurence-Moon-Bardet-Biedl and allied diseases in Switzerland. Clinical, genetic and epidemiological studies. *J Neurol Sci* 1969; 9:479.

Kopecky A, Seemanova E, Salichova J. Alstromuv syndrome u dvou sester. *Cas Lek Cesk* 1978; 117:921.

Koray F, Dorter C, Benderli Y, Satman I, Yilmaz T, Dinccag N, Karsidag K. Alström syndrome: a case report. *J Oral Sci* 2001; 43(3):221-4.

Mac Dermott S. Urological involvement in Alström syndrome. 2001 (Accessed August 4, 2004, at *http://www.jax.org/alstrom/documents/urology_abstract.pdf*).

Mac Dermott S. Renal and Urology. 2003 (Accessed August 4, 2004, at *http://www.jax.org/alstrom/documents/padua_congress.html*).

Macari F, Lautier C, Girardet A, Dadoun F, Darmon P, Dutor P, Dutor A, Renard E, Bouvagnet P, Claustres M, Oliver C, Grigorescu F. Refinement of genetic localization of the Alström syndrome on chromosome 2p12-13 by linkage analysis in a North African family. *Hum Genet* 1998; 103(6):658-661.

Maffei P, Imi M, Marshall JD, Martini C, DeCarlo E, Nishina PM, Collin GB, Sicolo N. *Endocrinological evaluation of Alström syndrome: Case report.* Program and Abstract Book 81st Annual Meeting of the Endocrine Society 1999; P1-61 (abs).

Maffei P, Munno V, Marshall JD, Scandellari C, Sicolo N. The Alström syndrome: is it a rare or unknown disease? *Ann Ital Med Int* 2002; 17(4):221-228. (a)

Maffei P, Marshall JD, Munno V, Collin GB, Martini C, Mioni R, De Carlo E, Vettor R, Nishina PM, Scandellari C, Sicolo N. *Growth hormone deficiency in Alström syndrome.* Abstract Book of the Endocrine Society's 84th Annual Meeting, San Francisco, June 19-22, 2002, P3-566 (abs) (b).

Maffei P, Boschetti M, Orsi I, Marshall JD, Paisey R, Beck S, Munno V, Minuto F, Barreca A, Sicolo N. The IGF system in Alström syndrome. *Growth Hormone and IGF Research* 2002; 12(4):291 (abs) ©.

Makaryus AN, Popowski B, Kort S, Paris Y, Mangion J. A rare case of Alström syndrome presenting with rapidly progressive severe dilated cardiomyopathy diagnosed by echocardiography. *J Am Soc Echocardiogr* 2003; 16 (2):194-196.

Marshall JD, Ludman MD, Shea SE, Salisbury SR, Willi SM, LaRoche RG, Nishina PM. Genealogy, natural history, and phenotype features of Alström syndrome in a large Acadian Kindred and three additional families. *Am J Med Genet* 1997; 73(2):150-161.

Marshall JD, Bronson RT, Paisey RB, Carey C, MacDermott S, Collin GB, Nordstrom AD, Maffei P, Beck S, Davis J, Pozzan GB, Shea SE, Shatirishvili G, Fiscuci C, McEwan L, Nishina PM. New phenotypes and diagnostic criteria for Alström syndrome based on evaluation of 134 cases. *Am J Hum Genet* 2002; Suppl to n° 71(4):550(abs).

Marshall JD. Research in Alström syndrome: history and future directions. 2003 (Accessed August 4, 2004, at *http://www.jax.org/alstrom/documents/padua_congress.html*).

Marshall JD, Bronson RT, Collin GB, Nordstrom AD, Maffei P, Paisey RB, Carey C, Macdermott S, Russell-Eggitt I, Shea SE, Davis J, Beck S, Shatirishvili G, Mihai CM, Hoeltzenbein M, Pozzan GB, Hopkinson I, Sicolo N, Naggert JK, Nishina PM. 2005; New Alström syndrome phenotypes based on the evaluation of 182 cases. *Arch Intern Med.* 165(6): 675-83

Michaud JL, Héon E, Guilbert F, Weill J, Puech B, Benson L, Smallhorn F, Shuman CT, Buncic JR, Levin AV, Weksberg R, Brevière G-M. Natural history of Alström syndrome in early childhood: onset with dilated cardiomyopathy. *Journal of Pediatrics* 1996; 128(2): 225-229.

Millay RH, Weleber RG, Heckenlively JR. Ophthalmology and systemic manifestations of Alström's disease. *American Journal of Ophthalmology* 1986; 102(4): 482-490.

Mykytyn K, Mullins RF, Andrews M, Chiang AP, Swiderski RE, Yang B, Braun T, Casavant T, Stone EM, Sheffield VC. Bardet-Biedl syndrome type 4 (BBS4)-null mice implicate Bbs-4 in flagella formation but not global cilia assembly. *Proc. Soc. Natl. Acad. Sci USA* 2004; 101:8664-8669.

Paisey RB, Carey CM, Bower L, Marshall J, Taylor P, Maffei P, Mansell P. Hypertriglyceridemia in Alström's syndrome: causes and association in 37 cases. *Clin Endocrinol* 2004; 60(2): 228-231.

Porcellini M.G. Successful kidney transplantation in a case of Alström syndrome. 2003 (Accessed August 4, 2004, at ttp://www.jax.org/alstrom/documents/padua_congress.html).

Quiros-Tejeira RE, Vargas J, Ament ME. Early-onset liver disease complicated with acute liver failure in Alström syndrome. *Am J Med Genet* 2001; 101(1): 9-11.

Richardson D, Shires M, Davidson AM. Renal diagnosis without renal biopsy. Nephritis and sensorineural deafness. *Nephrology Dial Transplant* 2001; 16:1291-1294.

Rudiger HW, Ahrens P, Dreyer M, Frorath B, Loffel C, Schmidt-Preuss U. Impaired insulin-induced RNA synthesis secondary to a genetically defective insulin receptor. *Hum Genet* 1985; 69(1): 76-78.

Russell-Eggitt IM, Taylor DSI, Clayton PT, Garner A, Kriss A, Taylor JFN. Leber's congenital amaurosis -a new syndrome with a cardiomyopathy. *British Journal of Ophthalmology* 1989; 73(4): 250-254.

Russell-Eggitt IM, Clayton PT, Coffey R, Kriss A, Taylor DS, Taylor JF. Alström syndrome. Report of 22 cases and literature review. *Ophthalmology* 1998;105(7):1274-1280.

Russell-Eggitt I. Ophthalmic manifestations of the Alström syndrome. 2003 (Accessed August 4, 2004, at *http://www.jax.org/alstrom/documents/padua_congress.html*).

Satman I, Yilmaz MT, Gursoy N, Karsidag K, Dinccag N, Ovali T, Karadeniz S, Uysal V, Bugra Z, Okten A, Devrim S. Evaluation of insulin resistant diabetes mellitus in Alström syndrome: a long-term prospective follow-up of three siblings. *Diabetes Res Clin Pract* 2002; 56(3):189-196.

Sebag J, Albert DM, Craft JL. The Alström syndrome: ophthalmic histopathology and renal ultrastructure. *British Journal of Ophthalmology* 1984; 68(7):494-501.

't Hart LM, Maassen JA, Dekker JM, Heine RJ, Maassen JA. Lack of association between gene variants in the ALMS1 gene and type 2 diabetes mellitus. *Diabetologia* 2003; 46(7):1023-1024.

Tai TS, Lin SY, Sheu WH. Metabolic effects of growth hormone therapy in an Alström syndrome patient. *Horm Res* 2003; 60(6):297-301.

Ten S, Matthews N, Anhalt H, Maclaren NK. *Effect of Metformin and Rosiglitazone in a prebubertal boy with Alström syndrome.* Abstract Book of the Endocrine Society's 86th Annual Meeting, New Orleans, June 16-19, 2004.

Titomanlio L, De Basi B, Buoninconti A, Sperandeo MP, Pepe A, Andria G, Sebastio G. Alström sindrome: intrafamilial phenotypic variability in sibs with a novel nonsense mutation of the ALMS1 gene. *Clin Genet* 2004; 65(2):156-157.

Tremblay F, LaRoche RG, Shea SE, Ludman MD. Longitudinal study of the early electroretinographic changes in Alström's syndrome. *American Journal of Ophthalmology* 1993; 115(5):657-665.

Van den Abeele K, Craen M, Schuil J, Meire FM. Ophthalmologic and systemic features of the Alström syndrome: report of 9 cases. *Bull Soc Belge Ophthalmol* 2001; 281:67-72.

Warren SE, Schnitth SJ, Bauman AJ, Gianelly RE, Landsberg L, Baim DS. Late onset dilated cardiomyopathy in a unique familial syndrome of hypogonadism and metabolic abnormalities. *American Heart Journal* 1987; 114(6):1522-1524.

Weichenhain B, Stemplinger J, Ziegler AG, Rabl W, Standl E, Stiegler H. Alström syndrome -a rare disease of diabetic association. *Med Klin* 1997; 92(3):175-178.

Weinstein RL, Kliman B, Scully RE. Familial syndrome of primary testicular insufficiency with normal virilization, blindness, deafness and metabolic abnormalities. *New England Journal of Med* 1969; 281(18):969-977.

Worthley MI, Zeitz CJ. Case of Alström syndrome with late presentation dilated cardiomyopathy. *Intern Med J* 2001; 31(9):569-570.

Wu WC, Chen SC, Dia CY, Yu ML, Hsieh MY, Lin ZY, Wang LY, Tsai JF, Chang WY, Chuang WL. Alström syndrome with acute pancreatitis: a case report. *Kaohsiung J Med Sci* 2003; 19(7):358-361.

Zumsteg U, Muller PY, Miserez AR. Alström syndrome: confirmation of linkage to chromosome 2p12-13 and phenotypic heterogeneity in three affected sibs. *J Med Genet* 2000; 37(7):E8.

In: New Developments in Obesity Research
Editor: Lawrence F. Ditmier, pp. 203-223
ISBN 1-60021-296-4

Chapter 8

IN VITRO, ANIMAL AND HUMAN APPROACHES IN NUTRITIONAL STUDIES AND INVESTIGATIONS

J. Alfredo Martinez and Miguel A. Martinez-Gonzalez
Department of Physiology and Nutrition and Department of Preventive Medicine
University of Navarra. Pamplona. Spain

ABSTRACT

Nutrition research concerns the study of those processes whereby cells, tissues and the whole body obtain and utilise substances contained in foods and drinks in order to maintain their structures and functions. Thus, nutrition oriented scientists are interested in understanding the balance between the supply and demand of substrates to ensure the functional integrity of the organisms, including homeostasis, growth and reproduction. Therefore, nutrition research involves the advances in knowledge concerning not only food composition, dietary intake and food/nutrient utilisation by the organism, but also and the short or long-term influences of food/nutrient consumption on health and nutrient functions, which can be achieved through "in vitro" and "in vivo" studies with different experimental models. Thus, processes involved in nutrient utilisation such as digestion, absorption, metabolism and excretion as well as nutrient functions can be evaluated and measured by using "in vitro" approaches. These tests involve a laboratory replication of an "in vivo" function. On the other hand, whole animal systems have been used in measuring nutrient utilization and fate as well as to assess nutrient functions. The selection of laboratory tests to be used in different models should take into consideration their precision, accuracy, analytical specificity and sensitivity, predictive value and validity. Most of the nutritional experiments and research methodology described for animals can be also applied to human beings. However, in some circumstances animal studies are preferred for practical reasons, but in others it is necessary to carry out the experimentation directly in man due to difficulties for data extrapolation. The impact of food intake on health and nutritional outcome is conveniently assessed thought dietary intake estimations, anthropometry, biochemical markers and clinical indications of the nutritional states. The selection of research tests to be used in different models should take into consideration their precision, accuracy, analytical specificity and sensitivity, predictive value and validity. In any case, adequate ethical guidelines and regulations must be followed and accomplished in all experiments carried out in animal and human beings.

Key words: In vivo, in vitro, nutrition studies and research.

INTRODUCTION

Any research approach in nutrition should be based upon the scientific method with the following steps: hypothesis formulation, experimentation under controlled conditions, results interpretation and theory formulation. The stages of research are cyclic and should be constantly tested by investigators in order to confirm or modify the existing theory. In general terms, experimental studies are devised to provide evidence for the effect of a nutritional impact (exposure) on a measurable variable (outcome). One of the first written examples about the causal relationship between an exposure and a health outcome can be found in the Bible (Dan 1:3 - 1:15). Thus, in this "research", the exposure to an intake of food (vegetables) and drinks (water) an health outcome (appearance and fatter flesh) was tested.

(Dan 1:3) Then the king commanded Ashpenaz, his chief eunuch, to bring some of the People of Israel, both of the royal family and of the nobility, (Dan 1:4) Youths without blemish, handsome and skilful in all wisdom, endowed with knowledge, understanding learning, and competent to serve in the king's palace, and to Teach them the letters and language of the Chaldeans. (Dan 1:5) The King assigned them a daily portion of the rich food which the king ate, and of the wine which he drank. They were to be educated for three years, and at the end of that Time they were to stand before the king. (Dan 1:6) Among these were Daniel, Hananiah, Misha-el, and Azariah of the tribe of Judah. (Dan 1:7) And the chief of the eunuchs gave them names: Daniel he called Belteshazzar, Hananiah he called Shadrach, Misha-el he called Meshach, and Azariah he called Abednego. (Dan 1:8) But Daniel resolved that he would not defile himself with the kings rich food, or with the wine which he drank; therefore he asked the chief of the eunuchs to allow him not to defile himself. (Dan 1: 9) And God gave Daniel favor and compassion in the sight of the chief of the eunuchs; (Dan 1:10) And the chief of the eunuchs said to Daniel, fear lest my lord king, who appointed your food and your drink, should see that you were in poorer condition than the youths who are of your own age. So you would endanger my head with the king. (Dan 1:11) Then Daniel said to the steward whom the chief of the eunuchs had appointed over Daniel, Hananiah, Misha-el, and Azariah. (Dan 1:12) "Test your servants for ten days; let us be given vegetables to eat and water to drink. (Dan 1:13) Then let our appearance and the appearance of the youths who eat the king's rich food be observed by you, and according to what you see deal with your servants. "(Dan 1:14) So he hearkened to them in this matter, and tested them for ten days. (Dan 1:15) At the end of ten days it was seen that they were better in appearance and fatter in flesh than all the youths who ate the king's rich food.

Nutrition research methodologies can be classified according to many criteria, for example, according to the aim (basic or applied), the strategy of data collection (historical or retrospective and prospective), the degree of control of the investigator over the exposures under study (experimental or observational), the epidemiological and timing design (case-control, cohort or cross-sectional), the setting (laboratory or field), or even the type of data (qualitative or quantitative), the kind of procedures (in vitro or in vivo) and the experimental model (animal or human beings).

The design of a nutritionally oriented investigation involves the selection of the research topic accompanied by the formulation of hypotheses and the aims, the preparation of a research protocol accompanied by the definition of required measurements and testing of the methods and, eventually the conduction of a pilot study. Other aspects concerning the study design are the selection of statistical analyses as well as ethical issues and commitments. In this context, the interpretation of research information should consider the role of chance and the occurrence of possible bias or results deviation due to systematic errors in selection or information gathering as well as the effects of other influences such as confounding factors or effect modifiers. An optimal study design must also take into account the sampling procedures, the study size, the statistical power, the duration of the follow period in addition to the research model (hypothesis and theory formulation).

"In vitro" Techniques for Use in Nutrition Research

Processes involved in nutrient utilization such as digestion, absorption, metabolism and excretion as well as nutrient functions can be evaluated and measured by using "in vitro" approaches. Biological outcome can be predicted by "in vitro" procedures, which have some advantages such as saving animal lives, time, space and expenses. These tests involve a laboratory replication of an "in vivo" function. A shortcoming of "in vitro" measurements is that they are relatively unphysiological and that the values usually are lower that those obtained in "in vivo" conditions, which has been mainly attributed to problems in the delivery of oxygen and substrates. However, despite the wide array of potential problems associated with "in vitro" measuring utilization of nutrients, a great deal of valuable data and meaningful conclusions regarding nutrient functions and regulatory mechanisms have been obtained from such methods, affording in some cases single aspects without interference from different types of cells or other factors.

The application of "in vitro" enzymatic digestion procedures, despite of some limitations, has provided useful information about macronutrient and dietary fibre availability by using acids or alkaline treatments or different proteases, lipases and amylases. A classical "in vitro" experiment to test the role of digestive juices on carbohydrates digestibility can be easily performed with saliva as the source of enzymes. Also, there are methods for determining apparent ileal protein and amino acid digestibility, which are complemented by mobile nylon bag techniques. Specific peptidase, disaccaridase, and lipase activities have been also quantified "in vitro" by applying mucosal scrapes on adequate substrates.

The publication of accurate and simple methods for amino acid analysis has made feasible the development of reliable protocols to anticipate protein quality and utilization. These procedures, based upon a food chemical analysis for their essential amino acids yield results, which are well correlated with biological values. Thus, the chemical score judge the nutritive adequacy and utilization by comparison of its amino acid composition with that of a reference protein (usually whole egg protein). Other classical chemical methods to evaluate protein quality are based on nitrogen analysis, protein solubility tests, microbiological assays, etc. Additional "in vitro" procedures to evaluate nutrient utilization are based on the assessment of nutrient absorption by means of everted intestine preparations, intestinal rings, Using´s type chambers or isolated enterocytes, which have been applied successfully for measuring sugar, amino acids, minerals and some vitamins transport and with more uncertainties for fat-soluble materials.

The study of nutrient fate and metabolism has been also benefited by the development of elegant and clever "in vitro" approaches, which have yielded information about the turnover of glucose, lipid, nitrogen compounds, minerals, vitamins and water as well as about energy metabolism in different cells and organs by means of appropriate markers and tracers (labelled compounds, enzymatic changes,...). Thus, the kinetics of different metabolic processes (glucolysis, glyconeogenesis, protein synthesis and breakdown, lipolysis and lipogenesis, etc.) have been evaluated under different physiological and nutritional circumstances in liver, muscle, blood cells, etc "in vitro" preparations... The application of radiolabelled markers, pulse change labelling, double-label methodologies and microinjection techniques as well as non-isotopic nutrient release protocols have provided valuable information about nutrient turnover and functions. As an example, cells preparations such as isolated adipocytes have been able to demonstrate the role of specific nutrients in lipolysis or lipogenesis, while the involvement of glucose precursors such a lactate in glyconeogenesis have been addressed in isolated hepatocytes. The availability of cells such as 3T3-L1 pre-adipocytes, Caco-2 cells obtained from colon carcinoma preparations, HepG2 human hepatoblastoma cell lines, etc has facilitated the standardisation of many studies concerning nutrient functions and metabolism.

Futhermore, the estimation of macronutrient synthesis and degradation has been afforded by investigators using a wide variety of "in vitro" strategies including perfused tissues, incubated intact tissues, organ cultures, cell-free systems, etc., changes in enzyme concentrations and the rates of enzyme formation and breakdown as well as the enzymatic activities of key molecules have produced valuable data concerning nutrient roles and utilization at the tissue level. Most studies have utilized tissues from rats, although biological samples from other animals have been used for specific purposes. The diversity of preparations provides flexibility in measuring "in vitro" metabolism processes. Furthermore, the effects of specific inhibitors, hormones, or other substances can be tested readily in addition to nutrients in "in vitro" preparations. In many instances, such effects can be studied by using matching in the design of experiments and applying paired statistical tests, providing a higher statistical efficiency and better of reliability in the measurements.

Assays related to host-defense and immunocompetence are other widely used "in vitro" techniques to assess the involvement of some nutrients in body homeostasis. Thus, a number of tests have been developed to identify and count different lymphocyte subtypes by standardised techniques to isolate inmune cells. Furthermore, lymphocyte function as an indicator of nutrient roles is assessed by the "in vitro" responses of lymphocytes to selected mitogens such as concanavalin A, phytohemaglutinin, lipopolisaccarides, etc., or nutrients (amino acids, fatty acids, minerals,...). A control group should also be included where possible, because genetic factors regulate many aspects of the immune response.

The light microscope is a powerful tool even in today's laboratory. Thus, visual examination of tissue subjected to various nutritional conditions can lead to novel discoveries on structure-function relationships. The use of antibodies to label or tag particular compounds, is commonly applied. Cell localization of particular proteins as affected by nutrient manipulation is often used as well. In addition to the use of immunohistochemistry with specific antibodies, "in situ" hybridization techniques are available in which a labeled cDNA can be hybridized to a tissue. Of course, this can provide quantitative information, as well as localization within tissues as to where the change in potential transcript expression may be occurring under different nutritional circumstances. The electron microscope allows the view in detail of cell organelles, such as mitochondria, that are not normally seen by a light microscope. However, the number of nutritionists using this powerful tool is limited.

The involvement of specific receptors in the nutritional functions of Vit B_{12}, retinol derivatives, etc., can be determined by using Scatchart plotts providing information about the impact of nutrition in receptor number and affinity. Additionally, the characterisation of second messangers and mediators at the nuclear or cytosolic level is of interest to identify both beneficial and detrimental influences and mechanisms of specific nutrients on nutritionally related aspects.

The application of molecular biology techniques to tissue and cell culture "in vitro" systems has provided researchers with powerful tools to evaluate and establish metabolic pathways and regulatory roles of nutrients and non-nutrient components of food. Thus, the use of Northern, Southern and Western blotting to quantitate specific RNA, DNA, and proteins in tissues in response to nutrients are a common tool in the nutrition laboratory. The influence of some nutrients or nutritional conditions on ribosomal dynamics as well as on cell hyperplasia or hypertrophy processes have been estimated through RNA, DNA or protein/ DNA values, respectively.

Subcells systems such as mitochondria, lysosomes, nuclei, etc.. have been specifically selected to characterise individual functions of differents minerals, vitamins and other nutrients. (fatty acids, amino acids, etc.), since some dietary constituents and signals can govern the expression of proteins at a number of transcriptional and posttranscriptional points. For example, fatty acids, retinoic acid and cholecalciferol have specific nuclear receptors. Following ligand activation, the receptors interact with specific DNA recognition sequences within a particular gene and either up or down regulate the gene's rate of transcription. Recent data also suggest that carbohydrates and polyunsaturated fatty acids may control the cytosolic content of specific transcripts by regulating the processing events involved in mRNA maturation. Another example is ferritin mRNA translation and transferrin receptor mRNA stability, which have now been shown to be regulated in response to changes in cytosolic iron. Similarly, other areas of molecular research often used by investigators applying molecular techniques to nutritional and metabolic research are measures of mRNA abundance, characterization of nuclear transcription factors and the DNA recognition sequences involved in the etiology of nutritionally related diseases.

Furthermore, molecular biology approaches have allowed numerous "in vitro" discoveries concerning certain "in vivo" nutrient functions at the gene level. Thus, the polymerase chain reaction (PCR) produces DNA amplification using "thermocyclers", transfection studies are involved in DNA insertion into cell to examine nutrient functions and metabolism. Thus, specific cell lines that lack the expression of a particular gene can be transfected with DNA containing the gene promoter as well as the gene to study the interactions of various nutrients with the expression of a particular gene. In either case, nutrient function at the cell level and the cell-gene level may be studied and provide definitive results. Gene regulation by nutrients has been assessed in different isolated cell and tissues using appropriate indicators and markers of gene expression RNA levels.

ANIMAL MODELS IN NUTRITION RESEARCH

Animals are used as providers of biological products such as hormones, antibodies, etc., but also are models for the study of biological responses and processes. Thus, whole animal systems have been used in measuring nutrient utilization and fate as well as to assess nutrient functions. A part of our knowledge regarding nutrition concepts stems from animal experiments, which are often extrapolated to man. Therefore, in many situations the animal serves as a substitute for man and it is referred to as an animal model.

The significance of results from animal experiments depends upon the selection of a suitable animal model, which have been categorised into four groups: individual animal models (e.g. administration or induced lack of a nutrient), spontaneous animal models (e.g. genetic models of nutrient intolerance), negative models (e.g. animal insensitive for a vitamin such as Vit.C) or orphan models (e.g. animals with a particular genotype serve to search for the human counterpart). In any case, data extrapolation from animals to man should always be carried out with reservation as should be the selection of animals on the basis of scientific, practical and ethical considerations. Intrinsic variability in animal responses can be reduced by using isogenic strains, genetic standardisation and specific pathogen free animals as well as by environmental and nutrition control, selection for uniformity, acclimatisation, unbiased selection and appropriate sampling.

The use of animals should be made following adequate regulations for animal experimentation according to local, national or international guidelines and legislation (Cruelty to Animals Act, Animal Welfare Act, Standards for the laboratory animals Act, EU directives...), which involve facilities, accommodation and animal care, competence, alternatives to animal experimentation anaesthesia and euthanasia procedures, registration, supply of animals as well as the involvement of the ethical committee.

Weanling or sexually nature male rat of different strains is the most widely used animal model, although mice, chickens, guinea pigs, dogs, sheeps, pigs, trouts, steers, primates among others have been bred for use in nutrition research throughout the world. The selection of specific animals for use in whole animal assays depends on the availability of the species, its ability to breed under captive conditions, the ease to be maintained, the previous use of the animal as a model and the possible extrapolation of the results to other species.

Food composition or nutrient content is measured by means of the animals response to dietary intake in body weight, organ or tissue growth and/or other appropriate criteria, such as the prevention or cure of some deficiency symptoms, the levels of specific nutrients in blood, urine, etc. The usefulness of animal bioassays is limited because their high cost, labor and time, although the fundamental advantage is that they give proof of biological response. In designing and conducting food analysis assays with animals, it is important to ensure that experimental influences act equally on all the animals. Therefore, the endocrine balance, maturation, age and previous nutritional status should be taken into consideration in addition to random sampling, size sample criteria and the characteristics of the animal model.

Food intake regulation is another relevant information that can be obtained from animal trials, which may help to explain the outcome under different experimental designs. Thus, metabolic balance is the result of intake minus output and reflects the amount of a given nutrient retained by the organisms, but also possible changes in body composition, which can be assessed by direct chemical methods or by using indirect approaches (i.e.: electric or energetic conductivity).

In nutritionally oriented studies, diets of laboratory animals are classified according to the degree of refinement of the ingredients as natural-ingredient diets, (formulated with corn, legume meal, fish, etc..), chemically defined diets (with pure chemicals), and purified diets (combination of natural ingredients, pure chemicals and refined nutrients). Variations in diet composition are often the key for the design of nutrition experiments. In this context, different feeding regimes can be applied to laboratory animal depending on scientific criteria: *Ad libitum*, in which the animal have free access to food or controlled feeding in which animals are offered a limited amount of food (restricted feeding) or in which the animal receive as much food as they can eat. (forced feeding). A specific form of restricted feeding is pair feeding, which involves the measurement of food consumed by some animals to match or equalise the intake in the following day of a test group. The intake or supply of nutrients may be administered orally, intravenously, intraperitoneally or by means of some specific tools (gavage, stereotaxis, etc.). Also, specific feeding with a selected combination of foods or ingredients, may lead to specific diet-induced models of obesity (cafeteria diets), hypercholesterolemia (highly saturated fat diets, etc.).

Digestibility assays in animal experiments aim to assess the portion of food or nutrient consumed, which is not excreted in the faeces and it is assumed to be digested by the animal. In a digestibility trial, an accurate measurement of the food intake and faeces/urine output is very important. For these trial, animals should be confined in metabolic cages or other devices to ensure the validity of the trial. For experiments conducted in poultry, the determination of digestibility is complicated by the fact that faeces and urine are voided together. In some circumstances, due to the lack of suitable equipment or the nature of the trial, indigestible markers could be used.

Growth and body composition are also widely used as criteria to assess nutrient utilization. A growth curve plots live body weight changes against time, supplying information about the general outcome from a nutritional exposure, which can be complemented by the estimation of the average absolute or relative growth rates, age equivalence in growth, etc.

Other approaches seek essentially at determining the changes in the content of the nutrient in the animal body in response to feeding a diet or meal containing that nutrient or food. While the nutrient levels in animal carcasses can be measured directly, virtually all measurements of nutrient levels in living animals depend largely on indirect strategies. There are basically two approaches to estimating changes in the nutrient content of the animal body. The first approach involves the measurement of weight changes (gain/loss), both direct and indirect, made on the animal body itself. Whole animal assays based on this approach for analysis of food protein and energy include protein efficiency ratio (PER), net protein ratio (NPR) and multilevel assays such as slope-ratio assays and net energy (NE). The second approach to estimating changes in the nutrient content of the body involves measurements of the intake and output of nutrients from the body through balance studies. Whole animal assays based on this approach for analysis of protein and energy metabolism include nitrogen or energy balance, biological value (BV), net protein utilization (NPU), and net energy (NE). Other strategies, known as depletion-repletion assays are based on the induction of a deficit, which is followed by monitoring the recovery to the basal level.

Whole blood or a fraction (leukocytes, erythrocytes, etc) are biopsy materials for nutritional assessment frequently used to estimate the level of a nutrient or its metabolite, although other fluids such as urine, saliva and the amniotic fluid or samples of hair, skin, nails, mucosa, etc., can be analysed. Laboratory tests of this specimens can provide specific and valuable information about nutrient utilization and function. An alternative procedure is to assess the urinary excretion rate, which reflects the body content or the size of the tissue store. The urinary outputs of some amino acids such as 3-methylhistidine or hydroxyproline are indicators of myofibrillar protein breakdown or collagen degradation, respectively under different nutritional circumstances.

Some functional tests inform about the impact of host nutriture to permit cells, tissues, or organs; to perform the nutrient-dependent biological functions. Plasma levels of minerals, vitamins and their metabolite are often good indictors of nutrient intake as well as the urinary output. Moreover, the circulating levels of haemoglobin, hematocrit, plasma proteins, insulin, glucose, ketonic bodies and other molecules may contribute to obtain information about nutrient fate. Other tests involve the measurement of abnormal metabolite production or changes in the activities of certain enzymes/blood components dependent on a specific nutrient. These assays may measure physiological functions as affected by specific nutrients and contribute to the quantitative and qualitative knowledge of specific metabolic pathways in animal models. Some factors that may confound the interpretation of laboratory tests are homeostatic regulation, diurnal variations, diseases, age, sex and nutrient interaction as well as some of the characteristics of the analytical method.

Nutrient turnover and deposition in animals have been conveniently measured by using isotopic tracers administered in labeled nutrients and by body composition techniques with simple or sophisticated devices (ultrasound, bioimpedance, etc), while immunocompetence approaches give information about the functional properties of some nutrients, particularly in depleted or deficient animals. Oxygen consumption and CO2 production values provide useful data about energy metabolism (basal metabolic rates, food induced thermogenesis, etc.), if adequate equipment and facilities are available. Thus, macronutrient oxidation (fat, carbohydrates, etc.) can be assessed by the calculation of the respiratory quotient ($CR=CO_2/O_2$ consumption) as well as by the application of specific equations for each energy-yielding substrates. Body temperature and heat dissipation measurements may provide also valuable information about nutrient metabolism and nutritional status in situations of under or over-nutrition as well as about energy expenditure.

Animal species that have been altered genetically are becoming more common in nutrition research. Either the presence of a powerful promoter linked to a gene of interest or the deletion of a gene by blocking an exon sequence, through transgenic models offer new insights into nutrients function. A transgenic model in which the gene of interest is no longer expressed is called a knock-out model. For example, the peroxisomal proliferating activating receptor-alpha (PPARα) is not expressed in one knock-out mouse model, producing fat accumulation as the result. Another example of a transgenic mouse presents an overexpression of Cu/Zn-superoxide dismutase. Such models represent an interesting setting to find answers to fundamental questions on the entire organism. In addition to a model being a whole-body knock-out or overexpression of a gene, it is possible to create such models wherein the altered genetic events are isolated to one or more organs.

The selection of laboratory tests to be used in different models should take into consideration their precision, accuracy, analytical specificity and sensitivity, predictive value and validity. All techniques may be subjected to both random and systematic measurement errors. Staff performance there procedures should be trained to use standardised and validated techniques, which must be continuously monitored by appropriate quality control protocols. Dietary requirements and ethical guidelines should be always considered.

Human Nutrition Studies

Most of the nutritional experiments and research methodology described for animals can be also applied to human beings. However, in some circumstances animal studies are preferred for practical reasons, but in others it is necessary to carry out the experimentation directly in man due to difficulties for the extrapolation of data. The impact of food intake on health and nutritional outcome is conveniently assessed thought dietary intake estimations, anthropometry, biochemical markers and clinical indications of the nutritional states. In any case, adequate ethical guidelines and regulations must be followed and accomplished.

Nutrient Intake and Requirements Studies

Food consumption data are collected for a variety of aims including the estimation of diet adequacy of individuals or population groups and the investigation of relationships between diet, health and nutritional status and also for the evaluation and implementation of nutrition programmes and other public health issues.

Dietary intake estimation entails the collection of information on the quantity of foods eaten and, using food composition values, the computation of the energy and nutrient content of these foods. Therefore, dietary consumption by individuals or groups is converted to nutrient intake using a nutrient database and, eventually adequate computer software. In fact, in the last few years a number of programmes have been developed for personal computers. The accurate recording and evaluation of the dietary intake is a difficult task of nutritional assessment because of psychological influences, under or over-reporting, subjects memory and also because nutrient variability in foods. Available methods for assessing food consumption of individuals, are the 24-hours recall, the estimated food record, the weighed food record, the dietary history and the food frequency questionnaire, which have different uses and limitations.

There are several acceptable approaches in determining nutrient requirements depending on age, gender, activity level and health status, population surveys can be used in which the intake of nutrients and the prevalence of particular clinical deficiency signs of the nutrient under investigation is evaluated. Evaluation of serum levels or urinary excretion of some nutrients in conjunction with dietary intake may be useful in establishing nutrient requirements. More sophisticated studies include excretion of nutrients in sweat and hair as well as enzymatic activities as balance studies are another approach in which subjects in a metabolic ward may be fed graded levels of a particular nutrient. However, such studies are expensive to conduct and may not evaluate inter-organ distribution of the nutrient. In any case, studies using free-living volunteers are valuable because the subjects are able to participate in their usual activities and may provide proof to be more insightful in a practical sense.

Digestibility and nutrient balance studies should be carried out in humans beings to specifically evaluate food utilization and nutrient functions. Other experimental approaches aiming to identify nutrient intake and to quantify the rates of different metabolic pathways have been boasted by the growing application of stable isotopes as kinetic markers of nutrient fate. The majority of elements exist in a stable form at a low natural abundance such as ^{13}C, ^{15}N, ^{2}H, ^{18}O, ^{70}Zn, etc., which can be used as tracers when incorporated into molecules or salts at high enrichments. The obvious advantage of using stable isotopes include that non-invasive methods are seeked and that they do not produce ionising radiation, although their assessment require mass spectrometry equipments. Many stable isotopes are still expensive. The applications of stable isotopes have been exploited over recent years in studies of but also for vitamin and mineral consumption and utilization nutrient turnover, body composition and energy metabolism. In recent years, the doubly labelled methods for the estimation of energy expenditure in humans have received much attention to complement indirect and direct calorimetry measurements, particularly in free-living conditions to validated food consumption data. The use of metabolic wards and a careful control of metabolic indicators is required in some experimental situations. The accurate measurement of energy intake following different approaches is of great interest in metabolic studies.

The provision of experimental diets to selected volunteers or free-living populations require dietary advice by well-trained professionals, a normal diet supplies amounts of all essential nutrients taking into account the requirements of the stage of the life cycle. Modified diets are based on the normal diet, which are designed to meet the needs of a given situation and may involve changes in individual nutrients, energy value, consistency and preparation in order to evaluate requirements. Levels of specific nutrients provided by a modified diet may be labelled as high, low, controlled, restricted, or free, as compared with the normal diet. When a limitation must be set on the amount of a particular nutrient, the diet contents should be specifically defined (i.e.: 20g protein and 0,5g of sodium). Diets for research or therapeutic purposes should be described, whenever possible, in qualitative and quantitative terms that leave no room for misinterpretation.

Antropometry and Body Composition Assessments

Antropometry and body composition assessment is an important part of the nutritional examination. The lack of proper growth is usually an early sign informing that nutrition may be at fault. Others measurements, such as weight for height, head circumference and skinfold thickness reflect present nutritional status and are used to assess energy reserves. Height and weight are common measurements, but since their significance and importance are not appreciated, they are frequently measured sloppily, incorrectly or inconsistently. The height and weight measurements are evaluated by comparing them to various norms. For children, the height and weight are recorded as a percentile, which reflects the percentage of the total population of children of the same sex at or below that height or weight at that age, and the child´s growth at each age, or growth "curve" can be followed.

Other indicators such as arm span length and head circumference usually reflects non-nutritional alterations rather than nutritional abnormalities. Skinfold thickness is a measurement of subcutaneous fat that is increasingly coming into use as health practitioners become more concerned with assessing body fatness and muscle protein reserves. Additional anthropometric informations are obtained by the mid upper arm circumference. When the mid upper arm circumference is combined with the triceps skinfold thickness measurement, an even better assessment of protein and energy intake can be made. Several formulas have also been developed for determining body fatness as a percentage of body weight using anthropometric data. The creatinine: height index has been used to assess the amount of skeletal muscle mass, which indicates muscle protein reserves, but the interpretation of this parameter requires some cautions.

Laboratory assessment of body composition includes several methods such as chemical analysis of cadavers, ^{40}K counting, total body water content measured by isotope dilution or densitometry (underwater weighing, plethysmography,…), bioelectrical impedance, or near-read interactance, ultrasound conductivity, computerised tomography, neutron activation analysis, magnetic resonance imaging (MRI) spectroscopy of dual energy X Ray absorptiometry (DEXA), etc. Body fatness can be also indirectly estimated through the Quetelet index or Body Mass Index (BMI), which is calculated as weight (Kg)/height ^{2} (meters) and allow to classify individuals as underweight (BMI<20), overweight (25<BMI<30) or obese (BMI>30), according to the OMS criteria.

The procedures used to assess antropometry and body composition, have a number of limitations. Thus, the selection of any method depends on the objectives of the study, cost, convenience, equipment and technical expertise as well as the health of the subject and the precision and accuracy required. Methods with the lowest cost are often the most imprecise.

Methods based on the two-compartment model (e.g. body fat and fat-free mass) include total body water and total body potassium, densitometry, total body electrical conductivity and, bioelectrical impedance. Such methods are not suitable for patients with chronic nutritional deprivation, because in such cases, the basic assumptions of the two-compartment model (i.e. water, protein, minerals, and fat) are not appropriate. These techniques include neutron activation analysis for protein and minerals and isotope dilution for total body water. Body fat can be calculated by subtracting the protein, minerals, and water from body weight.

Biochemical Measurements

Biochemical data regarding nutritional status can be obtained from examination of plasma, red blood cells, white blood cells, urine or tissues such as liver, bone, hair and fingernails. There are some principles of biochemical evaluation upon which authorities agree. Thus, heredity probably influences the biochemical nutritional profile, which it is also influenced by dietary and other environmental factors. The level of a nutrient in the body is also affected by the gender and age of the person. However, often there is not a definite value for a biochemical test that clearly delineates the deficient person from the non-deficient person.

Furthermore, some biochemical values reflect immediate nutrient intake, whereas others reflect past or long-term intake. The sensitivity of a test to changes in the status of a particular nutrient is important since a biochemical value for one nutrient can be influenced by the intake or body level of another nutrient. In summary, a good biochemical test of nutritional status should be sensitive to nutritional changes and be minimally influenced by other factors such as exercise, infection, physical or emotional stress, or trauma.

Functional tests (Table 1) are able to determine the influence of host nutriture in the physiological performance of the organism. Data indicative of immunological function can also be valuable indicators of nutrient utilization. Thus, delayed cutaneous hypersensitivity (DCH) usually consists of the response to five recall antigens, usually Candida, streptokinase-streptodornase (SK-SD), and purified protein derivative (PPD). The size of the induration is measured to give the clinician a measure of the body´s ability to mount an immune response.

Table 1. Some important examples of functional tests

Tests	Examples
Measurement of abnormal metabolic products in blood or urine arising from suboptimal intakes of the nutrient	Increased excretion of xanthurenic acid in vitamin B-6 deficiency
Measurement of changes in blood components or enzymes activities that are dependent on a given nutrient	Whole blood hemoglobin for iron assessment; Erythrocyte glutathione peroxidase activity for selenium; erythocyte glutathione reductase activity for riboflavin; erythrocyte transketolase activity for thiamin
In "vitro tests" of in vivo functions	Leukocyte chemotaxis for protein energy, zinc and iron depletion, uridine suppresion test for vitamin B_{12} and folate
Induced responses and load tests in vivo	Delayed cutaneous hypersensitivity for the assessment of protein-energy and zinc; relative dose-response for vitamin A; histidine load test folate; tryptophan load test for vitamin B_6
Spontaneous in vivo responses	Dark adaptation and taste acuity for vitamin A and zinc; muscle function for protein-energy deprivation
Growth or developmental responses	Sexual maturation for zinc; growth velocity for protein-energy, zinc or iodine; cognitive performance for iron

Other relevant functional tests of nutritional status are based on structural integrity (erythocyte and capillary fragility), experimental wound-healing, lipoprotein peroxidation (breath ethane/pentane); transport (intestinal absorption, plasma/tissue transport), hemostasis (prothrombin time, platelet aggregation), reproduction (sperm count); nerve function (dark adaptation, olfactory acuity, taste acuity and sleep pattern) and Work Capacity/

Hemodynamics (task performance/endurance, VO2 max capacity, VO2 submax cumulative, heart rate, etc).

In this context, it has been postulated that the amount of essential nutrients and non-nutrients occurring in the diet is likely to be related to body biochemical, functional or other nutritional indicators. However, the study of biochemical markers as predictors of dietary intake of nutrients is at an early stage of evolution and require adequate validation in many cases. Thus, the assay procedures used must be sufficiently sensitive, robust and free from fluctuation or interference to be valid and precise.

Table 2. Biochemical Markers of nutrient intake

Nutrient	Choice of index
Vitamins	
Vitamin A	Plasma retinol and vitamin A (mainly esters)
Vitamin D	Plasma 25-hydroxycholecalciferol and Vitamin D
Vitamin E	Plasma Vitamin E:cholesterol ratio
Vitamin K	Blood clotting time and plasma vitamin K
Thiamin	Red cell transketolase or plasma thiamin levels
Riboflavin	Activation of erythrocyte glutathione reductase
Vitamin B_6	Plasma pyridoxal phosphate, urine pyridoxic acid and red cell transaminases
Folate	Serum or red cell folate
Vitamin B_{12}	Serum vitamin B_{12}
Vitamin C	Plasma or urinary vitamin C
Other B vitamins	Plasma and urinary vitamin levels and enzyme activation
Minerals:	
$Na^+$and K^+	Urinary levels
$Ca^{2+}$and Mg^{2+}	Urinary levels
Fe^{2+}	Serum ferritin, red cell protoporphyrin; transferrin saturation, haemoglobin, etc.
Zn^{2+}	Plasma Zn^{2+} and hair zinc
Cu^{2+}	Red cell superoxide dismutase
Mn^{2+} ,Cr^{2+}, etc.	Plasma levels
F^- and Cl^-	Urinary level
I^-	Plasma thyroid hormone levels
Se^{2+}	Urinary levels; Plasma, platelet, hair and nail Se, Red cell glutathione perosidase
S^{2-}	Urinary levels
$PO4^{3-}$	Urinary levels
Carbohydrate	
Glucose	Plasma and urinary glucose, glycemic index,...
Lactose	Breath Hydrogen
Lipids:	
Fatty acids	Plasma Cholesteryl esters, phospholipid from cell membranes, Adipose tissue (long term)
Cholesterol	Plasma levels, LDL/HDL Cholesterol ratio
Protein:	
Amino Acids	Urea urinary levels, plasma and urinary amino acids
Protein quality	Digestibility and biological assays

The use of reference materials and validation procedures as well as careful selection of aspects concerning temporal variation in the biochemical, functional and clinical states indices is convenient to yield accurate and reliable data. However, other factors should be recognised and defined such as physiological status, intestinal transport rates, threshold levels, feedback control, etc, in addition to between-subjects variation, interactions between nutrients, interferences with non-nutrients and specific physiopathological circumstances (pregnancy, lactation, disease, etc..). Biomarkers are commonly obtained from fecal, urinary and blood specimens, although predictive information about the nutrient intake can be also provided by breast milk, saliva, fat biopsies, hair, nails, etc. (Table 2).

For most minerals, short-term intakes appear to be quite well predicted by the measurements of urinary excretion levels, provided that the completeness of urinary collection can be properly validated and dietary factors are reasonably constant.There are few instances of very close agreement between intake and biochemical markers for the vitamins, while each vitamin must be considered separately. For lipids, markers of usual intake are limited to the analysis of the proportions of fatty acids in serum or plasma as well as in cell membranes from adipose tissue.

Additionally, there are numerous factors (e.g. dietary fibre intake, alcohol consumption, smoking) that may influence fat uptake and metabolism. The relationship between protein intake and urinary excretion of the marker (urea or amino acids) is closer than for virtually all other nutrients (except perhaps sodium), assuming that the subject is in balance. However, for use in epidemiological studies there is a need to develop cheaper biochemical markers for the assessment of dietary fibre and energy intake.

Clinical Examination and Other Indices

Nutritionally-oriented clinical assessment consists of a physical examination and a medical history with emphasis on nutritional issues. Special attention should be given to the skin, hair, teeth, gums, lips, tongue and eyes, and in men to the genitalia, since these areas show earlier signs of nutritional deficiencies. Hair, skin and mouth are susceptible for examination because of the rapid cell turnover of epithelial and mucosal tissue. Early manifestations may also be reflected in the gastrointestinal tract, such as diarrhea because of gastrointestinal o mode mucosal changes. Clinical signs of nutritional deficiency should always be confirmed with biochemical and dietary data. A clinical symptom can and usually does reflect the presence of more than one nutritional deficiency.

Through the history, any behavioural or functional change in neuromuscular or cardiovascular system that may not be apparent to the clinician, but is remarkable to the patient or the patient´s family can be assessed. The clinical examination for nutritional evaluation in some hospitals now includes assessment of energy expenditure of the patient though measurement of O_2 consumption and CO_2 production in addition to other routine measurements such as weight, height, etc.

Information for the medical history can be obtained by administering a questionnaire, or from medical records, and it can be used in the study of the aetiology of the nutrient deficiency. Observer inconsistencies may also be a source of error, but can be minimised by training examiners and standardising the criteria used to define the signs.

The evaluation of physical fitness and functional capacity is another strategy for the assessment of the nutritional status. Thus, the ability to perform physical work depends on the efficiency of the cardiovascular, respiratory and neuromuscular systems, which are affected by nutrient intake and by changes in body weigh. Physical performance can be estimated by O2 uptake (VO_2 max) under gradually increasing work-loads on a treadmill or bicycle ergonometers or alternatively from submaximal exercise tests. Muscle strength and other indicators of nutritional status are assessed with electric or electronic manometers, while motor performance has been beamed by using several batteries of motor tests such as pull-ups, sit-ups, shutter run, 50 yard dash, etc.

Other functional testing such as grip strength can be a useful component of the nutrition assessment, since changes in function can occur before alterations in body composition are detected by anthropometry. Electrical stimulation of muscle has the advantage that it does not depend on voluntary effort. Psychometrical tests have been used to evaluate the impact of nutrition on health through a large number of approaches and test such as intelligence, personality, specific abilities, developmental schedules, cognitive psychology and neuropsychological evaluation.

Attitudinal information is often of interest in nutritional surveys. The reliability, validity and standardisation of tests depends of a number of factors, which should be identified to obtain valuable data through personal training, development of pilot studies, etc. Furthermore, psychosocial information concerning education, religious practices, economic status, smoking habits, or other lifestyle characteristics may be useful to understand the interactions of nutrition with health. Estimation of sedentarism and physical activity either through validates questionnaires or with movement recorders may be of interest in the nutritional evaluation.

The ideal nutritional indices are those that are accurate markers of nutritional status, accurate predictors of outcome, and easily monitored during nutritional repletion. To increase their predictive value, multi-parameter indices have been developed. The prognostic nutritional index (PNI) that incorporates serum albumin, serum transferrin, triceps skinfold thickness, and delayed hypersensitivity appears adequate for predicting nutritional outcomes. As interpretation of many of the laboratory indices is confounded by the effects of the disease process, an approach based exclusively on a medical history and physical examination has been developed. This technique, called subjective global assessment, is able to provide a valid and reproducible prediction of morbidity in surgical patients. Some mini-questionnaires have been developed for specific populations (i.e.: elderly, pregnant women, etc).

The performance of nutritional studying should take into consideration the fundamental ethical rights of the human being and must count with a squad informed consult of the participants (Annexes I and II).

ANNEX I

Model

The proposal respect the fundamental ethical principles including human rights and will deal only with individuals adequately informed and willing to participate. Also all research data partners, will obtain national authorization from ethical committee or equivalent body before any intervention with subjects. The study does not involve any genetic manipulation.

- Requested specifications:
  - Human embryos or foetus No Yes
  - Use of human embryonic or foetal tissue No Yes

– Use of other human tissue	No	Yes
– Research on persons	No	Yes
– If yes, further specify if it involves:		
– children	No	Yes
– persons unable to consent	No	Yes
– pregnant women	No	Yes
– healthy volunteers	No	Yes
– Use of non-human primates	No	Yes
– Use of transgenic animals	No	Yes
– Use of other animals	No	Yes
– Genetic modification of animals	No	Yes
– Genetic modification of plants	No	Yes

- Other specifications:
  - The regulations, concerning human and medical research, will be respected with precise reference of the recommendations of the Helsinki (1964), Tokio (1975), Venice (1983) and Hong Kong (1989) committees as well as other EU regulations RD 561/1993, 65/65 CEE, 75/318 CEE, Directive 91/507 and 89/843 EN-C (ISBN 92-825-9612-2).

ANNEX II

INFORMED CONSENT (model)
This form will cover the following aspects:
I..................(name)
I have read the volunteer's information
I have felt free to make questions concerning the study
I have received enough information
I have talked to the following personnel / responsible (names…)
I understand that my participation is in a voluntary basis
I understand that I can withdraw from the study:

1. If I wish
2. Without further explanations

Therefore, I freely confirm my availability to be involved in the trial

Date
Signature

In addition, all the partners agree with the following statement

In implementing the proposed research I shall adhere most strictly to all national and international ethical and safety provisions applicable in the countries where the research is carried out.

I shall conform in particular to the relevant safety regulations concerning the deliberate release into the environment of genetically modified organisms.

REFERENCES

Albanese, A.A. (1972) *Newer Methods of Nutritional Biochemistry with Applications and Interpretations.* Academic Press, New York and London.

Alexander, RR, Griffiths, JM. (1993). *Basic Biochemical Methods.* 2nd Edition. Wiley-Liss, Inc New York, USA.

Asp, N.G. (1995) Classification and methodology of food carbohydrates as related to nutritional effects. *American Journal of Clinical Nutrition.* 61, 930-937.

Barnes, S. and Kim, H. (2004) Nutriproteomics: identifying the molecular targets of nutritive and non-nutritive components of the diet. *J Biochem Mol Biol.* 37, 59-74.

Barrocas, A. Belcher, D. Champagne, C. and Jastram, C. (1995) Nutrition assessment: practical approaches. *Clinical Geriatric Medical.* 11, 675-713.

Bean, W.B. (1950) Control in research in human nutrition. *Nutr Rev.* Apr;8 (4), 97-9.

Berdanier, C.D. (1996) *Nutrient and Gene Expression Clinical Aspects.* (Berdanier CD editor). Boca Raton: CRC Press.

Bingham, S.A. (2002) Biomarkers in nutritional epidemiology. *Public Health Nutr.* 5, 821-827.

Bowman, B., Russell, R.M. (2001) *Present knowledge in Nutrition.* (8 edition) Editors ILSI Press Washington, DC.

Campion, J., Milagro, F.I. and Martinez, J.A. (2004) Genetic manipulation in nutrition, metabolism, and obesity research. *Nutr Rev.* 62, 321-330.

Carpenter, K.J. (2003) A short history of nutritional science: part 1 (1785-1885). *J Nutr.*133, 638-645.

Crews, H.M., Ducros, V., Eagles, J., Mellon, F.A., Kastenmayer, P., Luten, J.B. and McGaw, B.A. (1994) Mass spectrometric methods for studying nutrient mineral and trace element absorption and metabolism in humans using stable isotopes. *A review. Analyst.* 119, 2491-2514.

Cunningham-Rundles, S. (1998) Analytical methods for evaluation of immune response in nutrient intervention. *Nutrition Reviews.* 27-37.

Dennis, L.K. Snetselaar, L.G. Nothwehr, F.K. and Stewart, R.E. (2003) Developing a scoring method for evaluating dietary methodology in reviews of epidemiologic studies. *J Am Diet Assoc.* 103, 483-487.

Eaton, S.B. and Cordain, L. (1997). Evolutionary aspects of diet: old genes, new fuels. Nutritional changes since agriculture. *World Rev Nutr Diet.* 81, 26-37.

Elia, M. Stratton, R. and Stubbs, J. (2003). Techniques for the study of energy balance in man. *Proc Nutr Soc.;*62, 529-537.

Ensminger, A. Konlande, M.E. and Robson, R.J.K. (1994). *Foods and Nutrition Encyclopaedia* (2nd Edition) CRC Press Boca Raton.

Fidanza, F. (1994). Nutritional status assessment in perspective. *Bibl. Nutrition Diet.* 51, 9-17.

German, J.B. Roberts, M.A. Watkins, S.M. (2003). Personal metabolomics as a next generation nutritional assessment. *J Nutr.*133, 4260-4266.

Gibney, M.J. Vorster, H.H. and Kok, F.J. (2002) *Introduction to human nutriton.* Blackwell Publishing. London.

Gibson, R.S. (1990). *Principles of Nutritional Assessment.* New York Oxford: Oxford University Press.

Gladyshev, V.N. Kryukov, G.V. Fomenko, D.E. and Hatfield, D.L. (2004). Identification of trace element-containing proteins in genomic databases. *Annu Rev Nutr;* 24, 579-596.

Groff, JL, Gropper S.S. (2000). *Advanced Nutrition and Human Metabolism.* Wadsworth Minneapolis.

Hellerstein, M.K.(2003). In vivo measurement of fluxes through metabolic pathways: the missing link in functional genomics and pharmaceutical research. *Annu Rev Nutr.* 23, 379-402.

Hensurd, D.D. (1999) Nutrition screening and assessment. *Medical Clinics of North America.* 83, 1525-1546.

Hopkins, (1929). *The earlier history of vitamin research.* In: Nobel Lectures: Physiology of Medicine. Amsterdam: Elsevier, 1965:1922-1941.

Hui, D.Y. (1998) Utility and importance of gene knockout animals for nutritional a metabolic research. *Journal of Nutrition.* 128, 2052-2057.

Jeejeebhoy, K.N. (1998) *Nutritional assessment.* Gastroenterol Clinical North of America. 8, 113-123.

Kim, H, Page, G.P. Barnes, S. (2004). Proteomics and mass spectrometry in nutrition research. *Nutrition.* 20, 155-65.

Krause, M.V. and Mahan, L.K. (2001). *Food, Nutrition and Diet Therapy.* Philadelphia: W. B. Saunders Company.

Kushmer, R.F. (1993) Body weight and mortality. *Nutrition Reviews.* 51, 127-136.

Leedy, P.D. (1980). *Practical research: Planning and designs.* 2 New York Macmillan.

Lloyd , L.E. McDonald, B.E. Crampton, E.W. (1978). *Fundamentals of Nutrition.* (Salisbury G.W, Editor) W.H. Freeman and Company London.

Margetts, B.M. and Nelson, M. (1998). *Design concepts in Nutritional Epidemiology.* Oxford. Oxford University Press.

Marti, A. Martinez, J.A. (2000). *Methodological approaches to assess body weight regulation and etiology of obesity.* Proceeding of the Nutrition Society. 59, 405-411.

Martínez, J.A. (2002) Fundamentos teorico prácticos de nutrición y dietética. McGraw-Hill Interamericana. Madrid.

Mellon, F.A. and Fairweather-Tait, S.J. (1997) Stable isotope methods for studying nutrient mineral metabolism in humans. *Endeavour* 21, 12-18

Meyers, R.A. (1997). *Encyclopaedia of Molecular Biology and Molecular Medicine.* Weirhein VCH Ed..

Monsen, E.R. and Cheney CL. (1988). Research methods in nutrition and dietetics: Design, data analysis, and presentation. *Journal of the American Dietetic Association.* 88, 1047-1065.

Moreno, M.J. Martí, A. García-Foncillas, J. and Martínez, J.A. (2001) DNA hybridization arrays: a powerful technology for nutritional and obesity research. *British Journal of Nutrition.* 86, 119-122.

Moustaïd-Moussa, N. and Berdanier, C.D. (2001). *Nutrient-Gene Interactions in Health and Disease* (Moustaïd-Moussa N, Berdanier CD editors). Boca Raton: CRC Press.

Nissen, S. (1991). *Moderm Methods in Protein Nutrition and Metabolism.* (S. Nissen editor). New York. Academic Press, Inc.

Page, G.P. Edwards , J.W. Barnes, S. Weindruch, R. and Allison D.B. (2003). A design and statistical perspective on microarray gene expression studies in nutrition: the need for playful creativity and scientific hard-mindedness. *Nutrition.* 19, 997-1000.

Palmquist, D.L. and Jenkins, T.C. (2003). Challenges with fats and fatty acid methods. *J Anim Sci.* 81, 3250-4.

Paul, A.A. and Soughtgate, D.AT. (1988). *McCance and Widdowson's The Composition of Foods.* London: Her Majesty's stationery office.

Prentice, R.L.. Sugar, E. Wang, C.Y. Neuhouser, M. and Patterson, R. (2002). Research strategies and the use of nutrient biomarkers in studies of diet and chronic disease. *Public Health Nutr.* 5, 977-84.

Ruiz-Canela, M. et al. (1999). Informal consent and Approval by institutional Review Boards in published reports on Clinical trials. *N. E. J. Med.* 340, 1114-1115.

Sadler, M.J, Strain, J.J, and Caballero, B. (1999) *Encyclopaedia of Human Nutrition.* Academic Press New York.

Sauberlich, H.E. (1999). Assessment of Nutritional Status. (I. Wolinsky editor) Boca Raton: CRC Press.

Scholler, D.A. (1999) Recent advances from application of doubly labeled water to measurement of human energy expenditure. *Journal of Nutrition.* 129, 1765-1768.

Shaw, R.L. Booth, A. and Sutton, A.J. Miller, T. Smith, J,A. Young, B. Jones, D,R. Dixon-Woods, M. (2004). *Finding qualitative research: an evaluation of search strategies.* BMC Med Res Methodol.;4, 5.

Shils ME, Olson JA, Shike M, Ross AC. (2001). *Modern Nutrition in Health and Disease. 9th Edition,* Lippincott Williams and Wilkins, Baltimore USA.

Stubbs RJ, Johnstone AM, O'Reilly LM, and Poppitt SD (1998) Methodological issues relating to the measurement of food, energy and nutrient intake in human laboratory-based studies. *Proceedings Nutrition Society 57,* 357-372.

Trayhurn P (1998) Molecular biology and nutrition: the quest for integration. *British Journal of Nutrition. 80,* 305-306.

Varela G. (2004) De la alimentación a la nutrición. Universidad San Pablo Ceu. Madrid.

Van Zutphen L.F:M., Baumans V. and Beynen A.C. (1993) *Principles of Laboratory Animals Science.* Elsevier Amsterdam.

Velazquez, A. (1997) Gene-nutrient interactions in single-gene defects and polygenic diseases: methodologic considerations. *World Review Nutrition Dietetics* 80, 145-164.

Viteri, F.E. and Warren. R. (2002). Considerations on the use of radioisotopes in human nutrition research. *Food Nutr Bull.;*23, 7-16.

Waitzberg, D.L, Correia. M.I. (2003). Nutritional assessment in the hospitalized patient. *Curr Opin Clin Nutr Metab Care.*6, 531-8.

Whitehead, R.G, and . Prentice, A.T. (1991). *New techniques in Nutritional Research..* Cambridge: Academic Press, Inc.

Wildman R. E. C. and Medeiros D. M. (2001). *Advanced Human Nutrition.* (I. Wolinsky editor) Boca Raton: CRC Press.

Willett WC. (2000). Nutritional epidemiology issues in chronic disease at the turn of the century. *Epidemiologic Reviews;* 22, 82-6.

Willett,W. (1998). *Nutritional Epidemiology*. Oxford. Oxford University Press.

Yanovski, J.A. and Janovski, S.Z. (1999) Recent Advances in basic obesity research. JAMA. 282, 1504-1506.

Zamel, B.S. Riley, E.M., and Stallings, V.A. (1997) Evaluation of methodology for nutritional assessment in children: anthropometry, body composition, and energy expenditure. *Annual Review Nutrition.* 17, 211-235.

Further Readings

Alexander, R.R. and Griffiths, J.M. (1993). Basic Biochemical Methods (Second Edition). New York, NY.: Wiley-Liss, Inc.

Elmadfa, I., Anklam, E., and König, J.S. (2003). Modern Aspects of Nutritio –Present Knowledge and Furure Perspectives. Basel (Switzerland): Karger, AG.

Forbes, G.B. (1987). Human Body Composition. Growth, Aging, Nutrition and Activity. New York, N.Y.: Springer-Verlag.

Kathleen Mahan, L., and Arlin, M.T. (1992) Krause's Food, Nutrition and Diet Therapy. (8th Edition) Philadelphia, Saunders Company.

Linder M.A. (1991). Nutritional Biochemistry and Metabolism. With Clinical Applications. (Second Edition) New York. N.Y. Elsevier

Miján de la Torre A. (2002). Técnicas y Métodos de Investigación en Nutrición Humana. Barcelona: España, Editorial Glosa

Moustaïd-Moussa, N., and Berdanier, C.D. (2001). Nutrient-Gene Interactions in Health and Disease:Washington, D.C. N.Y. CRC Pres.

Ross, D.W. (1996). Introduction to Molecular Medicine (Second Edition). New York: Springer-Verlag.

Zempleni, J. and Daniel, H. (2003). Molecular Nutrition. Cambridge, USA: CABI Publishing.

Ziegler, E.D. and Filer, L.J. (1996). Present Knowledge in Nutrition. (Seventh Edition). Washington, USA, ILSI, Press.

Web Sites

www.eurekalert.org EurekAlert; American Association for the Advancement of Science
www.ilsi.org/pubs.html Nutrition Reviews
www.sennutricion/ Nutrition Society of Spain (SEN)
www.nutsoc.org.uk The Nutrition Society
www.nutrition.org/ American Society of Nutrition Sciences
www.nutr.annualreviews.org/ Annual Reviews of Nutrition
www.iuns.org/ International Union of Nutrition Sciences
www.nal.usda.gov./fnic/foodcomp/ Nutrient Data Laboratory
www.faseb.org/ajcn American Journal of Clinical Nutrition
www.ncbi.nlm.nih.gov/PubMed/ National Library of Medicine

www.aicr.org/ American Institute for Cancer Research
www.hnrc.tufts.edu/ Jean Mayer USDA Human Nutrition Research Center
http://www.fensweb.org/ Federation of European Nutrition Societies

INDEX

A

B

C

D

E

F

G

H

I

J

K

L

M

N

Q

R

S

T

U

V

W

X

Y

Z